THE EXPERIENCE OF RETURN MIGRATION

The Experience of Return Migration
Caribbean Perspectives

Edited by

ROBERT B. POTTER
University of Reading, UK

DENNIS CONWAY
Indiana University, USA

JOAN PHILLIPS
University of Reading and Policy Studies Institute, UK

ASHGATE

Published by
Ashgate Publishing Limited
Gower House
Croft Road
Aldershot
Hants GU11 3HR
England

Ashgate Publishing Company
Suite 420
101 Cherry Street
Burlington, VT 05401-4405
USA

Ashgate website: http://www.ashgate.com

British Library Cataloguing in Publication Data
The experience of return migration : Caribbean perspectives
   1. Return migration - Caribbean Area 2. Caribbean Area -
   Emigration and immigration 3. Caribbean Area - Social
   conditions
   I. Potter, Robert B. II. Conway, Dennis, 1941- III. Phillips,
   Joan
   304.8'729

Library of Congress Control Number: 2005924871

ISBN 0 7546 4329 8

Printed and bound in Great Britain by Antony Rowe Ltd, Chippenham, Wiltshire.

Contents

List of Tables

List of Contributors

Stephanie Condon is a permanent member of the research staff at the National Demographic Institute (INED) in Paris. Since training in Modern Languages and Geography in Britain, and completing a PhD in historical population geography at Queen Mary and Westfield (University of London), in 1987, she has been working on French Caribbean migration and, since 1993, in a comparative perspective with the British case.

Dennis Conway is Professor of Geography and Latin American and Caribbean Studies at Indiana University, Bloomington, Indiana. He has written extensively on the roles of tourism and migration in Caribbean small island development. Since 1966, he has conducted research on urbanization, migration and development issues in Barbados, Trinidad and the Windward Islands. His current research is on transnational migration networks, return migration and remittance impacts in the contemporary Caribbean; specifically focusing on young and youthful return migrant experiences in Trinidad and Tobago and Grenada.

Roger-Mark De Souza is the Technical Director of the Population, Health and Environment Program at the Population Reference Bureau, Washington DC, a research organization that examines the policy implications of population trends. De Souza was born and raised in Trinidad. He holds a BA and a postgraduate degree from the University of the West Indies and a Master's degree from George Washington University. Recently, he has been conducting research on Caribbean population dynamics and socio-economic conditions.

David Timothy Duval is Senior Lecturer in the Department of Tourism at the University of Otago, New Zealand. Originally trained as an archaeologist, and specialising in the prehistory of the Eastern Caribbean, his interests gradually shifted to issues of migrant mobilities whilst pursuing his PhD at York University in Toronto. David is currently undertaking ethnographic work with Filipino migrants living in Wellington, New Zealand in an attempt to understand the role of social networks in transnational mobilities.

Heather A. Horst is an Honorary Research Fellow in the Department of Anthropology at University College London. She completed her PhD thesis on the topic of return migration and the material culture of the home among first generation Jamaicans who migrated to the UK. More recently, she has been involved in a project concerning the impact and consumption of new communication technologies in Jamaica.

Marina Lee-Cunin is a Visiting Fellow at the Risk Centre, Shiga University, Japan and gained her PhD from the University of Kent in Social Anthropology. Her research interests concern gender, ethnicity and youth, with particular focus on student communities and Caribbean diasporic social issues. She holds dual citizenship of the UK and Trinidad and Tobago and has personal experience of the return and re-return migration process.

Gina M. Pérez is Assistant Professor of Comparative American Studies at Oberlin College, and is a cultural anthropologist with training in Caribbean history and ethnography and Latina/o Studies.   Since 1995, her research has focused on experiences of displacement, migration, transnationalism and gender.   In Puerto Rico and Chicago, as well on the experiences of Caribbean and Latina/o communities throughout the United States.

Joan Phillips is a Research Fellow at the Policy Studies Institute and Part-Time Postdoctoral Research Fellow in the Department of Geography at The University of Reading.   She is a Barbadian national and gained BA and MPhil degrees in sociology at the Cave Hill Campus of the University of West Indies, before completing a PhD at Luton University.

Robert B. Potter is Professor of Human Geography and Director of Research in the School of Human and Environmental Sciences at the University of Reading. He has been researching in the Caribbean region since 1980 on a variety of topics, including physical development planning, urbanisation, housing conditions and socio-economic change. He acted as a consultant on housing and population issues in connection with the third Physical Development Plan for Barbados.   Since 1999-2000 he has been investigating second generation migration to the region.

Joseph Rodman is in the graduate program in the Department of Geography, Indiana University, Bloomington. He conducted research in Grenada for his Master's thesis, and his chapter in this volume, co-authored with his advisor Dennis Conway, is a result of that work.   He has considerable field experience in West Africa, where he studied as an undergraduate and later worked with NGOs on community development projects.

Godfrey St. Bernard is Senior Research Fellow at the Sir Arthur Lewis Institute of Social and Economic Studies at the St Augustine Campus of the University of the West Indies, Trinidad and Tobago. He has worked extensively on Caribbean demographic patterns, including work as a statistician with the Central Statistical Office of Trinidad and Tobago. He was born in Port of Spain, Trinidad.

Preface

The idea for this book emanated from a session that we jointly organized and chaired at the 99th Annual Meeting of the Association of American Geographers, held in New Orleans in March 2003. At the conclusion of the session, which comprised two panels of papers, we felt that we had a sufficiently worthwhile core of materials to warrant putting together a wider collection of essays on the experiences of return migrants to the Caribbean. Accordingly, we invited a further group of active scholars, who had not been able to attend the meeting, to report on their research in the field.

The aim of *The Experience of Return Migration* is to focus attention on the new migration experiences of Caribbean returnees in a range of island and community settings. Thus, while several of the chapters consider the more extensively documented return of older retiree cohorts, most of the essays in this collection also focus attention on the experiences of younger returnees of working age who have decided to move to the island homes of their parents, or the islands that they left as children. How do their experiences compare with those of older returnees? How easily do they fit into their new island homes and what socio-cultural adjustments do they face on a day-to-day and year-to-year basis? The issues of how they fare in the job market and what opportunities they have to contribute to the evolution of both economic and social capital in the region are also of salience. Do such returnees face fewer or more problems in the workplace in comparison with their experiences in the realm of social relations? And what about the experiences of men and women; to what extent are they similar, or do they differ? A further fascinating topic concerns the degree to which such returnees are living truly transnational lives, perhaps involving multiple movements between the islands and metropolitan societies, thereby rendering multi-local patterns of existence and day-to-day experience.

Given the remit outlined in the previous paragraph, our second aim was to bring together a collection of case studies that was varied enough to enable comparative insights to be gained. Realistically, the Caribbean locations in which researchers were currently working was the major influence in this regard. To this end, the chapters cover aspects of return migration to Barbados (two chapters), St Lucia, Grenada, Trinidad (three chapters) and Jamaica in the Commonwealth Caribbean, together with the case of Puerto Rico in the Hispanic Caribbean, and the French West Indies taken as a whole. In terms of the group of authors, we are pleased to have among our contributors, several writers who are currently living and working in the Caribbean, along with several Caribbean nationals who are currently living and working outside the region. In addition, the team is made up of other researchers who have a longstanding interest and experience in the region. Virtually all of the essays presented here were specially written for inclusion in this volume. The only exceptions are chapters 2 and 3, earlier versions of which appeared in the journal *Caribbean Geography* (volume 12, no. 1 and volume 12, no. 2 respectively). They are included here with the kind permission of the Editor and Publisher.

During the period from 2002 to 2006, Rob Potter and Joan Phillips were working together on a research project under the title 'The social dynamics of foreign-born and young returning nationals to the Caribbean'. This research was generously funded by The Leverhulme Trust, and chapters 1, 2, 3, 4 and 14 can be seen as emanating in total or in part from that funded research. In 2004 through to 2006, Dennis Conway, Rob Potter and Godfrey St. Bernard carried out work on 'Young returning migrants in Trinidad and Tobago', supported by a grant from the National Geographical Society's Committee for Research and Exploration, and some of the products of this research can be seen in chapters 1, 8 and 13 and 14. We should like to thank both of these funding bodies for making the production of this research monograph a possibility as a part of our endeavours in the field, in addition to the specialist journal articles emanating from these projects.

Finally, our sincere thanks are extended to Chris Holland who efficiently assisted with the management of the computer files, together with the printing of the final manuscript.

Rob Potter, Dennis Conway and Joan Phillips
2005

Chapter 1

The Experience of Return: Caribbean Return Migrants

Dennis Conway, Robert B. Potter and Joan Phillips

Since the incorporation of the small island nations of the Caribbean into the external spheres of influence of European mercantilism in the sixteenth century, the region has experienced successive waves of immigration, emigration and circulation (Conway, 1989a). Hence, it is a well-exemplified observation that migrations of all descriptions have been a fundamental force in the creation and maintenance of Caribbean societies (Conway, 1999/2000; Patterson, 1978). Mobility strategies came to be the strategic responses of many Caribbean islanders to the realities of island existence; including the environment's limits, small territorial size, and the vulnerabilities of such territories. As Bryce-Laporte (1982: xxxviii) reminds us: 'Migration and return migration, and often-times remigration and return, have been an institutionalized aspect of Caribbean societies.' 'Livelihood mobility' is another characterization of this embedded small island tradition (Richardson, 1983). The range of movements has been wide and varied, extending from permanent movements (both voluntary and coerced) to transient or circular, temporary migrations. But, in common with many migration traditions has been an intention to return (Byron, 1994; Rubenstein, 1979; Thomas-Hope, 1998 among many others). And, that 'intention to return' has manifested itself in the generation of return flows, even though permanent emigration and non-return also figure prominently in the experiences (Conway, 1988).

Commonly, the short-term, contractual nature of overseas employment that many impoverished islanders sought after Emancipation necessitated a return, whether the migrant wanted to, or not (Conway, 1989a, 1997; Richardson, 1985). More recently, the Caribbean's proximity to the United States, and its island dependencies, Puerto Rico and the US Virgin Islands, and the relatively inexpensive travel costs incurred, ensure that a return move is a logical (and actual) termination of a Caribbean-North American circulation (Stinner, et al., 1982). Strengthening the institutional forces affecting this return of Caribbean sojourners are the conventional institutional barriers to border-crossing – demand for pre-arranged bonds, contracts and round-trip tickets as temporary visa entry requirements – and the threats of incarceration and deportation to illegal 'visa-overstayers'. Most Caribbean return migrants are neither behaving as free, unfettered voluntary movers, nor as fully coerced, recruited and/or forcibly contracted repatriates. Rather, they are conditioned and constrained by structural forces, and move as 'satisficers' (Pryor, 1976; Wolpert, 1965), in a world that has (with the exception of their elite and wealthy compatriots) rarely given Caribbean people – migrants and non-migrants alike – the chance to be optimizers.

Who Are these Return Migrants?

But, what is a 'return migrant', who was he, who is she? Why should we focus specifically on them and their experiences in this collection? One immediate answer springs to mind – namely, the relative lack of research, until recently, on the multi-faceted processes of return migration and on the selectivity and diversity of return migrants, their changing character, their experiences, and societal interactions. Return migrants are demographically selective, behaviorally diverse, they possess differing stocks of human capital, they have divergent attitudes, divergent images of their island homelands, divergent backgrounds, and consequently their experiences, adaptations, and behaviours will rarely be commonly shared, so they could mobilize and provide a 'solidarity' movement to initiate societal change. Bovenkerk's (1974, 1981) wishful thinking that such a lack of solidarity among Surinamese returnees contributed to their inability to be agents of societal change, unfortunately diverted the literature on this point so that the critical metaphor he coined in the title of his 1981 article, held sway for too long. Returnees were castigated as being unlikely to be agents of societal change, but rather, promoting intergenerational conflict, high levels of estrangement and discouragement characterized their experiences 'back home'.

On the other hand, when defined on the basis of the returnee's national identity, return migration becomes a manifestation of a basic human right. Part 2 of Article 13 of the 1948 *Universal Declaration of Human Rights* states that: 'everyone has the right to leave any country, including their own, and to return to their country' (UN General Assembly Resolution 217 A (III), United Nations Secretariat 1986: 88). And, although there has been a common tendency to view international migration as a one-way process of emigration (leaving) and immigration (arriving), where there is no return (of the successful) but only the return of failures, one scholar at least has recently written that '(r)eturn migration is the great unwritten chapter in the history of migration' (King 2000: 7).

The Diversity of Return Migrants

The diversity among return migrants has been categorized according to their patterns of mobility, which, as a first cut, yields some helpful insights. Gmelch (1980) reviews a host of typologies of return migration which deal with two dimensions of considerable diversity: the length of time migrants intended to remain abroad and their reasons for returning (Cerase, 1974; King, 1977; Rhoades, 1978; Richmond, 1968). Basic to these typologies is the useful distinction between those migrants who intend their emigration to be temporary and those who intended it to be permanent. Simply, but essentially, emigration behaviour is distinguished from circulation behaviour (see Conway, 1988). Gmelch (1980: 138) abstracts the core features of these typologies to compose three types of return migrants:

1. Returnees who intend temporary migration, or circulation. The time of their return is determined by the objectives they set out to achieve at the time of emigration.
2. Returnees who intended permanent emigration but were forced to return. Their preference was to stay abroad but because of external factors they were required to return.
3. Returnees who intended permanent emigration but chose to return. Failure to adjust and /or homesickness led to their decision to return.

Thomas-Hope (1985, 1986) also provides a typology of return migrants, with her primary distinction revolving around the time variable which distinguishes between long-stay and short-stay migrants, and with further sub-categories characterized by migrant's skill transformation and occupational change and migration intentions.

Four types of *settlers* or *long-stay migrants* are identified:

1. Long-term workers, who, for varying time periods, establish their households overseas before returning. They return either on retirement, or earlier in their life course when they fully anticipate continuing to work 'back home'.
2. Dependents, who leave as children, or who are born overseas and become accultured to their new homeland, and who are seldom among the permanent returnees.
3. Students, who study abroad, some of whom return after their acquisition of academic skills, while others remain abroad to maximize the opportunities their higher education qualifications have afforded them.
4. Long-term circulators are individuals who repeatedly go overseas, usually to the same country, re-entering the job market each circuit, often becoming a permanent resident of that country, or even taking out citizenship.

Four types of *transients* or *shuttle migrants* constitute Thomas-Hope's alternative *short-stay* grouping:

1. International vendors, who are mainly women 'higglers' and small-scale 'suitcase' traders.
2. Contract workers, who are usually agricultural or construction workers, domestic workers, restaurant waiters and barmen. These contracted workers

obtain visas for specified periods and for specified purposes defined by the contract.

3. Other itinerate migrants are principally men and include casual farm and construction workers, domestic servants, restaurant and entertainment workers who travel seasonally (and often illegally overstay their visas), but who do not gain entry under formal contractual agreements.

4. Business commuters are usually middle-class entrepreneurs, who travel within the region, or to Europe and North America on a regular basis, often several times a year.

Thomas-Hope's categorizations largely build upon her examination of Jamaican returnees and those of others who have analyzed that country's return migration flows (Nutter, 1986; Patterson, 1968: Taylor, 1976). It is more comprehensive and inclusive than its predecessors, but it is locationally and contextually limited, dealing mainly with Jamaican returnee experiences since the late 1940s through to the 1980s. Since that time, Jamaica's disappointing (and depressing) economic performance under the International Monetary Fund's Structural Adjustment Programmes, and the commensurate decline in living standards, has changed the island's situation quite dramatically. New forms of transnational networks have formed, return visiting – return migration strategies may very well have emerged as viable substitutes and evidence in the chapters of this collection points to much more variety in today's return strategies than in these earlier times. Her typology, however, remains a useful building board.

More recently, De Souza (1998) examined Trinidadian returnees and elaborated on the distinctions between returnees' determinants. He suggests there are four patterns of return: 1) seasonal, mobile livelihood circulation driven primarily by economic factors; 2) return visiting which categorizes repetitive visits of Caribbean emigrants; 3) long-term return migrants who are making the final return decision; and 4) the transnational movement of Caribbean circulators living 'between two worlds'. Segal (1998) has referred to these latter repetitive transnational movers as 'swallows'.

Interestingly, all of these categorical schemas omit one of Cerase's 1974 categories – namely *return of innovation*, which was referring to returnee 'brain gains' as opposed to emigré 'brain drains'. Cerase (1974) was, as it happens, referring to the returning 30 per cent, who left the United States after emigrating there during the period 1908 and 1957. But, its omission from extant categorical schemes for Caribbean return begs the question, why? Today's 'brain gains' are of considerable importance, especially since innovation, human skills and experiences are sought-for qualities. More needs to be detailed and conceptualized with regard to the growing volume and importance of transnational movers, their networks, their backward and forward linkages and the transfer of capital, information, culture and people that ebb and flow through them. Brain gains might counter brain drains, but of course, the relative authority and social power of these 'collectivities of human and social capital' need to be factored into any assessment of the consequential development outcomes (Olesen 2002). For now, the comparative frame of reference built upon extant literature is the goal.

Motivations for Return

Of course, there are unlikely to be single or simple reasons for return migration. After all, the migrant is overseas or abroad. She has, or has not, retained an intention to return. His experiences are influenced by nostalgic memories of the island home, by a range of critical assessments of his present situation, by comparisons between places, between societal milieus, between social networks, work experiences, among many other complexities. Her personal goals for emigration may have been met, not yet met, were unrealistic in the first place, or may have had to be drastically altered. His plans to return may have had to be changed because of changes in circumstances. Conditionality reigns!

Despite the above acknowledgement of motivational dissonance, an early review of Caribbean return migration and its underlying processes, abstracted the main reasons returnees offered into five generic domains (after Stinner, *et al.*, 1982). These domains included: 1) socioeconomic maladaptation; 2) life course transitions; 3) expiration, termination, or violation of contractual agreements; 4) 'homeland' linkages; and 5) societal socioeconomic situations. In addition to these five generic motivational determinants, a residual 'other' category contained epiphenomenal 'push' factors such as severe weather, natural disasters, health crises, family/personal-conflicts, run-ins with law, legal wrangling, even enforced deportation, and 'pull' factors such as the balmy weather, slow-life, ambience and extended family ties. Other factors include family home inheritance, inheritance of the family business, estate or real estate, and personal encounters, loves and marriages.

One of the authors, together with Bailey and Ellis, examined the reasons Puerto Rican women offered for undertaking circular migration between their island home and the US mainland, and documented 93 possible motivations (Bailey and Ellis, 1993; Conway, *et al.*, 1990; Ellis, *et al.*, 2001). Four broad categories differentiated these reasons – and comparisons between outward and return moves, and between first and second circuits, confirmed there were significant gendered differences in motivation. Labor market reasons which dominated outward moves, were supplanted by 'place utility' reasons for the return move. Tied moves, reflecting both patriarchal and family connections, were also significant categories among returning women. Proportionately, a lot of women circulators returned home either because they missed Puerto Rican life and culture or to improve their living environment, for themselves and their children. In total, approximately 90 per cent of the return moves of our sample of Puerto Rican women were made for reasons related directly to fulfilling gender roles, life course transitions and for improvements in place utility unrelated to employment and labor market opportunities (Ellis, *et al.*, 2001).

Bailey and Ellis (1993) pursued the issue further, by analyzing the relative importance of structural and behavioural factors in these Puerto Rican women's length of sojourn on the mainland and their return behaviours. Their findings were as follows: (1) single mothers leaving the island return sooner than other women; (2) women married on the mainland stay longer than women arriving already married; (3) women who give birth on the mainland are the least likely of any group of Puerto Rican-born women to return to the island, whereas women who return and give birth within six months will, however, have shorter mainland sojourns; (4) age influences the duration of sojourn – younger women return sooner, while older women have longer durations; and (5) college-educated women (upon arrival) have the speediest return behaviour. The overall conclusion of this collective work is that return behaviour and length of sojourns away from home are determined not so much by job or economic-structural factors as they are by the gendered nature of women's responsibilities, particularly family obligations at home on the island (Conway, *et al.*, 1990; Ellis, *et al.*, 2001). Leaving the island is very much tied to the male partner's migration calculus, in contrast, so this research definitely indicates the need for a gendered approach to understanding Caribbean return migrant behaviour, not only that of Puerto Rican women.

The Ideology of Return

From the days of the earliest 'off-the-island' sorties, many retained a 'return ideology' in which they had every intention of not permanently severing ties with their Caribbean homes (Philpott, 1973). Conceiving migration 'over there' – to Britain or North America, or to nearby more prosperous islands and more fruitful labor markets – as a temporary sojourn, was rationalized by harboring a 'return ideology', or a 'myth of return', in which the goals and objectives of the move were couched in terms of investing in a return, accumulating wealth for an eventual return, sending back remittances in anticipation of a return, and so on, as much as they were an international move to avail themselves of better employment opportunities. Common adherence to such a return ideology eventually became institutionalized as a multi-generational phenomenon within Caribbean societies and among all classes of the small island societies therein (Rubenstein, 1982).

Providing some useful insights for this 'ideology of return' among Caribbean migrants are two theoretical explanations of Caribbean migration behaviour; the first provided by Elizabeth Thomas-Hope (1992), the second by Mary Chamberlain (1995, 1997). Thomas-Hope's (1992) conceptual framework recognizes that the micro environments of localities and communities are the most important contexts which directly and indirectly influence a household's and individual's perception of, and ties to, 'home' and the 'home away from home'. Most importantly, Thomas-Hope focuses her lens on the role of individual perception and image, since the nature of these cognitive lenses represent the ways in which (Caribbean) individuals perceive their 'home' and 'away from home' situations and respond to the combined nexus of international, national and household circumstances – including constraints, expectations and opportunities – which impact on the individual.

Two theoretical observations are pertinent to our deliberations here:

1. People constitute complex information-processing systems, involving information-receipt and the translation of information into meaning which is determined, in the first place, by value, reference groups, aspirations and goals. The translation of information into meaning also bears relation to attitudes, including beliefs and the disposition toward certain action. Attitudes, often developed with reference to levels of risk assessment, take the image closer to the behavioral intention and thereby prompt the actual migration behaviour; whether it is emigration, circulation or return migration;

2. Emigration, circulation and return migration behaviours are determined by the (subjective) perceived environment, but are carried out in the (objective) actual environments at home and away. In addition, objective consequences result from such actions. Thus, there is cyclical feedback, whereby linkages occur and recur between the perceived and actual worlds. From this perspective, local, national, and transnational spaces are evaluated and take on significance in decision-making relating to emigration and return, the formation of return migration potential and the association of emigration and return migration objectives with other social, familial, economic and cultural objectives (Thomas-Hope, 1992:35-36).

Maintaining an 'intention to return', is part of the calculus of risk aversion, both helping to prompt the migration, as well as ensuring that the individual's 'strategic flexibility' in life-course decision-making is optimized (Carnegie, 1982). The embeddedness of 'migration and return', of 'remigration and return', and of 'return and remigration' experiences are therefore socio-cultural reinforcements of Caribbean peoples' vulnerabilities and livelihood options; an observation that finds common currency with Conway's (1988, 2004) categorizations of contemporary Caribbean mobility processes.

Mary Chamberlain's contribution to this explanatory construct is her insightful assessments of the role of family ties and family obligations in the maintenance of return ideologies among emigrating Barbadians/West Indians living in Britain (see also Philpott, 1973). Focusing on West Indian (Barbadian) migration to Britain and back, Chamberlain (1995: 255-256) notes that (t)he global dimension of migration, played out in international labor markets, and mediated by maneuvers of the host politics, engages with home-based social and cultural history which has furnished and continues to furnish Caribbean migrants with their own agenda. In this agenda, the family can be seen to play a significant role as both the end goal, and the means to achieve it'. Several important dimensions of the social process are suggested. 'First, the existence of a migration dynamic as a family dynamic which determined behavior and gave it meaning; second, the interplay between this migration dynamic and other family dynamics (such as colour), and family goals (such as social mobility); third, the importance of the family in approving and enabling migration; and fourth, an ethos reflecting and reproducing a broader culture of migration which perhaps ran parallel with, but did not necessarily conform to, the vagaries of international labor demands' (Chamberlain,1995: 256).

Furthermore, an increasing 'hybridity' in identity formation, and the development and adaptation of Caribbean/Black British cultural patterns and social formations, occurs as migrants go about their day-to-day lives. There appears to be a 'creolization' (Foner, 1998) of these Caribbean transnational communities that is best articulated in their material lives, within families, their workplaces, and their leisure spaces (Bailey, 2001). It is within these multi-local life-worlds that 'the points of similarity and difference, conformity and conflict are negotiated and resolved, where family values and cultural practices are transmitted, contested and transformed, and where national, transnational and/or hybrid identities evolve' (Chamberlain 1998: 8).

Maintaining an 'intention to return' is not incongruous to the 'creolization' of Caribbean peoples' identities away from home in Britain, or North America. 'Hybridity' and living between two worlds, or in transnational multi-local worlds, is built into the narratives of exile, and the family-traditions, family-tales and nostalgic reminiscences of island pasts that the first emigrant generation passes on to successive generations (Chamberlain, 1997). Affiliation of migrants with fellow nationals in sporting associations, neighborhood associations, marriage partnering, and a sharing of common interests and experiences, all revolve around the maintenance of ties to the family home and the intention to return 'someday' (Philpott, 1973; Rubenstein, 1979). Of course, the opposite occurs, with some emigrants deliberately avoiding fellow nationals, severing their ties to home, and eschewing their social obligations to dependents left behind, or to the wider extended family they intend to leave behind 'for good'. Social/segmented assimilation and absorption into their new metropolitan social milieu is, of course, an emigration option for many, but by no means all (Conway, 1988).

Return Migrants as Human Agents

We firmly believe that a focus on *return migrants as human agents* rather than return migration the process, the mass movement, or the numerical account, is the most appropriate viewpoint. Migration and circulation flows, population concentrations, rural depopulation exoduses, and transnational networks or diasporas in their spatial and temporal frameworks, might inform us of the complex character of the mobility patterns. However, examining their experiences and adjustments to their changing circumstances, assessing their accumulation of human capital and their contributions to social capital stocks, and examining their socio-spatial adjustments necessitates such a focus on people, not the process. In addition, their direct and indirect influences on local, regional or national development, on building community, creating transnational businesses, sustaining island enterprises and the rest, or their participation in communal efforts to achieve environmental sustainability, sustainable urbanism, or better quality lives for themselves and their families, requires us to investigate migrants (and non-migrants) – as individuals, in families and in communities – as human agents.

In a socially democratic world that we should hope continues to be the geo-political reality in the Caribbean, it is people's influential behaviors, after all, that determine societal decisions and public and private management practices, that influence the quality of residential environments, decisions on how remittances are disbursed, and the level of community activism that ultimately influences policy. Treating personal narratives as interactive texts (Miles and Crush, 1993) and listening more attentively to 'migration stories' (Lawson, 2000) will help us better understand the world's return migrants find themselves in, and better understand their daily, weekly and monthly experiences, their 'every-day material life', no less (Bailey, 2001).

A focus on return migrants' relations with their environments and their interactions and behaviors with their home societies is going to realize more informative answers to the questions we raise about how the complex interactions between migration and sustainable development play out in real geographic time and space (Conway, 2004). For example, it may well be the case that a relatively small number of return migrants, rather than the existence of a numerically large return migration flow, effectively mobilize their human and social capital to contribute to, and participate in, local development, which is environmentally friendly as well as sustainable (Conway, 1999/2000; Gmelch, 1992). A single returnee, or a return migrant family, may pave the way to changes in farm management and land use that instigate widespread environmental changes; both positive and negative impacts being possible (Gmelch, 1992). Or, return migrants may very well energize a community into activism to thwart an environmental threat to their homes (Berman Santana, 1996). Though unlikely to ever amount to a very large influx, those returning graduates and professionals who do come back are not only endowed with university and technical proficiencies, they still possess considerable local knowledge, education and expertise. Their repatriated human and social capital can make a difference, despite the inevitable recruitment leakage of this cohort to overseas opportunities – the 'brain drain'. For example, the recent flourishing of local NGOs in some Caribbean islands, and their efforts to undertake progressive, and successful, conservation and environmentally-sensitive initiatives, is prompted by the active involvement of a relatively small number of young professionally-trained returnees (Conway and Lorah, 1995). In addition, Gmelch (1987) in his Barbados study of return migrants' impacts on local community development found there were notable investments in small business ventures, and further that the private sector experiences of returnees were often progressive and productive in contrast to the re-assimilation-problems of returnees in public sector institutions.

Going beyond assigning significance to numerical majorities alone, we need to recognize the influence of groups of people, even a select set of individual returnees, on changes in today's Caribbean societies. We need to recognize the power and creativity of human agency of people; from the highly influential to the relatively powerless. At the scale of the local community and the family network, returnee influences can be significant, even if the impact at the national level is difficult to gauge. Focusing on such structure-agency relationships is important if we are seeking answers to questions about the complex interrelationships between return migration, transnationalism and sustainable development in 'home-places', or even 'away-places'. This re-focusing of attention on relatively small groups of returnees or even individuals, countermands, or at least qualifies, the recommendations made by Stinner *et al* (1982), who argued that the 'societal level consequences ... are dependent upon a number of factors including (1) the size of the return migration stream relative to the population base of the society; (2) the selectivity of the return migration stream; (3) societal receptivity; (4) the degree of solidarity among returnees; and (5) the prevalence of remigration' (Stinner, *et al.*, 1982: lviii).

Return Migration's Perceived Failings

Although return migration has been a subject of considerable interest in the recent past and there is a substantial literature on it (Bovenkerk, 1974; ICM, 1986; King, 1986; Kubat, 1984; United Nations Secretariat, 1986), only a few Caribbean societies have been the locus for research into the phenomena. West Indian returnees and their adaptation experiences received some attention in the late 1960s and early 1980s, but most of these studies concentrated either on returnee adjustment problems (Bovenkerk, 1981; Davidson, 1969; Patterson, 1968; Nutter, 1986; Taylor, 1976) or on the development implications of return migrants and retirees (Babcock and Conway, 2000; Conway and Glesne, 1986; Gmelch, 1987; Stinner, *et al.*, 1982; Thomas-Hope, 1985).

One common conceptualization of return migration viewed it as a relatively rapid response to incorrect information. Another stance viewed return migration as a failure to adapt to the new situation or to achieve advantages from the initial international move (Davidson, 1969; Dumon, 1986; Peach, 1968). Such returns home were to be expected in any long-distance movement, and accord with Ravenstein's (1885) long-held idea that every migration flow has its counter-flow. Alternatively, several European labor migration models included the assumption that migrants would return to their home countries on retirement, or when they ceased to contribute to the productive capacity of their destination (Castles and Kosack, 1985; Castles and Miller, 2003; Petras, 1980a). A more recent variant adds the proviso that when the inequalities of opportunity between the richer destination countries and the poorer countries of origin decrease, then the outflow lessens, and return movement increases (Stalker, 2000). Also, skilled workers were expected to behave differently to the unskilled, and the emigration of the more educated led to charges that international migration's main problem was that it caused a 'brain drain' of much-needed human capital. Indeed, a preoccupation with this 'brain drain' problem prompted much of the early research on international migration from less developed countries to the industrialized nations of Europe and North America (ICM, 1986; UNITAR, 1971; United Nations Secretariat, 1986).

The 'brain-drain' problem accompanying extra-regional emigration was also viewed as a major policy concern for Caribbean development, and several commentators addressed the issue in terms of how island governments might provide incentives to 're-drain the brains', or attract emigrants back home (Bascom,1990; Henry, 1990; ICM, 1986; North and Whitehead, 1990). This interest was, in part, stimulated by the growing significance of remittances from abroad, but it was also prompted by the shortages of skilled labor at home in the health and services sectors, brought about by increases in labor recruitment from abroad (Henry, 1990; Strachan, 1980, 1983; UNECLAC, 2003). Not withstanding these sallies into the domain of international migration, and its consequences for small island societies, by far the most interest to date has focused on returnees who have either returned to the Caribbean on retirement or during late middle-age (Thomas-Hope, 1985, 1992; Gmelch, 1985a, 1985b, 1985c, 1987; Byron and Condon, 1996; Nutter, 1986).

With respect to much of the developing world and southern Europe, return migration's impacts on housing and agricultural landscapes in national home-lands have been evaluated in terms of flows of remittances or repatriated gifts in kind (see King, 1986). While abroad, migrants remit large amounts of capital back to their countries of origin. Indeed, one of the major reasons migrants give for leaving home is to work hard in a foreign labor market to make a certain target sum, which could be remitted home (Conway and Glesne, 1986; Durand, *et al.*, 1996; Russell, 1986). However, some social commentators have argued that these remittances do not always serve positive ends (Brana-Shute and Brana-Shute, 1982; Rubenstein, 1983a, 1983b). Thus, it has been claimed that these monies from overseas have generally served to fuel conspicuous consumption, especially of imported goods, elaborate and luxury housing, along with investment in what some have referred to as marginally productive enterprises, such as shops and taxis (Lowenthal and Clarke, 1982; Richardson, 1975). The argument has been forwarded, therefore, that remittance investments have not always served to rejuvenate the rural economic sector. Indeed, in many circumstances, it has led to the ownership of land by those living overseas, and thereby, the creation of what are referred to as 'idle lands' (see Potter and Welch, 1996; Brierley, 1985). This situation has given rise to the suggestion that small Caribbean countries have become 'remittance-dependent' and that returnees are not significant agents of change (see Bovenkerk, 1981; Brana-Shute and Brana-Shute, 1982; Richardson, 1975; Stinner, *et al.*, 1982).

In fact, these arguments run parallel with those concerning the social impact of returning nationals themselves. Thus, local populations frequently regard returning nationals as interlopers who have 'had it all too easy abroad'. Such negative reactions to returnees have to be viewed in the light of a situation where Caribbean migrants faced a tough time in metropolitan societies, having to overcome racism and poverty in their efforts to start new lives (Essed, 1990; James, 1994). Notably, most West Indians who migrated to Britain in the late 1950s and early 1960s formed a replacement population, which undertook jobs that the bulk of the resident population chose to avoid (Peach, 1967, 1968; Brooks, 1975). In addition, many Black-British (West Indians) appear to have not done as well in the United Kingdom, when compared with the ultimate career placements of other compatriots who moved to North America (Foner, 1978, 1979, 1997a). In the latter case, active recruitment of educated West Indians in the public health and service sectors of New York City in the 1960s and 1970s meant that Jamaican, Barbadian, Trinidadian and Guyanese women embarked on career paths at higher salaries than their British equivalents who had emigrated to their mother country in the 1950s and early 1960s (Conway and Cooke, 1996; Conway and Walcott, 1999; Kasinitz, 1988, 1992; Kasintz and Vickerman, 2001).

In Toronto, West Indians faced similar exclusive practices and racism to that experienced in London (and elsewhere in England) and New York. Here too the emigrant experiences were bi-polar – the middle classes achieving some success and stability in public sector employment, while the lower classes gravitated into the domestic services and manual trades (Henry 1994). As migrants and newcomers and as a response to their common experiences of host-society racism, 'dual-allegiances' and the maintenance of dual/multiple identities among West Indians whether in London, New York City or Toronto was widespread (Vickerman, 2001; Waters, 1994). Clearly, this 'transnational trait' was well developed among West Indian families and communities in these metropolitan societies by the 1950s and 1960s (Chamberlain, 2001; Conway, 2000; Foner, 1997b; Sutton, 1992; Sutton and Mackiesky, 1975).

Despite the fact that elderly retirees are coming to the end of their working lives, Gmelch (1980, 1987, 1992) and Conway (1985, 1993) have argued convincingly that returning nationals and remittances should be looked upon in a much more favorable light than hitherto.[1] As such phenomena have long been an integral part of the social and economic fabric of the region, they argue that returnees have in the past, and are likely in the future, to play extremely significant roles in the region's development. For example, such groups are able to contribute wider experience and skills, developed over a lifetime of work. Many of them are parents and grandparents who add to the human and social capital of the nation in a wide variety of ways, and add to the reciprocal linkages which serve to bind members of the society together. And in addition, they have wider recreational and other skills and experiences from which island societies can potentially benefit (Conway, 1999/2000; Potter and Phillips, 2002).

In recent times, dramatic increases in the flow of remittances 'back home' have been observed globally, as well as in the Caribbean. This upsurge has lead many to not only re-evaluate their consumptive or productive roles, but also to treat such return flows of capital as potential investment pools (Oroczo, 2000, 2002). Obviously, return migrants as donors, and their families as dependent recipients of remittances, factor into any assessment of remittances' persistence, their investment portfolios and influences on family budgets, life-styles, and investment decisions. And, migrants who return while still in their active years can pool their remittance investments, using them as another source of finance capital, or even venture capital, if accumulation beyond immediate expenditures allows such micro-business practice (Connell and Conway, 2000; Conway and Cohen, 2003). Suffice to conclude that remittances are now viewed in a much more positive light than in times past.

New Roles for Return Migrants in Multi-local, Transnational Networks

Today, transnational networks are linking North American and European enclave communities and local 'home' communities together in a complex set of micro-economic and non-economic relationships at the family/household and individual levels (Bailey, 2001; Chamberlain, 1998). People, capital, information, even cultural ideas are circulating globally, and the consequences of these long- and short-term movements are important for island societies in this contemporary era of regional crisis accompanying global restructuring (Conway, 1999/2000; Watson, 1994, 1995). In addition, such international circuits are no longer bi-polar; more are developing as multi-local networks of movement (Fog Olwig, 2001), and the consequences are bringing about *new migrant experiences* (Smith, 1999; Vertovek and Cohen, 1999).

Not only are more and more Caribbean migrant families building transnational networks, but also the nature of the participants' material lives is undergoing transnational re-structuring, as well as re-adaptation and dynamic change. Migrant and non-migrant lives are being negotiated, organized (and disorganized) in multi-local transnational spaces and life-worlds, in which today's return migrants are quite a different group than those Thomas-Hope (1985, 1986) identified, and we noted earlier. Migrant and non-migrant behaviours in the contemporary Caribbean, in overseas metropolitan enclaves and among diaspora sub-populations, and in today's globalizing world, are being conditioned by the social, cultural, economic and technological changes underway, as were earlier behaviours determined and conditioned by their colonial and post-colonial contexts of yesteryear. Given the rapidly changing contexts of today, it is to be expected that contemporary migrants, as first, second and third generations, will respond differently to their different circumstances, different opportunity sets, different obstacles and barriers and the new sets of constraints they face at home and abroad, within their transnational networks, within and between transnational communities in which local power-dynamics present new challenges.

We should expect to find new migrant experiences, new cohorts of emigrants, new cohorts of circulatory migrants, new cohorts of return migrants, and also find that migrants and non-migrants alike are facing new circumstances as we enter the 21st century. Strategic flexibility, risk minimization, impromptu livelihood strategies, opportunistic 'satisficing' behaviour, individualism, familial-bonds, gender relations, are all likely parts of the Caribbean migrant's calculus in adapting and accommodating to today's ever-changing, but inherently unequal, world (not to mention its 'vulnerable' and 'dangerous' dimensions). Past lessons, inter-generational transfers of knowledge, know-how and expertise, family and community support structures – local, national and transnational – migrant family cultural traditions and strengths, will factor in to the calculus as continuing conditional influences, but new migrant types will emerge out of the traditional cohorts, and new adaptive strategies and new migrant experiences will logically follow.

It should not be surprising, then, that there has been a renewed interest in return migration from Britain to the Caribbean, in large part because the cohort of Black British who left the Caribbean in the 1950s and 1960s are now reaching retirement age and returning (Abenaty 2001; Byron, 2000; Byron and Condon, 1996; UNECLAC, 1998). In addition, there is a renewed interest in the social and economic impacts of return migrants, in part because international circulation and short-term movements appear on the increase, and in part because island governments have begun to reconsider the important effects these more transitory movers appear to be having on the growth of self-employed businesses and the tourism and service-sector industries (Brown, 1997; Conway, 1999/2000; Chevannes and Rickets, 1997; de Souza, 1998; Potter, 2003; Potter and Phillips, 2002; Thomas-Hope, 1999).

Return migrants are best viewed as people endowed with social capital, potential and realized, in which social and cultural knowledge as well as economic human capital comprise a significant bundle of re-investments (Conway, 1985, 1994, Conway and Cohen, 1998, 2003). 'Return of innovation', and return of the skilled, the experienced and professionally seasoned, is especially important for Caribbean communities, where emigration and brain drains have taken their toll on middle class cohesion, professional life in the service sectors, and the whole gamut of societal contributions – political party involvement, community services, education and vocational contributions and the like. Remittances invested, or spent, by return migrants are flows and stocks of financial capital that add to the bundle that is repatriated (Connell and Conway, 2000). The *quality* of the social and human capital returned, is likely to be of equal importance to the quantity, unless of course, the numbers are too small relative to the resident population, so that the effects of such a miniscule number of returnees are swamped by the accumulated deficits in human and social capital that mass emigration has caused.

On the other hand, the accumulated impacts of fixed capital investments of such relatively small numbers of returnees is unlikely to create inflationary pressures on land or housing markets, although it might very well be a societal charge leveled at their potentially harmful influence. If we are to ascertain the potential of these new cohorts of return migrant as agents of change and development, we need to ascertain the transnational character and gendered differences of experience, as well as their transnational roles and consequences. In particular, the consequences of return for the 'real-life' economics of Caribbean people need closer examination (Ekins and Max-Neef, 1992). Specifically, what role(s) do today's returning nationals play? What are their adaptation experiences? What social and economic roles are they undertaking? How do the experiences of returning young men and women differ? How are these returnees' transnational networks contributing to their adjustments, changing their situations, making them more mobile or more settled in their island homes?

Structure of the Collection

The chapters that follow this introductory review of the literature on return migration and its changing nature focus more specifically on the new migration experiences of Caribbean returnees in a range of different settings, islands and communities. It would be difficult to provide comprehensive regional coverage because the Caribbean is so diverse, multi-cultural and heterogeneous (see Boswell and Conway 1992: Potter *et al.*, 2004). Instead, we have gathered together a set of case studies and interrogations of return migration experiences in which the island and cultural contexts are divergent enough for comparative insights to be possible.

Most of the contributions started as papers in two panels convened at the Annual Meeting of the Association of American Geographers held in New Orleans in March 2003, with others invited subsequently. The particular return migrant populations under our lens differ from returning retirees who commonly have been the focus of earlier research. We do not duplicate that effort, but rather examine the adaptation experiences of younger and youthful cohorts who are choosing to return to the island homes of their parents, or the island homes they left as children. Our emphasis is on the current, contemporary experiences of returnees, who have undertaken their return to the Caribbean in the 1980s, 1990s and the first half-decade of the 21st century. We should expect some substantiation of previous generalizations made about return migrants' experiences, but we should also look for new experiences, and new transnational behaviors.

Providing a region-wide comparative sample of research studies, the chapters focus on topics such as the socio-cultural adjustments faced by returning transnational migrants, the degree to which they are 'othered' (treated as outsiders, or newcomers) and the gendered-character of their experiences. The consequences for the 'real-life' economics of Caribbean people are also examined. Specifically, what economic and social roles are returnees undertaking? How easy has return been in the work place as compared to adjustments to island society? Do the experiences of returning men and women differ, and in what ways might that be a consequence of gendered divisions in island society? How are the transnational networks of these returnees contributing to their adjustment experiences? Are they changing their situations, making them more mobile, more unwilling to settle permanently, or more favorably disposed to settle back home? Is there evidence that transnational business incubation is underway, that return is part of the entrepreneurial strategy? How are the island communities responding to returnees? How are the returnees responding to difficulties in adjustment?

These are a selection of questions, the authors address in their chapters, and we should not expect regional conformity, or common answers. We expect diversity, but seek generalizations, or at least comparative experiences which highlight the significant features of today's more youthful returning cohorts, who, to date, have not been the focus of attention in the extant literature.

Note

1. This more positive assessment of the effect of remittances on home communities in the Caribbean is now much more widely accepted, in large part because research on Mexican 'migradollar' remittance flows and their impacts by Massey and associated (Durand, *et al.*, 1996; Massey, 1988), and similar comparative research in remittances investments in Central American, have also concluded that the positive aspects of remittances outweigh the problems (Funkhauser, 1992; Lopex and Seligson, 1991).

References

Abenaty, F. (2001), 'The Dynamics of Return Migration to St. Lucia', in H. Goulbourne and M. Chamberlain (eds), *Caribbean Families in Britain and the Trans-Atlantic World*, London and Oxford: Macmillan Press, Caribbean Studies, 170-187.

Babcock, E.C. and D. Conway (2000), 'Why international migration has important consequences for the development of Belize', *Yearbook, Conference of Latin Americanist Geographers, 2000*, Austin, TX: University of Texas Press and Conference of Latin Americanist Geographers, 26:71-86.

Bailey, A.J. (2001), 'Turning transnational: Notes on the theorisation of international migration', *International Journal of Population Geography*, 7: 413- 428.

Bailey, A.J. and M. Ellis (1993), 'Going home: The migration of Puerto Rican-born women from the United States to Puerto Rico', *The Professional Geographer*, 45(2):148-158.

Bascom, W.O. (1990), 'Remittance inflows and economic development in selected Anglophone Caribbean countries', Commission for the Study of International Migration and Co-operative Economic Development, Working Paper, No. 58. Washington, D.C. July.

Berman Santana, D. (1996), *Kicking off the Bootstraps: Environment, Development and Community Power in Puerto Rico*, Tucson, AZ: University of Arizona Press.

Boswell, T.D. and D. Conway (1992), *The Caribbean Islands: Endless Geographical Variety*, New Brunswick, NJ: Rutgers University Press, August.

Bovenkerk, F. (1974) *The Sociology of return migration: A bibliographic essay,* The Hague: Martinus Nijhoff.

Bovenkerk, F. (1981), 'Why returnees generally do not turn out to be "agents of change": the case of Suriname', *Nieuwe West Indische Gids*, 55, 154-173.

Brana-Shute, R. and Brana-Shute, G. (1982), 'The magnitude and impact of remittances in the eastern Caribbean: a research note', in W.F. Stinner, K. de Albuquerque and R. Bryce-Laporte (eds), *Return Migration and Remittances: Developing a Caribbean Perspective*, Washington DC: The Smithsonian Institute, Research Institute on Immigration and Ethnic Studies, Occasional Paper no 3, 267-289.

Brierley, J. (1985), 'A review of development strategies and programmes of the Peoples' Revolutionary Government of Grenada, 1979-1983', *Geographical Journal*, 151: 40-52.

Brooks, D. (1975), *Race and labour in London Transport*, Oxford: Oxford University Press.

Brown, D.A. (1997), 'Workforce losses and return migration to the Caribbean', in P.R. Pessar (ed), *Caribbean Circuits: New Directions in the Study of Caribbean Migration*, Staten Island, New York: Center for Migration Studies, 197-223.

Bryce-Laporte, R.S. (1982), 'Preface' in W.F. Stinner, K. de Albuquerque and R.S. Bryce-Laporte (eds), *Return Migration and Remittances: Developing a Caribbean Perspective* (Washington, D.C.: The Smithsonian Institute, Research Institute on Immigration and Ethnic Studies, Occasional Paper, No. 3), ix-xxix.

Byron, M. (1994), *Post-War Caribbean migration to Britain: The unfinished cycle*, Aldershot, Hong Kong, Singapore, Sydney: Avebury.

Byron, M. (1999), 'The Caribbean-born population in 1990s Britain: Who will return?', *Journal of Ethnic and Migration Studies*, 25(2):285-301.

Byron, M. (2000), 'Return migration to the Eastern Caribbean: Comparative experiences and policy implications', *Social and Economic Studies*, 49(4)155-188.

Byron, M. and Condon, S. (1996), 'A comparative study of Caribbean return migrants from Britain and France: towards a context-dependent explanation', *Transactions of the Institute of British Geographers*, New Series, 21, 91-104.

Carnegie, C.V. (1982), 'Strategic flexibility in the West Indies', *Caribbean Review*, 11(1): 11-13, 54.

Castles, S. and G. Kosack (1985), *Immigrant Workers and Class Structure in Western Europe*, Second Edition, Oxford: Oxford University Press.

Castles, S. and M.J. Miller (2003), *The Age of Migration: International Population Movements in the Modern World*, 3rd *Edition*. New York: Guildford.

Cerase, F.P. (1974), 'Migration and social change: Expectations and reality: A case study of return migration from the United States to Italy', *International Migration Review*, 8: 245-262.

Chamberlain, M. (1995), 'Family narratives and migration dynamics: Barbadians in Britain', *Nieuwe West-Indische Gids*, 69, 253-275.

Chamberlain, M. (1997), *Narratives of exile and return*, London and Basingstoke: Macmillan Press, Caribbean Studies.

Chamberlain, M. (1998), *Caribbean Migration: Globalised Identities*, London and New York: Routledge.

Chamberlain, M. (2001), 'Migration, the Caribbean and the family', in H. Goulbourne and M. Chamberlain (eds) *Caribbean Families in Britain and the Trans-Atlantic World*, London and Oxford: Macmillan Caribbean Studies, 32-47.

Chevannes, B. and H. Ricketts (1997), 'Return Migration and Small Business Development in Jamaica', in P.R. Pessar (ed), *Caribbean Circuits: New Directions in the Study of Caribbean Migration*, Staten Island, NY: Center for Migration Studies, 161-196.

Connell, J. and D. Conway (2000), 'Migration and Remittances in Island Microstates: A Comparative Perspective on the South Pacific and the Caribbean', *International Journal of Urban and Regional Research*, 24.1: 52-78.

Conway, D. (1985), 'Remittance impacts on development in the eastern Caribbean', *Bulletin of Eastern Caribbean Affairs*, 11, 31-40.

Conway, D. (1988), 'Conceptualizing contemporary patterns of Caribbean international mobility', *Caribbean Geography* (2):145-163.

Conway, D. (1989a), 'Caribbean international mobility traditions', *Boletin Latino Americanos y del Caribe*, 46(2): 17-47.

Conway, D. (1993), 'Rethinking the consequences of remittances for eastern Caribbean Development', *Caribbean Geography*, 4:116-130.

Conway, D. (1994), 'The complexity of Caribbean migration', *Caribbean Affairs*, 7: 96-119.

Conway, D. (1997), 'Why Barbados has exported people: International mobility as a fundamental force in the creation of small island society', in J. Manuel Carrion (ed), *Ethnicity, Race and Nationality in the Caribbean*, Rio Pedras, Puerto Rico: University of Puerto Rico, Institute of Caribbean Studies, 274-308.

Conway, D. (1999/2000), 'The Importance of Migration for Caribbean Development', *Global Development Studies*, Winter 1999-Spring 2000, 2(1-2): 73-105.

Conway, D. (2000), 'Notions unbounded: A critical (re)read of transnationalism suggests that U.S.-Caribbean circuits tell the story better', in Biko Agozino (ed), *Theoretical and Methodological Issues in Migration Research: Interdisciplinary, Intergenerational and International Perspectives*, Ashgate Publishers, Aldershot, UK, and Brookfield, USA, 203-226.

Conway, D. (2002), 'Gettin' there, despite the odds: Caribbean migration to the U.S. in the 1990s', *Journal of Eastern Caribbean Studies*, 27(4): 100-134.

Conway, D. (2004), 'On being part of Population Geography's future: Population-environment relations and inter-science initiatives', *Population, Space and Place*, 10(4): 295-302. Online http://www3.interscience.wiley.com/cgi-bin/jhome/.

Conway, D. and J.H. Cohen (1998), 'Consequences of Migration and Remittances for Mexican Transnational Communities', *Economic Geography*, 74(1): 26-44.

Conway, D. and J.H. Cohen (2003), 'Local dynamics in multi-local, transnational spaces of rural Mexico: Oaxacan experiences', *International Journal of Population Geography*, 9(1): 141-161.

Conway, D. and T.J. Cooke (1996), 'New York City: Caribbean immigration and residential segregation in a restructured global city', in J. O'Loughlin and J. Friedrichs (eds), *Social polarization in post-Industrial metropolises*, Berlin and New York: de Gruyter and Aldine Press, 235-258.

Conway, D. and C. Glesne (1986), 'Rural livelihood, return migration and remittances in St. Vincent', *CLAG 1986 Yearbook, Volume 12*, Muncie, in: Conference of Latin Americanist Geographers, Ball State University, 3-11.

Conway, D. and P. Lorah (1995), 'Environmental Protection Policies in Caribbean Small Islands: Some St. Lucian Examples', *Caribbean Geography* 6(1): 16-27.

Conway, D. and S. Walcott (1999), 'Gendered Caribbean and Latin American Employment Experiences in New York City', *WADABAGEI*, 2(1):53-112.

Conway, D., Ellis, M. and N. Shiwdhan (1990), 'Caribbean international circulation: Are Puerto Rican women tied-circulators?', *Geoforum*, 21(1): 51-66.

Davison, R.B. (1966), *Black British: Immigrants to England*, London: the Institute of Race Relations and Oxford University Press.

Davidson, B. (1969), '"No Place Back Home": A Study of Jamaicans Returning to Kingston, Jamaica', *Race*, 9(4): 499-509.

De Souza, R-M (1998), 'The spell of the Cascadura: West Indian return migration', in Klak T. (ed), *Globalisation and Neoliberalism: the Caribbean Context*, London: Rowman and Littlefield, 227-253.

Dumon, W. (1986), 'Problems faced by migrants and their family members, particularly second generation migrants in returning to and reintegrating into their countries of origin', *International Migration*, 24(1): 113-128.

Durand, J., Parrado, E.A. and D.S. Massey (1996), 'Migradollars and development: A reconsideration of the Mexican case', *International Migration Review*, 30(2):423-444.

Duval, D.T. (2002), 'The return visit – return migration connection', in C.M. Hall and A.M. Williams (eds), *Tourism and Migration*, Netherlands: Kluwer Academic Publishers, 257-276.

Ekins, P. and M. Max-Neef (1992), *Real-Life Economics: Understanding Wealth Creation*, London and New York: Routledge.

Ellis, M., D. Conway and A.J. Bailey (2001), 'The circular migration of Puerto Rican women: Towards a gendered explanation', in K. Willis and B. Yeoh (eds), *Gender and Migration,* The International Library of Studies on Migration, Volume 10, Edward Elgar, Northampton, Mass, 119-150.

Essed, P. (1990), *Everyday Racism: Reports from Women in Two Cultures*, Claremont: Hunter House.

Fog, Olwig K. (1998), 'Constructing lives: Migration narratives and life stories among Nevisians', in M. Chamberlain (ed), *Caribbean Migration*, London: Routledge, 63-80.

Fog, Olwig K. (2001), 'New York as a locality in a global family network', in N. Foner (ed), *Islands in the City: West Indian Migration to New York*, Berkeley, Los Angeles and London: University of California Press, 142-160.

Foner, N. (1978), *Jamaica farewell: Jamaican migrants in London*, Berkeley and Los Angeles: University of California Press.

Foner, N. (1979), 'West Indians in New York City and London: A comparative analysis', *International Migration Review*, 13(2): 284-97.

Foner, N. (1997a), 'The immigrant family: Cultural legacies and cultural changes', *International Migration Review*, 31(4): 961-974.

Foner, N. (1997b), 'What's new about Transnationalism? New York immigrants today and at the turn of the century', *Diaspora*, 6(3): 355-375.

Foner, N. (1998), 'West Indian identity in the Diaspora: Comparative and historical perspectives', *Latin American Perspectives*, 25(3): 173-188.

Foner, N. (2001), *Islands in the City: West Indian migration to New York*, Berkeley, Los Angeles and London: University of California Press.

Funkhauser, E. (1992), 'Mass emigration, remittances, and economic adjustment: The case of El Salvador in the 1980s', in G. Borjas and R. Freeman (eds), *Immigration and the Work Force: Economic Consequences for the United States and Source Areas*, Chicago: University of Chicago Press, 135-175.

Gmelch, G. (1980), 'Return migration', *Annals, Review of Anthropology*, 9: 135-159.

Gmelch, G. (1985a), 'Emigrants who come back, Part I', *Bajan*, April 1985, 8-9.

Gmelch, G. (1985b), 'Emigrants who return, Part II', *Bajan*, May-June 1985, 4-5.

Grnelch, G. (1985c), 'The impact of return migration, Part III', *Bajan*, July-August, 4-6.

Gmelch, G. (1987), 'Work, innovation and investment: The impact of return migrants in Barbados', *Human Organisation*, 46: 131-140.

Gmelch, G. (1992), *Double Passage: the Lives of Caribbean Migrants and Back Home*, Ann Arbor: University of Michigan Press.

Goulbourne, H. (2001), 'The Socio-political Context of Caribbean Families in the Atlantic World', and 'Trans-Atlantic Caribbean Futures', in H. Goulbourne and M. Chamberlain (eds) *Caribbean Families in Britain and the Trans-Atlantic World*, London and Oxford: Macmillan Press, Caribbean Studies, 12-31 and 170-187.

Goulbourne, H. and M. Chamberlain (2001) *Caribbean Families in Britain and the Trans-Atlantic World*, London and Oxford: Macmillan Press, Caribbean Studies.

Henry, F. (1994) *The Caribbean Diaspora in Toronto: Learning to Live with Racism*, Toronto, Buffalo and London: University of Toronto Press.

Henry, RM. (1990), *A Reinterpretation of Labor Services of the Commonwealth Caribbean*, Working Paper, No. 61, Commission for the Study of International Migration and Co-operative Economic Development, Washington, D.C., July.

Ho, C.T.G. (1999), 'Caribbean Transnationalism as a Gendered Process', *Latin American Perspectives*, 26(5): 34-55.

ICM (1986), 'Seventh seminar on adaptation and integration of migrants: Economic and social aspects of voluntary return migration', Special Issue of *International Migration*, 24(1).

James, W. (1994), 'Migration, racism and identity formation: The Caribbean experience in Britain', in W. James and C. Harris (eds), *Inside Babylon: The Caribbean Diaspora in Britain*, London and New York: Verso, 231-287.

Kasinitz, P. (1988), 'From Ghetto Elite to Service Sector: A Comparison of the Role of Two Waves of West Indian Immigrants in New York City', *Ethnic Groups*, 7:173-203.

Kasinitz, P. (1992), *Caribbean New York: Black Immigrants and the Politics of Race*, Ithaca and London: Cornell University Press.

Kasinitz, P. and M. Vickerman (2001), 'Ethnic Niches and Racial Traps: Jamaicans in the New York Regional Economy', in H.R. Cordero-Guzman, R.C. Smith and R. Grosfuguel (eds), *Migration, Transnationalization, and Race in a Changing New York*, Philadelphia: Temple University Press, 191-211.

Keely, C. (1986), 'Return of talent programs: Rationale and evaluation criteria for programs to ameliorate a "Brain Drain"', *International Migration*, 24(1): 179-189.

King, R.L. (1977), 'Problems of return migration: Case study of Italians returning from Britain', *Tidschrift voor Economische Geografie,* 68(4): 241-245.

King, R. (1986), *Return migration and regional economic problems*, London, Sydney, Dover, New Hampshire: Croom Helm.

King, R. (2000), 'Generalizations from the History of Return Migration', in B. Gosh (ed), *Return Migration: Journey of Hope or Despair*, Geneva, Switzerland: International Organization for Migration and the United Nations, 7-56.

Kubat, D. (1984), *The politics of return: International return migration in Europe*, Staten Island, NY and Rome: Center for Migration Studies.

Lawson, V.A. (2000), 'Arguments within geographies of movement: The theoretical potential of migrants' stories', *Progress in Human Geography*, 24: 173-189.

Lopez, J.R. and M.E. Seligson (1991), 'Small business development in El Salvador: The impact of remittances', in S. Diaz-Briquets and S. Weintraub (eds), *Migration, Remittances and Small Business Development: Mexico and Caribbean Basin Countries*, Boulder: Westview, 175-206.

Lowenthal, D. and Clarke, C. (1982), 'Caribbean small island sovereignty: chimera or convention?', in O. Fanger (ed), *Problems of Caribbean Development*, Munich: WF Verlag.

Massey, D.S. (1988), 'Economic development and international migration in comparative perspective', *Population and Development Review*, 14(3):383-413.

Miles, M. and J. Crush (1993), 'Personal narratives as interactive texts: Collecting and interpreting migrant life-histories', *Professional Geographer*, 45(1): 95-129.

North, D.S. and J.A. Whitehead (1990), 'Policy Recommendations for Improving the Utilization of Emigrant Resources in Eastern Caribbean Nations', Washington, D.C.: Commission for the Study of International Migration and Co-operative Economic Development, Working Paper, No. 25.

Nutter, R. (1986), 'Implications of return migration from the United Kingdom for urban employment in Kingston, Jamaica', in R. King (ed), *Return migration and regional development problems*, London: Croom Helm, 198-212.

Olesen, H. (2002), 'Migration, Return and Development: An Institutional Perspective', *International Migration*, 40(5): 125-150.

Orozco, M. (2000), *Remittances and Markets: New Players and Practices.* Washington, D.C.: Inter-American Dialogue and the Tomas Rivera Policy Institute.

Orozco, M. (2002), 'Globalization and Migration: The Impact of Family Remittances in Latin America', *Latin American Politics and Society*, 44(2): 41-67.

Patterson, H.O. (1968), 'West Indian migrants returning home: Some observations', *Race*, 10(1): 69-77.

Patterson, O. (1978), 'Migration in Caribbean societies: Socioeconomic and symbolic resource', in W.H. McNeill and R.S. Adams (eds), *Human migration: Patterns and policies*, Bloomington, in: Indiana University Press, 106-145.

Peach, C. (1967), 'West Indians as a replacement population in England and Wales', *Social and Economic Studies*, 16, 289-294.

Peach, C. (1968), *West Indian Migration to Britain: A Social Geography*, Oxford University Press for the Institute of Race Relations.

Peach, C. (1984), 'The force of West Indian island identity in Britain', in C. Clarke, D. Lay and C. Peach (eds) *Geography and Ethnic Pluralism*, London: George Allen and Unwin, 214-230.

Petras, E. (1980a), 'The role of national boundaries in a cross-national labour market', *International Journal of Urban and Regional Research*, 4(2): 157-195.

Petras, E. (1980b), 'Towards a theory of international migration: The new division of labor', in R.S. Bryce-Laporte and D. Mortimor (eds), *Sourcebook on the New Immigration*, Brunswick, NJ: Transaction Books, 439-449.

Phillips, J. and R.B. Potter (2003), 'Social Dynamics of "Foreign-born" and "Young" returning nationals to the Caribbean: A review of the literature', University of Reading, Department of Geography: Geographical Paper, No. 167.

Philpott, S. (1968), 'Remittance obligation, social networks and choice among Montserratian Migrants in Britain', *MAN*, 3(1): 465-476.

Philpott, S. (1973), *West Indian Migration: The Montserrat Case*, London: Athlone Press.

Philpott, S. (1977), 'The Montserratians: Migration dependency and maintenance of island ties in England', in J.L. Watson (ed), *Between two cultures: Migrants and minorities in Britain*, Oxford: Basil Blackwell, 90-119.

Portes, A. and L.E. Guarnizo (1991), 'Tropical capitalists: U.S.-bound immigration and small-enterprise development in the Dominican Republic', in S. Diaz-Briquets and S. Weintraub (eds), *Migration, remittances and small business development: Mexico and Caribbean Basin countries*, Boulder: Westview, 101-131.

Potter, R. B. (2003), '"Foreign-born" and "Young" returning nationals to Barbados: A pilot study', University of Reading, Department of Geography: Geographical Paper, No. 166.

Potter, R.B. and J. Phillips (2002), 'The Social Dynamics of Young and Foreign-born Returning Nationals to the Caribbean: Outline of a Research Project', *Centre for Developing Areas Research Paper*, No. 37, 30 pp.

Potter, R.B. and Welch, B. (1996), 'Indigenization and development in the eastern Caribbean: reflections on culture, diet and agriculture', *Caribbean Week*, 8, 13-14.

Potter, RB., D. Barker, D. Conway and T. Klak (2004), *The Contemporary Caribbean*, London: Pearson Education/Prentice Hall.

Rhoades, R. (1978), 'Intra-European return migration and rural development: Lessons from the Spanish case', *Human Organization*, 37(2): 136-147.

Pryor, R.J. (1976), 'Conceptualizing Migration Behavior: A Problem in Micro-Demographic Analysis', in L.A. Kosinski and J.W. Webb (eds), *Population at Micro Scale*, Hamilton, New Zealand: IGU Commission on Population Geography and New Zealand Geographical Society, 105-119.

Richardson, B.C. (1975), 'The overdevelopment of Carriacou', *Geographical Review*, 65, 290-299.

Richardson, B.C. (1983), *Caribbean migrants: Environment and human survival on St. Kitts and Nevis*, Knoxville, TN: University of Tennessee Press.

Richardson, B.C. (1985), *Panana Money in Barbados, 1900-1920*, Knoxville TN: University of Tennessee Press.

Richmond, A. (1968), 'Return migration from Canada to Britain', *Population Studies*, 22: 263-271.

Rubenstein, H. (1979), 'The return ideology in West Indian migration', in R.E. Rhoades (ed), *The Anthropology of Return Migration*, Papers in Anthropology, 20: 330-337.

Rubenstein, H. (1982), 'The Impact of Remittances in the Rural English Speaking Caribbean: Notes on the Literature', in Stinner et al. (eds), *Return Migration and Remittances: Developing a Caribbean Perspective*, Washington D.C.: Smithsonian Institute, Research Institute on Immigration and Ethnic Studies, Occasional Papers No.3, 237-266.

Rubenstein, H. (1983a), 'Remittances and rural underdevelopment in the English-speaking Caribbean', *Human Organization*, 42(4): 295-306.

Rubenstein, H. (1983b), 'Migration and underdevelopment: The Caribbean', *Cultural Survival Quarterly*, 7(4): 30-32.

Russell, S.S. (1986), 'Remittances from International Migration: A Review in Perspective', *World Development*, 14(6): 677-696.

Segal, A. (1998), 'The Political Economy of Contemporary Migration', in T. Klak (ed), *Globalization and Neoliberalism: the Caribbean Context*, Lanham, Maryland: Rowan & Littlefield, 211-226.

Smith, R. (1999), 'Reflections on Migration, the State and the Construction, Durability and Newness of Transnational Life', in L. Pries (ed), *Migration and Transnational Social Spaces*, Aldershot UK, Brookfield USA, Singapore and Sydney, Australia: Ashgate, 187-219.

Stalker, P. (2000) *Workers without frontiers: The impact of globalization on international migration*, Boulder: Lynne Rienner.

Stinner, W.F., de Albuquerque, K. and Bryce-Laporte, R.S. (1983), *Return migration and remittances: Developing a Caribbean perspective*, Washington DC: Smithsonian Institute, Research Institute on Immigration and Ethnic Studies, Occasional paper, no 3.

Strachan, A.J. (1980), 'Government sponsored return migration to Guyana', *Area,* 12(2): 165-169.

Strachan, A.J. (1983), 'Return migration to Guyana', *Social and Economic Studies*, 32(3): 121-142.

Sutton, C.R. (1992), 'Some Thoughts on Gendering and Internationalizing our Thinking about Transnational Migrations', in *Towards a Transnational Perspective on Migration: Race, Class, Ethnicity, and Nationalism*, Annals of the New York Academy of Sciences, Volume 645, edited by N. Glick Schiller, L. Basch and C. Blanc-Szanton. New York: The New York Academy of Sciences, 241-249.

Sutton, C. and S.R. Makiesky (1975), 'Migration and West Indian Racial and Ethnic Consciousness', in H.I. Safa and B. du Toit (eds), *Migration and Development: Implications for Ethnic Identity and Political Conflict*, The Hague, Netherlands: Mouton Press, 113-144.

Swyngedouw, E. (1997), 'Neither Global or Local: 'Glocalization' and the Politics of Scale', in K.R. Cox (ed) *Spaces of Globalization*, New York: Guildford, 137-166.

Taylor, E. (1976), 'The social adjustment of returned migrants to Jamaica', in F. Henry (ed), *Ethnicity in the Americas*, The Hague: Mouton, 213-230.

Thomas-Hope, E.M. (1985), 'Return migration and its implications for Caribbean development', in Pastor R.A. (ed), *Migration and Development in the Caribbean: the Unexpected Connection*, Westview Press: Boulder Colorado, 157-177.

Thomas-Hope, E.M. (1986), 'Transients and settlers: Varieties of Caribbean migrants and the socio-economic implications of their return', *International Migration (Geneva)*, 24(3): 559-571.

Thomas-Hope, E.M. (1988), 'Caribbean skilled international migration and the transnational household', *Geoforum*, 19(4): 423-432.

Thomas-Hope, E.M. (1992), *Explanation in Caribbean migration: Perception and the image: Jamaica, Barbados and St Vincent*, Basingstoke: Macmillan Caribbean.

Thomas-Hope, E.M. (1998), 'Globalization and the development of a Caribbean migration culture', in M. Chamberlain (ed), *Caribbean migration: Globalised identities*, London and New York: Routledge, 188-202.

Thomas-Hope, E.M. (1999), 'Return migration to Jamaica and its development potential', *International Migration*, 37(1): 183-205.

Thomas-Hope, E.M. and R.D. Nutter (1989), 'Occupation and status in the ideology of Caribbean return', in R. Appleyard (ed), *The impact of international migration on developing countries*, Paris: OECD, 287-300.

UNECLAC (1998), *A Study of Return Migration to the Organization of Eastern Caribbean States (OECS) Territories and the British Virgin Islands in the Closing Years of the Twentieth Century: Implications for Social Policy*, Port of Spain, Trinidad and Tobago: UN Economic Commission for Latin America and the Caribbean, LC/CAR/G.550, December 1998.

UNECLAC (2003), *Emigration of nurses from the Caribbean: Causes and consequences for the socio-economic welfare of the country – Trinidad and Tobago – A case study*, Port of Spain, Trinidad and Tobago: UN Economic Commission for Latin America and the Caribbean, LC/CAR/G.748, 12 August 2003.

UNITAR (1971), *The Brain Drain from five developing countries*, New York: UN Institute for Training and Research.

United Nations Secretariat (1986), 'The meaning, modalities and consequences of return migration', *International Migration (Geneva)*, 24(1): 77-93.

Vertovek, S. and R. Cohen (1999), *Migration, Diasporas and Transnationalism*. Cheltenham, UK, Northampton, MA, USA: An Elgar Reference Collection – The International Library of Studies on Migration.

Vickerman, M. (2001), 'Tweaking the Monolith: West Indian Immigrant Encounter with "Blackness"', in N. Foner (ed), *Islands in the City: West Indian Migration to New York*, Berkeley, Los Angeles and London: University of California Press, 237-256.

Waters, M.C. (1994), 'Ethnic and Racial Identities of Second-Generation Black Immigrants in New York City', *International Migration Review*, 28(4): 795-820.

Watson, H.A. (1994), *The Caribbean in the Global Political Economy*, Boulder & London: Lynne Reiner.

Watson, H.A. (1995), 'Global Powershift and the Techno-Paradigm Shift: the End of Geography, World Market Blocs, and the Caribbean', in M. Aponte Garcia and

C. Gautier Mayoral (eds), *Postintegration Development in the Caribbean*, Rio Pedras, Puerto Rico: Social Science Research Center, 74-146.

Wolpert, J. (1965), 'Behavioral Aspects of the Decision to Migrate', *Papers of the Regional Science Association*, 15: 159-169.

Chapter 2

The Socio-Demographic Characteristics of Second Generation Return Migrants to St Lucia and Barbados

Robert B. Potter

Migration and Caribbean Development

Migrations have been a fundamental correlate of Caribbean social and economic development and change since the age of discovery (Marshall, 1982; Conway, 1994). Indeed, some writers have referred to the 'uprootedness' of Caribbean peoples. This expression is helpful only in so far as it serves to stress the fact that migrations, both great and small, permanent and transitory, have been an integral component of Caribbean social and economic change throughout time. However, any suggestion that Caribbean migration can somehow be characterised as uncoordinated and essentially chaotic, has to be firmly rejected at the outset.

The creation of the contemporary Caribbean has been premised on the largest enforced migration, that of black West African slaves by Europeans. On the abolition of the slave trade and emancipation in 1834, this inhumane movement, reckoned to have accounted for 6-10 million people, was followed by the migration of indentured labourers, in particular, those from India. Subsequently, in the nineteenth and early twentieth centuries, this movement was followed by the strong intra-regional migration of West Indians, most conspicuously in relation to the building of the Panama Canal between 1879 and 1914 (Richardson, 1985, 1992; Newton, 1984).

In the twentieth century, there has been an almost non-stop movement of Caribbean denizens to major metropolitan regions in North America and Europe, although due to various immigration acts, the former has become much more important than the latter. As the twenty-first century has opened, the mass movements to the 'North' in the immediate post war period have been complemented by the return migration of a proportion of those who made the journey from the Caribbean to London, Birmingham, Reading, New York, Boston and elsewhere.

Thus, as Conway (1998) has observed, since the incorporation of the small island nations of the Caribbean into the external spheres of influence of European mercantilism in the sixteenth century, the region has experienced successive waves of immigration, emigration and circulation.   As implied above, the range of migratory movements has been broad and varied, extending from enforced permanent movements to transient or shuttle migrations.   Such has been their diversity, they have consisted of short-term, circular and recurrent migration.

This chapter deals with a relatively new migration stream to the Caribbean region, which up to now remains entirely uncharted and unstudied.   This is the movement of comparatively young returning nationals to the Caribbean drawn from the progeny of the twentieth century migrants who came to Europe, North America and elsewhere, in the wake of the Second World War.   But before examining the characteristics of this new cohort of migrants, the better-known and studied phenomenon of retiree return migration to the Caribbean is briefly reviewed.

Return Migration to the Caribbean

There has been much interest in that cohort of men and women who first migrated from the Caribbean to the United Kingdom in search of gainful employment and the chance to better their lives in the 1950s and 1960s. By far the most interest has focused on that sub-group of the displaced population that has either returned to the Caribbean on reaching retirement or during late middle-age. This situation has been true of the mass-media coverage of such *Return Migrants*, or *Returning Nationals*, as they are now frequently known. In addition, a mounting, although by no means extensive volume of academic research since the late 1970s has also focused on older return migrants to the Caribbean (see, for example, Thomas-Hope, 1985, 1992; Gmelch, 1985a, 1985b, 1985c, 1987; Byron and Condon, 1996; Nutter, 1986; Byron, 2000).

With respect to much of the developing world and southern Europe, an extensive literature has developed documenting return migration to national homelands and its impacts on housing and agricultural landscapes (see King, 1986). The salient point is that whilst abroad, such nationals have remitted large amounts of capital back to their countries of origin. However, some writers have argued that these remittances do not always serve positive ends. Thus, it has been argued that these monies from overseas have generally served to fuel conspicuous consumption, especially of imported goods, elaborate and luxury housing, along with investment in what some have referred to as marginally productive enterprises, such as shops, bars and taxis (Gmelch, 1980; Gmelch, 1985c; Lowenthal and Clarke, 1982; Richardson 1975).

The argument has been advanced, therefore, that such capital has not always served to rejuvenate the rural economic sector. Indeed, in many circumstances, it has led to the ownership of land by those living overseas, and thereby, the creation of what are referred to as 'idle lands' (see Potter and Welch, 1996; Brierley, 1985). This situation has prompted the suggestion that some small Caribbean countries have become 'remittance-dependent' and that returnees cannot be regarded as significant agents of change (see Bovenkerk, 1981; Brana-Shute and Brana-Shute, 1982; Stinner *et al.*, 1983). However, in the Caribbean context, Gmelch (1980;

1987; 1992), De Souza (1998) and Conway (1993) have disputed this argument, and have written strongly in favour of the developmental efficacy of remittances from overseas. In this connection, it is salient to emphasise that when settled back into the country of their birth, such returnees continue to bring in substantial amounts of money in the form of pension payments and investment income. For example, in the case of Barbados, it was estimated by the Central Bank that in 1996 remittances from overseas and returned nationals amounted to $US 62.5 millions. For the same year, the estimated revenue lost as the result of the duty free importation of the household effects of returnees was just in excess of $Bds 9 millions (FURN, Ministry of Foreign Affairs, ca 1997).

In fact, these arguments run parallel with those concerning the social impact of returning nationals themselves. Thus, local populations frequently regard returning nationals as interlopers who have 'had it all too easy abroad'. Such feelings of resentment and suspicion have been well-documented by Gmelch (1980) in the case of returning nationals as a whole. Such negative reactions to returnees have to be viewed in the light of a situation where Caribbean migrants faced a tough time in metropolitan societies, having to overcome poverty and racism in their efforts to earn a better living. As is well-documented, most West Indians formed a replacement population, that is they frequently undertook jobs that the bulk of the indigenous UK population was largely unprepared to tackle (Peach, 1967, 1968; Brooks, 1975). In addition, there is evidence that West Indians have not fared as well in the United Kingdom, when compared with the ultimate career placements of essentially the same population that moved to North America.

One of the issues which is considered as contributing to resentment against returning nationals is that, as alluded to above, returnees are eligible to bring their household contents, including cars and consumer durables, into the country without having to pay import duties (Ministry of Foreign Affairs, Barbados, 1996). Further, there is a view that returnees are relatively well-off, and this has led to accusations that they are routinely over-charged by traders and craftspeople.

Despite such problems, and notwithstanding the fact that elderly retirees are now reaching the end of their working lives, Gmelch (1980, 1987, 1992) and Conway (1985, 1993) have taken up the argument that returning nationals and remittances should be viewed in a much more favourable light than hitherto. As such phenomena have long been an integral part of the social and economic fabric of the region, they argue that returnees have in the past, and are likely in the future, to play extremely significant roles in the region's development. This is in addition to the considerable amounts of foreign currency that is being brought into the country. For example, such groups should be able to contribute wider experience and skills, developed over a lifetime of work and domestic experience.

Young Returning Nationals to the Caribbean: An Uncharted Path

As explained, all of these studies have primarily focused on returning nationals of retirement age. On research trips to the eastern Caribbean in the late 1990s the author met a growing number of '*Young Returning Migrants*' to the Caribbean, most of whom could be described as '*Foreign-Born Returning Nationals*' (see below for a precise definition). These are second and third generation West Indians, that is those born in the United Kingdom, United States, Canada or elsewhere of first generation West Indian immigrants, who for a variety of reasons, have decided to 'return to' the countries that they themselves had not come from, but from which their parents have.

However, this cohort of migrants has never been the focus of a specific study, and although aware of the existence of such migrants, the respective Ministries of Foreign Affairs and High Commissions in London do not have precise details as to the number of such migrants, their migration histories, their employment and wider socio-economic characteristics. As noted over twenty years ago, while nations collect precise statistics concerning 'aliens', the same seldom applies for returning citizens (Rhoades, 1979; Gmelch, 1980).

An exploratory research project, the first of its kind, was therefore carried out, from 1999 to 2000 by the author. The aim was to provide a first analysis of the nature of this new movement, its duration and socio-economic impacts.   In particular, by means of open-ended semi-structured interviews, the project sought to investigate the experiences and attitudes of such young returnees to Barbados and St Lucia. As such, the project had important policy implications, including the possible contribution of young returnees to social and human capital and the skills base of small developing nations. The project sought to establish whether this new cohort of young returnees is regarded in a different light to the better-known category of older returnees. What motivates them, and why are they prepared to leave the country of their birth? How well have they adjusted to their new homes and what have they faced in so doing?

'Tales of Two Societies': The Research Design

The principal target group for the study was what the present research refers to as *Foreign-Born Returning Nationals*, those who were born in the United Kingdom (or the United States etc) and who have decided to make Barbados or St Lucia their home. They might also be referred to as 'British Barbadians', 'American Barbadians' or 'British St Lucian' etc. All those who have a Caribbean parent can claim nationality by descent. Another group are those who were born in the Caribbean, but who later travelled to the United Kingdom (or elsewhere) with their parents. If after ten years or more they return to live in the Caribbean, they also qualify as returning nationals. In the present research, members of this group are described as *Young Returning Nationals*.

The research was carried out from October 1999 to February 2000, with the assistance of a grant from the British Academy and focused on both Barbados and St Lucia. Formal discussions were held with senior civil servants at the *Facilitation Unit for Returning Nationals* (FURN), within the Ministry of Foreign Affairs in Barbados, and the Ministry of Foreign Affairs, St Lucia. As already noted, the overall aim of the pilot project was to provide the first ever examination of this cohort of foreign-born and young returning nationals. The project sought to understand their motives for migrating, and their experiences on migration. What do they see as the advantages of such a move, and what adjustments do they feel that they have had to make? Do they feel that they have started to be assimilated within society?

Using background information provided by the respective Ministries, and snowballing contacts thereafter, 40 semi-structured interviews were carried out with foreign-born and young returning nationals – 25 in Barbados and 15 in St Lucia. Virtually all the informants were contacted by telephone in the initial instance. Where it was the wish of the informant, they were interviewed over the telephone. Reflecting the issues that such returnees feel they face, a number preferred the relative anonymity provided by this approach. However, where informants were happy to be interviewed face to face, this is what was done.

The Socio-Economic and Demographic Characteristics of the Young Return Migrants

Barbados

As shown in Table 2.1, of those interviewed in Barbados, the majority were females, 21 out of 25. Indeed, one of the four male informants specifically commented that he saw his move as temporary, and that he believed most males of West Indian origin would seek work in major metropolitan cities such as New York, Toronto or London. This suggests that Barbados is seen as a more appropriate migration destination for females. The majority of those interviewed were foreign-born, with 11 hailing from the UK, and one each from the USA, Hong Kong and Singapore (Table 2.1). Two informants had been born in other Caribbean territories (St Vincent and Trinidad), but had been brought up in the UK/USA. There were nine 'young' returning nationals who, having been born in Barbados, had been brought up overseas. Most of these had left Barbados before starting primary education. The majority had parents who were Barbadian (42 out of 50, see Table 2.1).

The foreign-born and young returnees interviewed ranged from 23 to 45 years old. Their mean age at interview was 33.72 years, with a median of 35, and clearly, the returnees were second generation West Indians. Reflecting their average age in their early 30s, ten had children. Further, the sample was split almost equally between those who had partners and those that did not (Table 2.1). Of those with partners, seven said their partners had been born in the UK, and four were married to Bajans.

The median age at which the informants had returned to Barbados was 30, with a mean of 29.65 years. At the time of interview, they had lived in Barbados from 4 months to 16 years. The modal duration of residence in Barbados was three years, with a median of three years and a mean of 3.97 years. Several informants said that they considered three years to be a critical period, during which time young return migrant had either acclimatised to living in Barbados or had decided to 're-return' to the UK or USA. The influence of first generation parents who had returned to Barbados attracting their offspring is implied by the fact that 23 out the total of 50 parents of the young returnees were living in Barbados at the time of interview (Table 2.1). Eight of the parents were still living in the UK, six in the USA and two in Canada.

Table 2.1   Basic socio-economic and demographic characteristics of the young return migrants to Barbados

(i) Sex:

Female	21	Male	4

(ii) Place of Birth:

United Kingdom	11	USA	1
Barbados	9	Hong Kong	1
Other Caribbean	2	Singapore	1

(iii) Parent's Place of Birth:

Barbados	42	United Kingdom	2
St Vincent	2	Jamaica	1
USA	2	Hong Kong	1

(iv) Age of Young Returning Nationals:

Range 23-45 years			
Median	35 yrs	In their 20s	7
Mean	33.72 yrs	In their 30s	13
		In their 40s	5

(v) Age at Returning to Barbados:

Range 15-39 years			
Median	30 yrs	In their teens	1
Mean	29.65 yrs	In their 20s	10
		In their 30s	12
		In their 40s	2

(vi) Period Lived in Barbados since Return:

Range 0.4-16 years			
Median	3 yrs	Over 10 years	3
Mean	3.97 yrs	4-6 years	3
		2-3 yrs	7
		1-2 yrs	3
		0-1 yr	6
		Not stated	3

Table 2.1 (continued)

(vii) Country of Domicile of Parents:

Barbados	23	Canada	2
United Kingdom	8	No data	8
United States	6	Dead	3

(viii) Family Status of Young and Foreign-Born Returnees:

With partner	12	Partner from UK	7
Without partner	13	Partner from Barbados	4
		Not stated	1

No children	15
1 child	8
2 children	2

Source: Author's survey 1999-2000.

The occupations of the returnees to Barbados at the time the interviews were conducted are listed in Table 2.3a. The majority were in employment, with only three stating that they were unemployed and one female describing herself as a 'housewife'. A high proportion of jobs requiring skills, training and qualifications are featured among the female respondents. Several of the informants in their 30s strongly emphasized the need to be well-qualified, preferably to degree level. They also spoke of younger returnees, those in their 20s without qualifications, who had 're-returned' to the UK. On the other hand, three of the four male informants were in manual occupations in the service sector (car mechanic, barman, waiter). The predominance of young female returnees to Barbados was clear, as was the professional nature of many of their occupations.

St Lucia

In contrast to Barbados, of those interviewed in St Lucia, the majority were males, amounting to 11 out of 15 (Table 2.2). The majority of those interviewed were foreign-born, with seven having been born in the UK, and one in Canada (Table 2.2). Six informants had been born in St Lucia and one in another Caribbean territory, but had subsequently been brought up in the UK. The majority of returnees had parents who were St Lucian, specifically amounting to 27 out of 30 biological parents (Table 2.2). One parent came from Guyana, another from Dominica, and one from China.

The foreign-born and young returnees ranged from 27 to 51 years of age at the time of interview. Their mean age was 38.93 years, with a median of 39 years. Reflecting the average age in their early thirties, eight stated they had children. Further, the sample was divided almost equally between those who had partners and those that did not (Table 2.2). Of those with partners, four said their partners had been born in St Lucia, and one was married to a UK national.

Table 2.2 Basic socio-economic and demographic characteristics of the young return migrants to St Lucia

(i) Sex:

| Female | 4 | Male | 11 |

(ii) Place of Birth:

| United Kingdom | 7 | Other Caribbean | 1 |
| St Lucia | 6 | Canada | 1 |

(iii) Parent's Place of Birth:

| St Lucia | 27 | Guyana | 1 |
| Dominica | 1 | China | 1 |

(iv) Age of Young Returning Nationals:

Range 27-51 years
Median	39 yrs	In their 20s	1
Mean	38.93 yrs	In their 30s	8
		In their 40s	4
		In their 50s	2

(v) Age on Returning to St Lucia:

Range 24-46 years
Median	35.50 yrs	In their 20s	10
Mean	34.93 yrs	In their 30s	12
		In their 40s	3

(vi) Period Lived in St Lucia since Return:

Range 0.17-10 years
Median	3.5 yrs	Over 8 years	4
Mean	3.94 yrs	4-5 years	3
		3-4 yrs	1
		2-3 yrs	1
		1-2 yrs	2
		Less than 1 year	4

Table 2.2 (continued)

(vii) Country of Domicile of Parents:

St Lucia	12	Deceased	2
United Kingdom	7	No data	9

(viii) Family Status of Young and Foreign-Born Returnees:

With partner	7	Partner from UK	1
Without partner	7	Partner from Barbados	4
Divorced	1	Not stated	2

No children	5
1 child	4
2 children	4
No data	2

Source: Author's survey 2000.

The median age at which the informants had 'returned to' St Lucia was 35.5 years, with a mean age of 34.93 years. At the time of interview, they had lived in St Lucia from less than a month to 10 years. The median duration was 3.5 years, with a mean of 3.94 years. The influence of first generation parents who had returned to St Lucia in attracting their offspring is again implied by the fact that 12 out the total of 30 parents of the young returnees were living in St Lucia at the time of the interview (Table 2.2). Seven of the parents were still living in the UK, two were deceased and data were not to hand for nine other interviewees.

The occupations of the returnees at the time the interviews were conducted are listed in Table 2.3b. As in the case of Barbados, the majority were in employment, with only one being unemployed. Also, most of the males interviewed were in relatively skilled jobs, including a civil engineer, an assistant manager of a hotel, self-employed businessmen, car sales executives and the head of watersports at a leading hotel. However, two were working as part-time barmen or waiters. The occupations of the four females interviewed are also listed in Table 2.3b, and ranged from a secretary/personal assistant to a Manager in the public sector.

Thus, the salient difference was that female young returnees dominated in Barbados, whilst males were most frequent in St Lucia. In terms of commonalities, although the returnees to St Lucia were on average a few years older, both groups were in their early 30s on migration and a majority of their parents had also returned to their Caribbean island of origin, suggesting the importance of family networks to migration. In both St Lucia and Barbados, at the time of interview, those interviewed had retuned just over three years previously.

Table 2.3 The occupations of the young return migrants by sex

(a) Barbados:

Females:

Legal secretary
Personnel Officer
Solicitor
Secretary
Psychiatric nurse
Secretary
Computer network administrator
Guest House manager
Secretary
Accounts manager
Qualified Chef
Unemployed
Airline ground staff
Housewife
Unemployed
Airline marketing manager
Pharmacist (currently unemployed)
Restaurant manager
Psychologist
Graphic designer
Not stated

Males:

Car mechanic
Barman at a club
Waiter
Fashion business

(b) St Lucia

Females:

Public sector Manager
Currently Unemployed
Secretary/Personal Assistant
Hotel Office Manager

Males:

Self-employed Businessman
Assistant Manager of a Hotel
Part-time barman
Civil Engineer
Barman/waiter
Self-employed Businessman
Head of watersports at a hotel
Car sales Executive
Car sales Executive
Office Manager
Self-employed businessman

Reasons for Leaving the Metropole and for Migrating to the Caribbean

Barbados

When asked why they had decided to migrate, relatively few major reasons were given by the informants, as shown in Table 2.4a. Indeed, three informants specifically noted that they had encountered no real problems, commenting:

'There were no negative reasons why I left the UK'.

Such comments suggest that most of the migrants left the UK more due to 'pull' than 'push' factors. This is also borne out by the fact that several said that they had reached a natural watershed in their lives and just needed a change, (for example,

Table 2.4 Reasons stated as to why the young return migrants had decided to leave the UK/USA or Canada

Reason stated	Number of citations
(a) Barbados	
No real problem	3
UK educational system bad for black people	3
'The UK race thing'/Barriers to black people	2
Wanted a change	2
Overlooked for promotion/'glass ceiling' at work	2
Made redundant and given a lump sum	1
On the dole/no job	1
'End of a long period of Conservative government'	1
Mundane way of life	1
Bad relationship	1
'I am not a city girl'	1
No commitments in the UK	1
To get away from the cold	1
To get away from constantly changing fashions	1
To get away from crime in the USA	1
(b) St Lucia	
No real negative reasons for leaving	4
Things bad for black people/racism	2
Never felt at home in the UK/US etc	1
Son's education	1
Wanted a change of lifestyle	1

'no commitments in the UK', 'the end of a bad relationship') (Table 2.4a). Of those that did mention a push factor, the view that the UK educational system is bad for black children, especially boys, was cited by three informants. One commented directly that:

> 'Black boys don't do well in the UK'.

Other push factors included what was described as 'the UK race thing/barriers to black people', being overlooked for promotion, and reaching a 'glass ceiling' at work. Following the Stephen Lawrence case in the UK, it was anticipated that issues of institutional racism might feature prominently as reasons for migrating, but in the event, the informants rarely emphasized such issues – rather where mentioned, this was seen as a common, almost taken-for-granted aspect of everyday life.

Turning to the 'pull' factors, the main reason stated as to why the returnees had decided to migrate to Barbados was that family and parents were living there (cited by ten informants), followed by statements to the effect that 'it's home, it is where you belong' (two informants) (Table 2.5a). The second most cited factor was the existence of better opportunities in terms of land, housing and living standards. Three others stated that they wanted to start their own businesses and that this is a more realisable goal in Barbados than in the UK or USA. Several informants stressed that although wages are much lower in Barbados, that it offered more potential for the future.

Several interviewees maintained that those with skills find it relatively easy to find a job in Barbados and to keep it. This was borne out by the fact that several informants explained that they had either migrated to try things out or had travelled to Barbados on a temporary basis and had stayed because they had found a job before their planned return date. Climatic factors were cited by four of the informants as a reason for migrating, with one respondent arguing that:

> 'The climate is in your blood...'.

Barbados prides itself on having an excellent educational system, with an official literacy rate of 97 per cent. Although the veracity of this statistic may be questioned (see Potter and Dann, 1987, for example), there is no doubting that there are some very good secondary schools in Barbados. Thus, not surprisingly, three of the informants emphasized that the education of their children/the quality of schools had been important factors in their decision to settle in Barbados.

Other motivating factors included the relatively slow pace of life, what was seen as an overall attractive life style, and the perception that Barbados is a relatively safe place to live. Another issue mentioned was the view that Barbados has recently become a very entrepreneurial society. It was also felt that in Barbados race is a far less important factor in terms of obtaining loans and finance than is the case in the UK.

St Lucia

As in the survey of Barbadian returnees, when asked why they had decided to migrate to St Lucia, the young returning nationals cited relatively few negative or 'push-oriented' reasons for their move (Table 2.4b). Indeed, four informants specifically noted that they could not cite specific negative reasons for leaving, implying that they were attracted to St Lucia by positive factors. One informant commented:

> 'No real negative reasons. I did not really experience racism in the UK. In fact, I miss England'.

However, two respondents were noticeably more equivocal, arguing that things are bad for black people in the UK, and they specifically mentioned issues of racism. Both of these respondents were males and both cited the activities of the National Front (NF):

> 'I had few direct problems. I am English, black English. I encountered racism. From every group – from the NF to "mods" in Croydon'.

> 'Things were pretty shitty for blacks when I grew up, due to the NF and the like'.

Another informant was more ambivalent, reporting that whilst he was relatively sanguine about issues of racism, he never felt totally comfortable living in the UK:

> 'I never felt fully at home in the UK. But racism was not the real issue. You sometimes got annoyed – you know, a tool used against you'.

The only other push factor was mentioned by a mother who emphasized the need to provide an appropriate education for her son. Another respondent observed that they wanted a change in lifestyle. Table 2.5b lists the migratory 'pull' factors cited by the young returnees to St Lucia. Clearly, the desire to start a small business was a critical motivation for many, along with the feeling that in St Lucia, this might be an achievable goal. This was cited by seven of the respondents and thereby came out in prime position. At the very least, the general feeling seemed to be that there was the chance to better things in the future. As might be expected, this was very similar to the Barbados sample.

Having family in St Lucia (seven informants), plus the climate (five informants) were the next most important factors cited as having promoted migration to St Lucia. But noticeably, a strong accent was also placed on enhancing the overall quality of life experienced ('to improve quality/way of life', open/out-doors way of life', 'beauty of the island', 'slower pace of life') (Table 2.5b). Typical comments made by the young returnees included:

> 'Life here is more relaxed – there's no rat race'.

Table 2.5 Reasons stated as to why the young return migrants had decided to live in the Caribbean

Reason stated	Number of citations

(a) Barbados

Family home/parents or partner there	10
Better opportunities (land, housing etc)	6
The weather/climate	4
Education of children/quality of schools	3
Ease of getting a job and keeping it	3
Wanted to start a business	3
'It's home/home is where you belong'	2
Slower pace of life/easy going place	2
Much safer/better security	2
Quality of life/attractive lifestyle	2
Availability of other part-time sources of income	1
Wages are lower but you can afford things	1
Infrastructure is good	1
Raised on romantic stories of the Caribbean	1
'Surfing, especially at Bathsheba'	1
Loans easier here for black people	1
Country very entrepreneurial for black people	1
'Men stand by their responsibilities to children'	1
Friendly people	1
More space/more like a rural area	1

(b) St Lucia

Start up a business/potential for small businesses	7
Family here/due to family circumstances	6
Weather/climate	5
To improve quality/way of life	5
Open way of life/outdoors way of life	2
Beauty of the island	1
Slower pace of life	1
Better education for son	1
Purchased land	1

The salient point highlighted in both Tables 2.4b and 2.5b is that in general, the informants cited relatively few push factors in explaining their migration, although where such factors were cited, they reflected issues related to racism. The overwhelming influence was attributed to the possibilities of starting up a business, and in the case of St Lucia, this was cited more frequently as a factor influencing relocation than the fact that parents and other family members were living on the island. For the returnees to Barbados, the fact that parents and other family were living in Barbados was mentioned more frequently.

Adjustments to Living in the Caribbean Region

Barbados

The major adjustments the young return migrants to Barbados felt they had to make are summarized in Table 2.6. Naturally, having already explained the reasons why they had decided to live in Barbados, overall reactions related to problems of adjustment. However, notably, a number of more positive reactions were cited. Three interviewees reported that they were generally satisfied, one mentioned the strong focus on education in Barbados, while another pinpointed the existence of genuine communities.

One issue, however, was mentioned by the majority of foreign-born and young returnees to Barbados, and this was difficulties in making new friends. This issue had a strong gender component. Thus, it was cited exclusively by female informants, and involved difficulties in making friends with Bajan women, but generally not Bajan men. Sexual and workplace competition were given as the reasons for such difficulties. Issues relating to accent and language, principally speaking with an English (as opposed to an American) accent, were the second most frequently mentioned. Several of the informants said that speaking with an English accent in a social context represented a problem, with this often being the subject of mockery. This factor was followed by the slow/relaxed pace of life (7 informants), feeling like an outsider (6), poor shopping/high prices (5), and aspects of a 'culture shock' (5). Although only mentioned by a few interviewees, the Americanisation of Bajan society, English people being regarded as 'mad', and perceived aspects of 'racism within society', were topics that were stressed forcefully by a small number of informants. Similarly, feelings of resentment, the unregulated 'macho' behaviour of bosses and having to accept things as they are without criticism, were also cited by more than two respondents.

Strong feelings were expressed by the informants in relation to many of these areas of adjustment. Indeed, in an early review of the literature on return migration, Gmelch (1980) noted that even retiree returnees who were originally raised in a country are ill prepared for return and suffer what he referred to as a 'reverse culture shock'. How much more of a problem is this for those who have never lived in a country, but have only visited it and seen it through their parents experiences in the past?

Table 2.6 The adjustments the young returnees to Barbados felt that they had to make

	Adjustment	Frequency of citation
1	Difficulties in making new friends/fixed friendship patterns	19
2	Issues relating to accent/language	8
3	Slow pace of life/relaxed way of life	7
4	Feeling like an outsider	6
5	Poor shopping opportunities/high prices	5
6	Aspects of a 'culture shock'	5
7	Americanisation of the society	4
8	English people being regarded as 'mad'	4
9	Racism in society	4
10	Feelings of resentment ('you have had it easy')	3
11	The 'macho' behaviour of bosses	3
12	Having to accept things as they are/not being able to change things	3
13	Generally satisfied	3
14	Relatively low salaries	2
15	Class-based society/colour-class system	2
16	Being expected to change at once to Bajan ways	2
17	Overcharging in shops and market stalls	2
18	Peoples' mindset and attitudes	2
19	Not making comparisons with the UK/USA etc	2
20	Poor application at work	1
21	Being cut off from world news/parochialism	1
22	Lack of respect for each other	1
23	Lack of provision for female sport (eg football)	1
24	'Sharp tongues'	1
25	Strong focus on education by parents	1
26	Money society/materialistic society	1
27	Loss of Black/Caribbean solidarity	1
28	Accepting mentality	1
29	Bajans not inviting you into their homes	1
30	'You can't change things'	1
31	Need to be polite at all times	1
32	The existence of genuine communities	1
33	Lack of confidentiality	1
34	Red tape and bureaucracy	1
35	Missing the seasons	1

Source: Author's survey, 1999-2000.

St Lucia

In the case of the young return migrants to St Lucia, the range of adjustments cited is shown in Table 2.7. Again, having already explained the push factors that had led them to migrate, overall reactions related primarily to problems of adjustment. However, several positive reactions were cited. These included three informants who commented on the slow pace and laid back way of life in St Lucia. In addition, one informant mentioned adjusting to the weather and another specifically mentioned that he experienced 'less hassle' from the police in St Lucia (Table 2.7).

Table 2.7 The adjustments the young returnees to St Lucia felt that they had to make

Adjustment	Frequency of citation
1 Colour-class system/racism in society	5
2 Difficulty of making 'local' friends	4
3 laid back way of life/slower pace of life	3
4 Competition with locals/resentment	3
5 Accent makes you stand out/gives an advantage	2
6 Narrowmindedness of the people	2
7 Poor quality of education	2
8 Work ethic being different	2
9 Being seen as an outsider/not St Lucian	1
10 Low wages	1
11 High land prices	1
12 Expensive place to live	1
13 Earnings lower	1
14 Need to moderate views	1
15 Large gender differences	1
16 The weather	1
17 People assuming you have money	1
18 Lack of a middle class	1
19 Difficulties in relationships with local women	1
20 Who you know in society being so important	1
21 Less hassle from the police in St Lucia	1
22 Americanisation of society	1
23 Dependency culture	1
24 Problem of having an English accent	1
25 Bosses not liking you exercising your rights	1
26 Need to speak patois to be fully accepted	1

Source: Author's survey, 2000.

The most frequently cited adjustment among the young returnees to St Lucia was the operation of the colour-class system, associated with issues of racism in society. The difficulty experienced in making friendships with local St Lucians was the second most frequently cited issue of adjustment, followed by the 'laid back way of life/slow pace of life'. Problems relating to competition with members of the local indigenous population, along with issues of resentment, were also cited.

Interestingly, the argument that an English or American accent marks returnees out was specifically addressed by two of the informants. However, this was referred to in a basically positive manner, suggesting that in the context of St Lucia, possessing such an accent bestowed specific advantages. This was in direct contrast to the case in Barbados, where having an English accent was frequently seen as a major problem in the field of social relations, although an accent was regarded as helpful in the workplace. The only other adjustments mentioned by at least two of the respondents were what was referred to as 'the narrowmindedness of the people'. The perceived poor quality of education, and inferred differences in work ethics between St Lucia and the UK/US (Table 2.7) were further factors mentioned by the informants. Interestingly, poor secondary education was cited by several young returnees as a reason that might promote re-return to the metropolitan society or onward migration to the Francophone Caribbean. This was another salient difference between the young returnees to St Lucia and Barbados.

Conclusion

Following a review of the relationships between migration and Caribbean development, the socio-economic and demographic characteristics of a new cohort of young return migrants to the Caribbean have been described. In both St Lucia and Barbados, on interview the young returnees had an average age in their mid- to late-30s, having decided to migrate to the Caribbean in their early- to mid-30s. At the time of interview, on average, the informants had lived in the Caribbean for a period between 3 and 4 years. In both countries, the majority of the parents of the young returnees were to be found living in their country of birth, exemplifying the importance of family in Caribbean migration. But the gender composition of the two groups differed markedly, with female returnees predominating in Barbados and males in St Lucia. The tendency for the female returnees to Barbados to occupy professional and semi-professional occupations was marked. When asked why they had decided to migrate to the Caribbean, more 'pull' factors were mentioned than 'push' factors. Reaching a natural watershed in their lives and needing to make a positive change were predominant. However, issues of poor education, especially for boys, and racism were cited as push factors.

The issues of adjustment cited by the interviewees suggested that first and foremost, in both St Lucia and Barbados, the young returnees occupy 'inbetween' and 'hybrid' positions. Inevitably, young return migrants carry markers from two societies. In the case of returnees to Barbados, this was experienced in terms of difficulties experienced in making friends, problems associated with having an English accent, feeling like an outsider, the strong perceived Americanisation of society, general feelings of resentment, and English returnees openly being branded as 'mad'. In St Lucia, the returnees also spoke of the difficulties experienced in making new friends, competition with indigenous St Lucians, resentment and being regarded as an outsider. However, in the St Lucian context, the operation of the colour-class system was regarded as the leading issue of adjustment. There appeared to be no parallel narrative concerning the madness of return migrants from the UK, while having an English accent in St Lucia was talked about in much more positive terms.

These different nuances of adjustment suggest that transnationalism has to be regarded as a relatively 'elastic' concept, in that the experience of youthful return varies somewhat from territory to territory – something that should come as no great surprise given the heterogeneity of Caribbean societies. These differences in experience on the part of young returnees with regard to issues such as race, gender, accent, friendship, plus the commonalities which they also experience, are the focus of a major comparative research project which is being carried out by the author in Barbados and St Lucia. It is hoped that subsequently, the research will be extended to other Caribbean territories, such as Jamaica and Trinidad.

Acknowledgements

The field work carried out from 1999 to 2000 in Barbados and St Lucia was funded by the British Academy (APN 29771). Ongoing investigations into the social dynamics and attitudes of foreign-born and young returning nationals are to be funded by The Leverhulme Trust. The assistance of the Facilitation Unit for Returning Nationals (FURN) of the Ministry of Foreign Affairs in Barbados is also gratefully acknowledged, especially in relation to the work carried out 1999-2000.

References

Bovenkerk, F. (1981), 'Why returnees generally do not turn out to be "agents of change": the case of Suriname', *Nieuwe West Indische Gids*, 55, 154-173.

Brana-Shute, R. and Brana-Shute, G. (1982), 'The magnitude and impact of remittances in the eastern Caribbean: a research note', in Stinner, W.F., de Albuquerque, K. and Bryce-Laporte (eds), *Return Migration and Remittances: Developing a Caribbean Perspective*, Washington D.C.: The Smithsonian Institute, ResearchInstitute on Immigration and Ethnic Studies, Occasional Paper no 3, 267-289.

Brooks, D. (1975), *Race and Labour in London Transport*, Oxford: Oxford University Press.

Brierley, J. (1985), 'A review of development strategies and programmes of the Peoples' Revolutionary Government of Grenada, 1979-1983', *Geographical Journal*, 151, 40-52.

Byron, M. (2000), 'Return migration to the eastern Caribbean: comparative experience and policy implications', *Social and Economic Studies*, 49, 155-188.

Byron, M. and Condon, S. (1996), 'A comparative study of Caribbean return migrants from Britain and France: towards a context-dependent explanation', *Transactions of the Institute of British Geographers*, New Series, 21, 91-104.

Clayton, A. and Potter, R.B. (1996), 'Industrial development and foreign direct investment in Barbados', *Geography*, 81, 176-180.

Conway, D. (1985), 'Remittance impacts on development in the eastern Caribbean', *Bulletin of Eastern Caribbean Affairs*, 11, 31-40.

Conway, D. (1993), 'Rethinking the consequences of remittances for eastern Caribbean development', *Caribbean Geography*, 4, 116-130.

Conway, D. (1994), 'The complexity of Caribbean migration', *Caribbean Affairs*, 7, 96-119.

Conway, D. (1998), 'Why migration is important for Caribbean development', *Paper presented at the 94th Annual Meeting of the Association of American Geographers*, Boston MA.

Dann, G. and Potter, R.B. (1997), 'Tourism in Barbados: rejuvenation or decline?', in Lockhart, D. and Drakakis-Smith, D. (eds), *Island Tourism: Problems and Perspectives*, London: Mansell.

Dann, G. and Potter, R.B. (2001), 'Supplanting the planters: new plantations for old in Barbados', *International Journal of Tourism and Hospitality Research* (in press).

De Souza, R-M. (1998), 'The spell of the Cascadura: West Indian return migration', in: Klak, T. (ed), *Globalisation and Neoliberalism: the Caribbean Context*, London: Rowman and Littlefield, 227-253.

FURN (Facilitation Unit for Returning Nationals) Ministry of Foreign Affairs (ca 1997), *First Annual Report of the Facilitation Unit for Returning Nationals (FURN)*, Government Printing Department.

Gmelch, G. (1980), 'Return migration', *Annual Review of Anthropology*, 9, 135-140.

Gmelch, G. (1985a), 'Emigrants who come back, Part I', *Bajan*, April 1985, 8-9.

Gmelch, G. (1985b), 'Emigrants who return, Part II', *Bajan*, May-June 1985, 4-5.

Gmelch, G. (1985c), 'The impact of return migration, Part III', *Bajan*, July-August, 4-6.

Gmelch, G. (1987), 'Work, innovation and investment: the impact of return migrants in Barbados', *Human Organisation*, 46, 131-140.

Gmelch, G. (1999), *Double Passage: the Lives of Caribbean Migrants and Back Home*, Ann Arbor: University of Michigan Press.

King, R. (1986) (ed), *Return Migration and Economic Development*, Bechenham: Croom Helm.

Lowenthal, D. (1972), *West Indian Societies*, Oxford: Oxford University Press.

Lowenthal, D. and Clarke, C. (1982), 'Caribbean small island sovereignty: chimera or convention?', in Fanger, O. (ed), *Problems of Caribbean Development*, Munich: W.F. Verlag.

Marshall, D. (1982), 'The history of Caribbean migrations', *Caribbean Review*, 11, 6-9.

McAslan, E. (2001), *Poverty reduction and social capital in Barbados*, Unpublished PhD thesis, Royal Holloway, University of London.

Ministry of Foreign Affairs of Barbados (1996), Returning Nationals Information Booklet, *Nation Publishing Co Ltd for the Ministry of Foreign Affairs of Barbados*.

Newton, V. (1984), *The Silver Men: West Indian Labour Migration to Panama 1850-1914*, Mona, Jamaica: Institute of Social and Economic Research.

Nutter, R. (1986), 'Implications of return migration for economic development in Jamaica', in King, R. (ed), *Return Migration and Economic Development*, Beckenham: Croom Helm.

Peach, C. (1967), 'West Indians as a replacement population in England and Wales', *Social and Economic Studies*, 16, 289-294.

Peach, C. (1968), *West Indian Migration to Britain: A Social Geography*, Oxford University Press for the Institute of Race Relations.

Peach, C. (1984), 'The force of West Indian island identity in Britain', chapter 9 in Clarke, C., Lay, D. and Peach, C. (eds), *Geography and Ethnic Pluralism*, London: George Allen and Unwin, 214-230.

Potter, R.B. and Dann, G. (1987), *Barbados: World Bibliographical Series*, Oxford: Clio Press.

Potter, R.B. and Welch, B. (1996), 'Indigenization and development in the eastern Caribbean: reflections on culture, diet and agriculture', *Caribbean Week*, 8, 13-14.

Rhoades, R. (1979), 'Toward an anthropology of return migration', *Papers in Anthropology*, 20, 1-111.

Richardson, B. (1975), 'The overdevelopment of Carriacou', *Geographical Review*, 65, 290-299.

Richardson, B. (1985), *Panana Money in Barbados, 1900-1920*, Knoxville: Tennessee University Press.

Richardson, B. (1992), *The Caribbean in the Wider World, 1492-1992*, Cambridge: Cambridge University Press.

Robotham, D. (1998), 'Transnationalism in the Caribbean: formal and informal', *American Ethnologist*, 25, 307-321.

Stinner, W.F., de Albuquerque, K., Bryce-Laporte, R.S. (1983) (eds), *Return Migration and remittances: Developing a Caribbean Perspective*, Washington DC: Smithsonian Institute, Research Institute on Immigration and Ethnic Studies, Occasional paper no 3.

Thomas-Hope, E. (1985), 'Return migration and its implications for Caribbean development', in Pastor, R.A. (ed), *Migration and Development in the Caribbean: the Unexpected Connection*, Westview Press: Boulder Colorado, 157-177.

Thomas-Hope, E. (1992), *Explanation in Caribbean Migration:Perception and the Image: Jamaica, Barbados and St Vincent*, Basingstoke: Macmillan Caribbean.

Vertovec, S. (1999), 'Conceiving and researching transnationalism', *Ethnic and Racial Studies*, 22, 447-462.

Watson, M. and Potter, R.B. (2001), *Low-Cost Housing in Barbados: Evolution or Social Revolution?*, Jamaica, Barbados and Trinidad: University of the West Indies Press.

Chapter 3

'Tales of Two Societies': Narratives of Adjustment among Second Generation Return Migrants to St Lucia and Barbados

Robert B. Potter

Introduction

This is the second of two chapters presenting the findings of a study of second generation overseas-born West Indians who have decided to migrate to their parents' country of birth. In the previous chapter, the socio-economic, demographic and family life-cycle characteristics of these young return migrants were examined, along with the reasons that lay behind their decision to migrate to the Caribbean. The chapter finished with a listing and brief overview of the major adjustments that the sample of young returnees reported they had had to face in 'returning to' the country in which they had not been born. In the present chapter, a detailed commentary concerning these socio-cultural adjustments is presented. This is achieved by means of a qualitative analysis of the narratives that were provided by the young returnees concerning the adjustments they have had to face and still face.

The overall research design involved in the study has been fully explained in Chapter 2. The target group for the study was foreign-born returning nations – that is British and American Barbadians and St Lucians. Using background information provided by the respective Ministries of Foreign Affairs, and snowballing contacts thereafter, forty semi-structured interviews were carried out with such foreign-born and young returning nationals – 25 in Barbados and 15 in St Lucia. In both Barbados and St Lucia, on interview, the young returnees had an average age in their mid- to late-30s, having decided to migrate to the Caribbean in their early- to mid-30s. Thus, at the time of interview, on average, the informants had lived in the Caribbean for a period of 3-4 years. The majority of the parents of the young returnees were also to be found living in their country of birth, stressing the salience of family in Caribbean migration. But there were two major differences between the young returnees in Barbados and St Lucia. In Barbados, female return migrants predominated (21 of 25), while males were more numerous in St Lucia (4 of 15). Further, the female returnees to Barbados showed a high proportion of professional occupations, while the males in St Lucia were more likely to be engaged in service occupations (see Chapter 2).

In the main body of the present chapter, the issues of adjustment mentioned by the young returnees are presented and analysed, first for Barbados and then for St Lucia. In both cases, the order of consideration of the issues raised by the young returnees follows the frequency with which the specific adjustments were mentioned by the interviewees, as shown in Tables 2.6 and 2.7 for Barbados and St Lucia respectively in Chapter 2. We now turn to the narratives that lie behind the quantitative listings of socio-cultural adjustments given in Tables 2.1 and 2.2. In the case of the young returnees to Barbados, we look broadly at the twelve most cited adjustments mentioned, all of these having been cited by more than three (or 12 per cent) of those interviewed. In respect of the returnees to St Lucia, the eight most frequently cited adjustments, those enumerated by more than 2 (or 13.33 per cent) of the informants are considered. In the present chapter, when issues of difference are encountered in the experiences of the two national sets of young returnees, the contrast is emphasized in the account concerning St Lucia.

Adjustments among the Young Return Migrants to Barbados

Difficulties in Making Friends

One issue was mentioned by the majority of young return migrants to Barbados, 19 out of 25, and that was the difficulties involved in making friends (Table 2.1). Making friends in Barbados was recurrently seen as very difficult, with a measure of relative alienation from mainstream Barbadian society being implied:

> 'In Barbados, friendship patterns are fixed'.

> 'It's the clique mentality of Barbadians. Nobody ever phones you up and asks you out'.

Notably, an important causal role was ascribed to friendship patterns in Barbados having been established via patterns of secondary school attendance. Barbados has a common entrance examination at the age eleven, and a very clear and strong pecking order of secondary schools exists, and it is apparent that these act as primary agents of socialisation. As one young returning national explained:

> 'In Barbados school friends and who you know are very important. It is tough if you do not know anyone'.

The same sentiment was expressed from a different perspective by a returnee born in the USA who had moved to the island with her Bajan-born husband:

> 'My husband has a lot of friends from High School and they helped a lot when we came down here. They made for a smooth network'.

The importance of other returning nationals and expatriates was stressed time and time again, and it was also implied that social relations are easier in the workplace than in the social realm:

> 'I would not see my work colleagues socially. There are only
> four people in Barbados I would phone. They are all English
> of Bajan parents and have lived here for less than four years'.

> 'My friends are all from outside Barbados'.

Naturally, having strong family ties, marriage and targeting friends were all considered to be factors that make matters easier, but virtually none of the informants claimed to have female Bajan friends, regardless of how long they had lived in Barbados:

> 'I was very lonely at first. All my friends came from the UK.
> But I married here'.

> 'I have not found friendship patterns too big a problem, but I
> have strong family support. Plus, my best friend is another
> UK-based returning national who left almost exactly the same
> time as me'.

A frequent observation was that Bajans go to work and then go straight home. Several young returnees said that they had suggested to their colleagues going for a drink after work on a Friday, but that this had been greeted by a measure of incredulity. Barbados is traditionally seen as a conservative and proud country, in which people keep their private lives and problems to themselves (Lowenthal, 1972). Lowenthal (1972) cites O.R. Marshall that: 'Jamaica has a difficulty for every solution, Trinidad a solution for every difficulty: Barbados has no difficulties!' An associated contention among some of the respondents was that Barbadians are 'standoffish' about outsiders.

But the principal difficulty expressed by the young female returning nationals was in making female friends:

> 'It is difficult to make friends with Bajan women'.

Several informants were unequivocal in attributing this to rivalry, both sexually, with respect to finding a male partner, as well as economically, in respect of the job market:

> 'It is difficult to make female friends. Bajan women think you
> are going to take their men and their jobs'.

> 'Friendship patterns were very difficult at first. Bajan girls are
> not very warm'.

The informant who had lived the longest in Barbados, for over 15 years, stated:

> 'I now have a few Bajan friends. Many think you are taking
> things away from them'.

Some informants explained the problems experienced in making female friends in terms of gender vistas, plus issues of social standing:

> 'There is some bigotry. Womens' vistas are more limited
> here...being secretaries, having children, being mothers. They
> are suspicious of outsiders. I expect they think we see
> ourselves as better than them. And in the workplace they are
> worried about their jobs'.

> 'My circle of friends is small. I have made more male friends
> than female friends. I feel I have very little in common with
> Bajan females'.

This situation may perhaps be exacerbated by the frequently mentioned fact that a good number of prominent Barbadian male politicians are married to non-nationals, including women from Guyana and Jamaica.

On the other hand, friendships with males were seen as much easier both to make and to regulate. The view that being different drew attention from Bajan men was cited on several occasions:

> 'Women here are not friendly. But the men are friendly!'.

> 'Making male friends is not a problem. But you have to be
> very clear what the score is. But this is accepted'.

Interestingly, it has been suggested in the past that too little attention has been given to female migrants (Gmelch, 1980). In the present research, as well as in respect of friendship patterns, the particular situation faced by women migrants is also addressed below in respect of 'macho bosses'.

Problems of Accent/Language

For several of the young returnees to Barbados, issues of accent and language (Table 2.1) represented a real problem. Typical comments included:

> 'They take the mickey out of your accent'.

> 'There is a good deal of teasing about my accent'.

Saliently, this was frequently commented upon as a particular issue for those with British accents, as opposed to those from North America:

> 'It is harder for those with a British accent'.

> 'People may try to overcharge you if you have an English
> accent, like my Mother when she visits the market'.

> 'An English accent is a double-edged sword really. There is some resentment obviously, a feeling perhaps that you are trying to be grand. But in truth there is some jealousy. Most would love to live in the UK/USA if they had the chance'.

There appeared to be a general suggestion that in the social domain an English accent is seen as suggesting arrogance, whilst in the job market an overtly English accent could be a passport to job opportunities. Thus, having an English accent might bestow a significant advantage if working in tourism or the property market. It was universally held that it is easier for those with a North American accent:

> 'A UK accent is more difficult than one from the USA/Canada'.

> 'Some people do tease me about my accent – it is better to have a US accent'.

> 'Those from the USA and Canada are more accepted'.

> 'I have no problem with my accent. I can slip from Bajan to American'.

While some informants said that they made a joke out of being teased about their UK accent, others felt that the situation was more malicious and derogatory. These interviewees argued that nobody would think of openly mocking a white tourist about their accent, so why should they be treated differently. It is tempting to conjecture as to whether the relative antipathy to an English accent in the social domain is part of a distinctly 'post-colonial turn', some 35 years after independence. On the other hand, an English accent can be an advantage in some areas of the job market. For example, a popular new female DJ on the radio has a marked English accent.

Slow Pace of Life

Almost inevitably, the slower and more relaxed pace of life experienced in Barbados as opposed to that experienced in a metropolitan context was regarded as an issue demanding adjustment:

> 'At first, I found the place very slow'.

> 'The slowness of Barbados has nearly driven me to distraction'.

As might well be expected, however, some had adapted to this and regarded it as a major advantage:

> 'The attitude to life is much more relaxed here'.

> 'The need to slow down – but why rush like in the UK? I no longer get frustrated in queues'.

> 'But now I have completely adjusted'.

For some, this was a context that continued to influence day-to-day patterns of living, both in the work place and more widely:

> 'It's the mindset here. Things are very slow'.

> 'The accepting mentality of the people is a problem'.

Feeling Like an Outsider

Inevitably, some returnees reported that they were regarded as total outsiders. The following was a typical comment, replete with inferences of resistance:

> 'I encountered lots of hostility from people at work. I was regarded as an outsider. But I decided that their opinions were of no value to me'.

For several, it was frustrating that they were constantly reminded of their national identity and the way in which it clashes with their racial identity:

> 'You may think that you are Bajan, but you are not really! I am referred to at work as the 'English person'. I am still the outsider at work. I join in with the laughing and joking and I am accepted to a point. You cannot expect to be wholly accepted'.

> 'You are constantly reminded that you are English and that you are not Bajan'.

The experience of the appellation 'English person' was a very frequent one. This clear identification as being English clashed with the previous identities of the informants as West Indian when they were living in the UK. In an extreme case, one informant lamented that:

> 'The worst things that have been said to me here have been almost as cruel and negative as those I faced in the UK'.

Poor Shopping/High Prices

Poor shopping opportunities were specifically mentioned by several of the informants. One aspect of this clearly related to the expense of goods:

> 'When you shop in Barbados you pick out what you need, not what you want'.

This also reflected the perceived difficulties involved in getting a good range of fresh fruit, vegetables and non-frozen meats in Barbados. The fact that many local clothing shops are old-fashioned was mentioned, along with the fact that there is little in the way of enforceable consumer legislation:

'Consumer protection is appalling. There is no real redress or refunds'.

Like those Bajans who have the resources to do so, young and foreign-born returning nationals tended to shop whilst on trips to the UK, the US, or elsewhere.

Aspects of a 'Culture Shock'

Several informants specifically invoked the idea of a 'culture shock' to explain their experience of living as foreign-born and young returning nationals in Barbados, stressing that it is 'a different culture' in a wide variety of ways:

'The cultural differences are so wide...it's a real culture shock'.

'It is a culture shock, the differences, but you have got to work at it; don't let people bother you'.

Several respondents alluded to differences in the general work ethic between themselves as English-born migrants and indigenous members of the workforce. Typical comments were that they were used to seeing jobs through to completion without supervision and would report back on progress as a matter of established practice, neither of which it was claimed inevitably happen in the workplace in Barbados. The principal argument was that English- and Amercian-born Barbadians do things faster at work than their indigenous counterparts. One informant bemoaned that because she did her work quickly, she was persistently allocated more work than her colleagues.

The Americanisation of Society

An important part of the culture shock experienced by the young returning nationals appeared to relate to the perception that Barbadian society has become highly Americanised. For long referred to as 'Little England' as a consequence of its singular British colonial history, it seemed that the trend toward Americanisation came as something of a surprise to many of those who had been brought up by parents whose primary recollections of Barbados emanated from several decades ago. Typical comments included:

'But Barbados is becoming Americanised'.

'Barbados is now *so* Americanised'.

'Barbados *is* Americanized'.

Some informants added the rider that this trend was particularly true of the younger generation, at the same time drawing specific attention to what they perceived as the existence of marked inter-generational contrasts:

> 'Younger people are very pro-USA'.

> 'Bajans adore Americans. The younger generation have adopted the American way of life. The older people find this uncomfortable'.

One interviewee took matters further, implicitly arguing that the age-old sobriquet 'Little England' is now wholly inappropriate, and thereby directly suggesting the operation of strong 'post-colonial' forces:

> 'This is Little America, desperate to free itself from its colonial past. Barbadians would change every name to an American one if they could! They really want to forget the past.

It is tempting to suggest that the rejection of English accents and the propensity to regard British returnees as 'mad' (explored in the next section), must both be seen as socio-cultural correlates of the exposure of the present generation of Bajans to strong forces of Americanisation. In respect of this simple working hypothesis on the one hand there is the rejection of colonialism and post-colonial forces; on the other, there are forces of Americanization and globalisation associated with the post-modern condition. It would indeed be surprising if young West Indians were not influenced by the globally hegemonic forces of music, fashions, videos/DVDs, films, CDs, and sports emanating from the USA, linked to American accents, speech and modes of dress.

English People being Regarded as 'Mad' and Issues of English Identity

A further salient issue complicated the fact that a primarily English identity was being ascribed to the returnees. Several of the respondents reported that English returning nationals are openly referred to as being 'mad':

> 'They think that the English are mad'.

> 'Bajans say that the English *are mad*'.

> 'The attitude to returnees from England is far from positive. Returning nationals from England are regarded as mad. Old English attitudes are seen as arrogant'.

The last comment raises an important issue that most foreign-born and young returnees seemed to find hard to explain. Is this frequently cited view of English returnee madness also part of the post-colonial turn, that is a further way of discounting the perceived arrogance of the British-dominated past? Or does this derogatory labeling reflect the stress hypothesis of migration – namely, that migration is a testing experience and is associated with higher rates of stress-related mental illness? Notably, Mallett (*pers. comm.*) argues that a substantial number of Caribbean migrants to the UK who developed mental illnesses went back to the Caribbean, thereby giving credence to the association of mental ill-health with return in the eyes of Caribbean denizens themselves. Certainly in general conversations that the author has had with Bajans, the saying 'English people have snow (or rain) on their brains' has been used quite frequently. It seemed that the appellation 'mad' was being used as a method of 'Othering' these young returnees.

Given such views regarding reactions to English accents, and the claim that English returnees are 'mad', plus the trend toward the Americanization of Bajan society, several informants seemed to be mounting strategies of resistance. Specifically, several argued forcefully that the path to successful adjustment was through the retention and fostering of an English identity:

> 'I like my English identity: a Bajan at heart and through heritage'.

> 'I need to keep my English identity'.

> 'I *am* English!'.

For several informants, such was the force of this counter identity that they stressed the irony that while they had generally felt Bajan or West Indian in the UK, now that they live in the Caribbean, they feel more English than they had ever done previously in their lives.

Issues of Race and Class

Further fundamental issues relating to racial and national identity were addressed by several of the returnees. Before addressing these issues, several important facts must be borne in mind concerning race in Barbados. First, as already noted, Barbados was from first European settlement a British colony, and unlike many other islands did not come under the control of any other European colonial power. Further, it had a low level of miscegenation during the colonial period. Although Barbados has a higher proportion of whites than most other Caribbean societies, standing at some 4 per cent, many have commented on the operation through time of an effective colour bar, with the longstanding existence of black and white cricket clubs and other organisations. Thus, influential white Barbadian families run all the renowned 'Big Six' companies that control much of the Barbadian economy. Finally, in common with the rest of the post-colonial Caribbean, the strong and continued operation of a colour-class system is openly acknowledged, where skin shade is strongly linked to occupation and inferred social standing (Dann and Potter, 1990).

It seemed clear that the operation of these facets of racial distinction in Barbados had come as something of a shock to several of the young returnees. For example, one interviewee commented that:

> 'The unity of black people in the UK is lost when you come
> here. For example, at lunchtime at work, there is a white-black
> split. It is just accepted. Barbados is the most racist society I
> have ever lived in'.

The argument that the relative unity of black people experienced in the UK regardless of country of origin is dissipated in Barbados was encountered several times. Similarly, another respondent reacted strongly to the operation of the colour-class system:

> 'Racism is horrendous in Barbados. Not black and white, but
> among themselves, by shades of colour. I am very saddened
> by this. I can't understand it'.

Whilst it is possible to enter into long debates as to whether the recognition of colour gradations represents a form of labeling or *de facto* racism, it is clear that the existence and operation of the colour-class system, and the economic hegemony of the white minority of Barbadians came as a real surprise to several of the foreign-born and young returning nationals:

> 'Racism here is a major issue. I was not really aware of this
> before I came here permanently'.

> '…a lack of respect for each other. I have developed a see and
> not see attitude'.

Reflecting matters of class and race, several of the informants stressed that they considered Barbados to be a very class-based society, and that this served to explain some of their wider socio-cultural experiences:

> 'It is a very class-based society. If you are seen as a gap-
> tenantry [low-income housing area] person, then you will be
> treated as such'.

Another informant clearly sought to link this to the generally centre-right political tendency of contemporary Barbadian society, in a manner that apparently linked this to the problems experienced in making Bajan friends:

> 'This is a money-society. People are very materialistic and
> concerned with external appearances. Generally, people do not
> invite you into their homes. People want a good car and a big
> house that looks good from the outside'.

Feelings of Resentment

Several of the young returning migrants said that they felt the same type of negative reactions as those experienced by their older counterparts, the retiree returning nationals. These generalized feelings of resentment did not appear to be related to the fact the returnees can bring household items and vehicles into the country without paying duties (Ministry of Foreign Affairs, 1996). The general sentiment appeared to be expressed by one respondent when she observed how:

> 'Some people feel that you have had it easy in the UK, with the dole and housing. You have to try hard. People are not going to make it easy for you. They have even said this to me'.

The Behaviour of 'Macho Bosses'

Another issue that was expressed in gender-specific terms was what informants described as the 'macho-style behaviour' of male bosses in Barbados. For example, female employees observed that:

> 'Bosses here think they can get away with murder. They treat staff bad, they curse and carry on at them. They couldn't work like this anywhere else'.

> 'Macho-bosses exist here, and complaints procedures rarely exist. Managers just turn a blind eye'.

> 'My first job was male dominated and the men felt that they could say whatever they liked to me'.

Several foreign-born returning nationals asserted that some bosses act in inappropriate ways in the workplace and that indigenous female employees do not feel that they can complain in any way about this. One of the returnees said that she had specifically told her boss that she would report him, and that as a result, he had stopped on her third and final warning. Another female informant related how her male boss had told her how as a young woman she should wear particular items of personal clothing. But on the other hand, the same interviewee said that she felt safer in general in Barbados. She reported that it was when she was in the UK that she felt more worried about the risks of sexual assault.

Having to Accept Things as They Are

Almost inevitably, the foreign-born and young returning migrants noted that they were expected to change to Bajan ways of doing things, rather than trying to change the way things are done locally. Typical was the comment that:

> 'You have to accept things as they are. You mustn't try to change things'.

In the workplace, several informants made it clear that allusions to the way in which things are done in the UK or North America are no-go, and that more subtle strategies involving demonstration effects have to be employed:

> 'Don't say this is how things are done in the UK!'.

One informant cogently summarised the pragmatic reality of the situation:

> 'You can't expect people to adjust to you. You have to adjust
> to the new situation'.

However, a further respondent complained that non-Bajan born migrants were expected to adjust to, and assimilate, Bajan ways almost instantaneously:

> 'On the other hand, you are expected to change immediately to
> the Bajan way of doing things'.

The early literature on older returning nationals tended to ask whether returnees could be seen as repositories of new ways of doing things within relatively conservative societies (Gmelch, 1980). It would appear that even for these young returnees, the degree to which they can contribute to the development of more flexible and adaptable societies is limited, at least in the short-term.

Adjustments among the Young Return Migrants to St Lucia

The Colour-Class System and Racism in Society

In the case of St Lucia, issues relating to race and the operation of the colour-class system were the most frequently cited among the young returnees, as attested by Table 2.2. Interestingly, this was more frequently cited than in the case of Barbados.

In common with the rest of the post-colonial Caribbean, the continued operation of a strong colour-class system is openly acknowledged in St Lucia, whereby skin shade is frequently directly correlated to occupational standing and inferred social status (Lowenthal, 1972). St Lucia has a mixed French and British colonial history, the island having changed hands between these two colonizing powers as many as 14 times. There has been a greater incidence of miscegenation, leading to a more racially diverse population than is found in Barbados, for example.

The interviews indicated that the operation of some facets of such racial distinction and labelling in St Lucia had come as something of a shock to several of the young returnees. Thus, one respondent who was living on the island, but who had been moving backwards and forwards in a peripatetic manner noted with some dismay how:

> 'The racism thing is bad here. It's this that keeps me away
> from moving here permanently'.

Another talked about the operational expression of this at work, whereby, when visiting his office, some St Lucians would tend to discount his presence, presumably on account of his black skin:

'Some St Lucians ignore me at work. They ask for my
partners who are white, and complain about 'nobody' being
around!'.

Interestingly, two of the informants, in relating their experiences of race in St
Lucia directly mentioned the operation of the colour-class system:

'There is a lot of racism in St Lucia – black and white judged
by the colour-class system'.

'There's an in-built class system in the islands: a colour-class
system … racialism is wherever you go'.

But there was a further interesting take on this. One light-skinned UK-born
returnee noted that whilst he had been seen as black in the UK, locally in St Lucia
he was regarded as white, and this labelling was the source of some annoyance to
him:

'There is racism here. I am seen as white in St Lucia. Local
people refer to "you people". That really bugs me!'.

Perhaps the most important interpretation is that the young returnees were
acutely aware of the existence and the operation of colour-class in the Caribbean
region, thus dispelling any simple notion that class alone has displaced the
combined influence of colour-class in the post-colonial era (see Lewis, 1990; Dann
and Potter, 1990; Beckles, 1996; de Albuquerque and McElroy, 1999; Potter,
Barker, Conway, Klak, 2004).

Difficulties in Making Local Friends

As with the returnees to Barbados, several of the informants in St Lucia
emphasized what they regarded as the problem they were having making friends
with indigenous Lucians. One reported that most of his friends were drawn from
those who had migrated from the UK and the same individual implied that a lack of
local identity was a fundamental obstacle:

'Most of my friends are from the UK. I tried to be St Lucian
for the first year, but it just did not work'.

The feeling that some local St Lucians regarded young returnees, like older
returnees, as relatively affluent, and the linked feeling that this might lead to
exploitation was clearly expressed by one individual. This seems to be a direct
parallel with the experiences of older returnees:

'I have made a few friends, but it isn't easy. It is difficult to be
friendly with local St Lucians. I made some local friends but
they wanted to use me. If you are from England, they think

you are rich … They think you feel you are better than them.
They give you a real hard time'.

One female interviewee emphasised that she found it particularly difficult to make female friends in St Lucia, and there appeared to be the suggestion that this reflected aspects of sexual rivalry and competition:

> 'I do not have St Lucian female friends. It's their way of thinking. They are insecure and there is much narrowmindedness. If you are friendly with a man, people immediately think something's going on. I keep myself to myself … the attitude is different'.

In the case of the interviews carried out in Barbados, difficulties in making friends were cited by some 76 per cent of the respondents. In particular, in the case of Barbados there was a strong gender component in that it was female informants who reported extreme difficulties in making local female friends. So, although expressed in very similar terms by this individual, the incidence of the problem seemed to be far less prevalent in the case of St Lucia. Indeed, one respondent specifically commented on the welcoming nature of St Lucians and implied the operation of reciprocity:

> 'A lot of the locals are very nice. They are happy to welcome you. You are going to give back'.

The 'Laid Back' Way of Life

Another similarity with the Barbadian informants was the feeling that it had been necessary to adjust to the generally slow and laid back pace of life experienced in St Lucia. In the case of Barbados, this was also the third most frequently cited issue of adjustment. Also, as in the case of Barbados, at the outset, some found the slow pace of life difficult to adjust to:

> 'Can I live in such a laid back society?'.

However, others had clearly come to terms with this, and saw it as one of the principal reasons why they wanted to live in St Lucia:

> 'In the beginning, I found it a long-winded process. As I slowly pieced together what was happening, I accepted things on their grounds and the laid back way of doing things'.

Competition and Feelings of Resentment

As already noted, in the Barbados research, having an English accent and identity was seen as being connected with a number of problems in the realm of social relations, although it was seen as an advantage in the workplace. In the interviews carried out in St Lucia, the issue of being English and a degree of anti-English feeling towards the foreign-born returnees only emerged when discussing the workplace. This appeared to be in almost direct contrast to what was found in Barbados. For example, one informant in St Lucia emphasized that:

'There is anti-English feeling. When they see the English come and work, they are resented'.

This connected to the wider issue of rivalry in the workplace, and for at least one respondent, it also connected with issues relating to differences in work ethics between employees in St Lucia and the UK:

'Some worry we are going to take their jobs. People go out of their way to give you stick ... Some would not be employed in the UK'.

Yet another young returnee in St Lucia pointed to what he clearly saw as the divisive nature of competition between the indigenous population and returnees, suggesting that on occasion, local people are all too inclined to attribute success to illegal activities:

'Competition can bring things to a head: we can be our own worst enemies. If you do well, some ask how. Was it through drugs?'.

Issues Relating to an English Accent

The fifth issue raised by the interviewees, which related to the fact that an English accent serves to make returnees stand out from the crowd, proved to be of some interest. The salient point was that most of those who mentioned their English accent argued that in St Lucia this posed no real problem. Indeed, several informants saw an English accent as a distinct advantage:

'You can get served faster in shops if you have a light skin and accent'.

'With an English accent you can get in anywhere. The accent makes you sound like a tourist'.

'English people can go into all the hotels, because of their accent. You can be a tourist and a local at the same time!'.

Such extracts seem to imply that the possession of an English accent appears to bestow '*symbolic*' or '*token whiteness*' on the returnees. In this way, in the St Lucian context, an English accent appeared to be associated with a 'fuzzy identity' which transcends both local and white-privileged-tourist identities. In this way, the returnees appear to occupy an essentially 'hybrid' and 'inbetween' position within society (Fanon, 1967; Loomba, 1998; Bhabha, 1994). In this way the returnees appear to be both black and symbolically white, advantaged and disadvantaged.

One young returnee who had been brought up in Canada even commented on the fact that her Canadian accent was sometimes confused for an English accent, and that this was frequently commented on favourably by St Lucians:

> 'They say, "what a nice accent – is it British"?'.

Such reactions to an English accent in a social context were in direct contrast to those expressed in Barbados, where although it was accepted that this might bestow certain advantages in the workplace and job market, it was generally perceived as a problem in the realm of day-to-day social relations.

One interviewee pointed to what he saw as a subtle nuance relating to the possession of a strong regional accent, together with the class connotations associated with this, in suggesting that he could easily make the transition from a St Lucian to an English accent:

> 'I can slip from St Lucian to English with ease. But a strong
> London accent might be a problem'.

Another informant suggested that whilst there might be some teasing regarding his accent, he found it easy enough to turn a blind eye to this:

> 'If people do take the mickey, they are people I don't want to
> know'.

Just as in the general social context there appeared to be a somewhat more positive view about an English accent in St Lucia, compared with Barbados, so two respondents seemed to be arguing against the proviso drawn in the context of Barbados that having an English accent might bestow advantages in the workplace. In this connection it was contended that:

> 'St Lucian employers will pass you up due to your English
> accent, especially in Government'.

Yet another respondent seemed to be widening this issue in suggesting that her former employee had said that he would not hire people hailing from the UK in the future as a direct result of their tendency to take action and seek redress in the workplace:

> 'My ex-boss said he would never employ another English
> person. He said this because English people exercise their
> rights'.

Although expressed primarily in relation to the issue of accent, the view expressed points a finger at the existence of macho bosses, an allegation also encountered in the Barbados pilot study.

The 'Narrowmindedness' of the People

Two of the respondents argued that they had to adjust to what they referred to as the 'narrowmindedness' of the local population. This accusation related mainly to those members of the indigenous population who had not traveled beyond the confines of the island. Typical was one informant who argued that:

> 'Local people are not used to discussing big issues. And if you
> disagree, people take it personally'.

The Caribbean islands are frequently characterised by their essentially insular nature, and so such attributions of parochialism are perhaps only to be expected. One well-connected female returnee stressed how she had been forced to moderate her social and political views since returning to St Lucia. In the metropolitan society in which she had lived before coming to St Lucia, she had actively campaigned in respect of gender issues. The inference was that her well-connected family had reminded her not to 'rock the boat' in St Lucia. She noted that gender differences are marked and summarized these as 'women in the kitchen, men on the terrace!'.

Another comment was that:

> 'There is a 1950s culture here'.

By which it was meant that social issues such as homosexuality were as hush-hush in today's St Lucia as they had been in Britain fifty years or more ago.

Poor Quality of Education

Two of the respondents were quick to stress that while state education up to the age of eleven was generally adequate in St Lucia, they regarded the standard of secondary education to be far from acceptable. In discussing her English-St Lucian friends who had returned to the United Kingdom, a female informant specifically stated how:

> 'Most have gone back due to schooling'.

Indeed, this particular mother was at the time of the interview actively considering moving to Martinique as part of the Francophone Caribbean, where she felt that her children would receive a better secondary education. The fact that her sister was currently living in Martinique encouraged her to consider taking this step.

A young returning St Lucian who had studied to degree level in the United Kingdom was generally condemnatory regarding the state of the education system in St Lucia, commenting that:

> 'Education is the big problem in St Lucia. The quality of
> teaching is poor, and bilingualism is a big problem. The

average St Lucian is poor at English. And there is a shortage
of vocational training'.

The idea that secondary school education was an issue that was promoting re-
return and onward migratory movements was of particular interest. In the case of
Barbados, what was generally perceived to be the excellent education system was
regarded as one of the factors promoting return, as well as serving to retain those
who had returned. This appeared to be an interesting difference.

Differences in Work Ethic

The argument that there are marked differences in the general work ethic revealed
in the UK and St Lucia was mentioned above in relation to revealed Englishness
via accent. Essentially, it was argued that those trained in England were likely to
see a job through to completion and would routinely report back on progress to line
managers, without the latter having to request such information. It was argued that
in the St Lucia workplace this was rarely something that happened spontaneously.
One respondent summarily concluded that:

'The work ethic is not the same here'.

Indeed, in several instances, this gave rise to the complaint that citizens by
descent were more likely to be given a larger workload than their indigenous
colleagues.

Conclusions

In this second of two closely-linked chapters, the qualitative narratives provided by
a sample of young return migrants to the eastern Caribbean have been considered.
The overall analysis demonstrates just how much the young transnational returnees
exist between two cultures and societies. They essentially occupy an 'inbetween'
and 'hybrid' liminal position within their adopted societies.

The returnees to both St Lucia and Barbados have had to adjust to the laid
back/slow pace of life they have encountered in the Caribbean. They have also had
to face marked difficulties in establishing new friendship patterns, although this
seems to have amounted to more of a problem in Barbados than in St Lucia.
Having an English accent featured for both national samples, but seemed to
represent much more of a problem for the returnees to Barbados. Indeed, in the
case of returnees to St Lucia, the possession of an English accent appeared to
bestow advantages, allowing access to both 'local' and 'tourist-expat' facilities.
Further, perceptions of competition and feeling like an outsider were mentioned in
both contexts, but again seemed to loom larger for the sub-group of returnees to
Barbados. Both national sub-samples reacted to issues of race relations and colour-
class, although these took pole position in the case of returnees to St Lucia. But the
accusation of 'madness' only occurred in the case of the young returnees to

Barbados. Indeed, several of the returnees to St Lucia stressed that they felt St Lucia to be a very tolerant society, where they are seldom refused assistance.

Thus, as in the previoius chapter, the study shows that both in St Lucia and Barbados, the young returnees occupy what can only be described as 'inbeween' and 'hybrid' positions within society. Inevitably they carry markers from two societies. The qualitative data show that transnationalism has to be regarded as a relatively elastic and fluid concept, in that the precise experience of youthful return varies somewhat from Caribbean territory to Caribbean territory.

Acknowledgements

The field work carried out from 1999 to 2000 in Barbados and St Lucia was funded by the British Academy (APN 29771). Ongoing investigations into the social dynamics and attitudes of foreign-born and young returning nationals are funded by The Leverhulme Trust. The assistance of the Facilitation Unit for Returning Nationals (FURN) of the Ministry of Foreign Affairs in Barbados is also gratefully acknowledged, especially in relation to the work carried out 1999-2000.

References

Beckles, H. (1996), 'Independence and the social crisis of nationalism in Barbados', in Beckles, H. and Shepherd, V. (eds), *Caribbean Freedom: Economy and Society from Emanicipation to the Present*, Princeton NJ: Marcus Weiners, 528-539.

Bhabha, H. (1994), *The Location of Culture*, London and New York: Routledge.

Dann, G. and Potter, R.B. (1990), 'Yellow man in the yellow pages: sex and race typing in the Barbados telephone directory', *Bulletin of Eastern Caribbean Affairs*, 15, 1-15.

deAlbuquerque, K. and McElroy, J. (1999), 'Race, ethnicity and social stratification in three Windward Islands', *Journal of Eastern Caribbean Studies*, 24, 1-28.

Fanon, F. (1967), *Black Skin, White Mask*, New York: Grove Press.

Gmelch, G. (1980), 'Return migration', *Annual Review of Anthropology*, 9, 135-140.

Lewis, L. (1990), 'The politics of race in Barbados', *Bulletin of Eastern Caribbean Studies*, 15, 32-45.

Lowenthal, D. (1972), *West Indian Societies*, Oxford: Oxford University Press.

Loomba, A. (1998), *Colonialism-Postcolonialism*, London: Routledge.

Mallett, R. (2002) Personal communication.

Ministry of Foreign Affairs of Barbados (1996), *Returning Nationals Information Booklet*, Nation Publishing Co Ltd for the Ministry of Foreign Affairs of Barbados.

Potter, R.B., Barker, D., Conway, D. and Klak, T. (2004) *The Contemporary Caribbean*, London and New York: Pearson/PrenticeHall.

Chapter 4

Incorporating Race and Gender into Caribbean Return Migration: The Example of Second Generation 'Bajan-Brits'

Joan Phillips and Robert B. Potter

Introduction

Studies of Caribbean return migration have tended to neglect issues relating to race and gender. This neglect is not altogether surprising viewed in the regional context. Indeed, it was only from the 1990s that researchers readily acknowledged that migration as a whole is a strongly gendered process (Grasmuck and Pessar, 1991; Pedrazza, 1991; Hondagneu-Sotelo, 1994; Grieco & Boyd, 1998; Pessar and Mayler, 2001; Ryan, 2002; Pratt and Yeoh, 2003). Although nascent research has suggested that Caribbean return migration must be viewed as a process that impacts significantly on the social position of women (Bueno, 1997), little detailed work has examined how exactly return migration is influenced by national and transnational constructions of gender. Further, race has also received limited attention as a significant variable affecting return migration. For instance, while post-colonial theories have articulated the 'inbetweenness' and 'hybridity' of migrant identities (Fanon, 1967; Gilroy, 1983; Hall, 1989; Bhabha, 1994), little is known about the impact and circumstances of such return 'hybrid' identities in particular post-colonial spaces (Phillips & Potter, 2003, Potter 2003a, 2003b).

The present chapter seeks to respond to this dearth of both gendered and racialised analyses with regard to Caribbean return migration. It examines how race and gender intersect and impact on the processes of second generation or relatively 'young' return migration to Barbados. Within the context of a post-colonial theoretical framework, the study interpretatively explores how young Bajan-Brits are characterised by gendered and racialised-hybrid identities in England, and consequently, how this inbetween identity provides the basis for, and motivation to, return to Barbados. Further, the research examines the myriad ways in which such hybrid identities negotiate, articulate and challenge the local construction of racial and gendered spaces in Barbados as a post-colonial setting.

The Research Design

The principal source of information here is 51 in-depth, qualitative interviews conducted by the authors with foreign-born and young returning nationals to Barbados. The overwhelming majority of those interviewed were females, amounting to 38 against 13 males. Nearly 63 per cent were foreign-born, with 29 out of 32 of those born outside Barbados having been born in the UK. The average age of the young returnees was just over 33 years when they migrated to Barbados, and they had been living in the country for an average of just over seven and a half years. The majority, some 44, were classified as black and nearly all had attended college; 35 per cent had studied at university.

The interviews were semi-structured in that they sought to cover all of the major life domains associated with migration that had previously been identified in a pilot study (see Potter, 2001). In all but one case, the interviewees were happy for the discussions to be taped. The research design also included detailed discussions with relevant policy-makers, searches of national newspaper archives and focus group discussions with members of the indigenous Barbadian public, concerning their attitudes to returnees. All of the interviews were fully transcribed and NUD-IST (Non-numerical Unstructured Data – Indexing Searching and Theorising) software was used to assist with the qualitative analysis of the data. The interviews rendered invaluable insights concerning migration histories, family and socio-economic standing, motives for migrating, employment and educational histories, and socio-cultural adjustments and experiences. The interviews were carried out between March and June 2002.

There are, of course, good socio-economic and development-oriented reasons why the progeny of Caribbean migrants to the UK and USA might consider 'returning' to the land of their parents. Quite simply, Barbados has achieved considerable economic success in the period since independence in 1966. Although only a small country (430 sq kms), Barbados now scores 'high' on the *United Nation's Human Development Index (HDI)*. Indeed, on this basis, in recent promotional literature, Barbados has described itself as the 'leading nation in the developing world'. Barbados recorded a per capita income of $US 14,353 in 2001, putting it ahead of several European nations. It boasts an excellent education system and according to national statistics at least, a 97 per cent adult literacy rate.

During the period of British colonial rule, extending from 1627 to 1966, Barbados was almost exclusively developed as a monocultural sugar economy, based on plantations and slavery. It is generally agreed that the conditions faced by black slaves in the Caribbean were far harsher than those faced by slaves in America (Lowenthal, 1972). The legacy of the colonial system is still witnessed in the highly skewed relationship that exists between race and ownership within the economy. Thus, much of the economy is in the hands of a small number of white families, although the majority of the population is described as black for census purposes (87.79 per cent in 1991). These families have in the past been referred to as owning the 'Big Six' companies. In the political arena, however, there has been the emergence of a black ruling elite, together with associated black professional and business groups within the population.

In the period since independence, manufacturing, tourism and latterly offshore data processing have become the mainstays of the Barbadian economy. Thus, Barbados now affords good employment prospects for migrants, especially via its well-developed tourism sector (Potter and Phillips, 2004), but also in the public and private service sectors more generally. For example, tourism currently accounts for 13,970 jobs directly and manufacturing 11,390; whilst 25,390 are employed in Government services and 18,190 in the general service category.

As noted by Potter and Dann (1987), Barbados has clearly come a long way since the colonial period. However, in the post-independence era, the people of Barbados have turned their attention northwards in the direction of the United States. As early as 1987, Potter and Dann suggested that 'Little America' might be a more appropriate description of this former British colony. This certainly seems increasingly true in terms of modernisation effects and the consumption of mass culture, including North American music and cable television, migration paths and the regular recreational travels of the relatively wealthy and mobile.

Incorporating Race and Gender in Caribbean Return Migration

While post-colonial theories of identity have articulated the essential 'inbetweenness' and hybridity of migrants living in the diaspora (Fanon, 1967; Gilroy, 1983; Hall, 1989; Bhabha; McLeod, 2000), according to Gilroy (1993) they have failed to recognise the importance of transnational connections, crossings and tensions between Africa, the Caribbean, America and Britain. Moreover, such theories of identity have tended to overlook how these hybrid identities are constructed and the role that is played by race and gender in such contexts.

In examining how Bajan-Brits construct their racialised and gendered identities in England, we make use of Fanon's (1967) paradigmatic framework to firstly, argue that an ingrained knowledge of racial inferiority overshadows and dominates the post-colonial experience of black people living in Europe. Moreover, this knowledge of racial inferiority provides the basis for a racial identity which is both dialectical and alienating (Loomba, 1998). This identity, borne of racial difference, is based on the adoption of a 'white mask' of perceived egalitarian identity, and with it, an associated disavowal of blackness. According to Fanon, this 'black skin-white mask' identity is ambivalent, uneasy and racially flawed. Further, ultimately it can never be entirely successful, since, the fact of blackness is forever fixed in white imaginings, in terms of difference.

In addition we add the conceptual framework of '*social location*' to show how the intersection of gender and race in a post-colonial context allows for a construction of identity based on both racial and gendered difference. We use the term social location to mean a 'person's position within power hierarchies created through historical, political, economic, geographic, kinship-based and other socially stratifying factors' (Pessar and Mayler, 2001:6). Thus a person's social location is affected and conditioned by hierarchies of class, race, sexuality, ethnicity, nationality and, of course, gender. In other words, according to Pessar and Mayler (2001), 'social location' confers certain advantages and disadvantages.

The nexus of these multiple dimensions of identity will influence how people think and act.

Thus, using the social location framework, prospective second generation Bajan-Brit returnees are viewed as a racially-hybrid, economically and socially disadvantaged group within post-colonial England. Deemed to be neither English nor West Indian, neither Black nor White, the Bajan-Brits inhabit an inbetween liminal space. Theirs is a social location/identity based on the experience of the post-colonised living in the world of the coloniser. Furthermore, this post-colonial existence is dominated by an ingrained understanding of prevailing racism and racial inferiority (Fanon, 1967). Moreover, we argue that it is this ingrained knowledge of racial inferiority and the disadvantages inherent in their particular social location that provides the impetus for young returnees' emigration in the first place.

The 'Inbetween Identity' of the Bajan-Brit and the Motivation to Return

Fanon (1967) maintained that the black skin-white mask identity is dialectically alienating and precarious, since it often involves the adoption and internalizing of a British 'white' identity and the disavowal of the black 'inferior' identity. The precarious nature of the black skin-white mask conceptualisation lies in the fact that it is never entirely successful within the post-colonial context. Thus, it results in a new identity, the state of flux of which is its inbetween nature, of being neither white nor black, neither English nor West Indian. This hybrid identity of the Bajan-Brit embodies features of both its white and black characteristics. Moreover, the need to affirm such an identity provided the basis to migrate to Barbados for many of our informants, see Potter and Phillips, 2006.

In order to understand the importance of the hybrid identity as a factor promoting return, an appreciation of how this identity is constructed is provided here. This examination is presented through an exploration of the early socialisation of young Bajan-Brits. Part of the early socialisation of the Bajan-Brits in the UK appeared to comprise a disavowal of blackness and a rejection of Caribbean identity. For example, this process of rejection by the Bajan-Brits is witnessed in these two narratives:

> 'When I grew up in England, I knew absolutely nothing about the Caribbean. The only concept that I had of the Caribbean was of Tarzan and Jane, from the programmes I saw. I knew nothing. There was no real discussion, you know, in our household ... this is where I came from; this is who you are ... My identity was mixed up. So I went through life wishing you had the white skin, the blond hair and blue eyes and the skinny figure. When in reality, no one grounded me to say, 'well this is where you are from'. So when I was young, growing up in school, I never considered myself Bajan'.

> 'It was only when I was sixteen that I realised that I was black, because at that age, I started going out and competing in the open world with other workers'.

The wearing of the white mask is further entrenched with the internalization of a British identity. Initially, this stance does not prove problematic, until that is racial difference begins to be actualised and pointed out by white peers. When this happens, the white mask superimposed on the black skin starts to prove an uneasy fit. It had thus proved alienating for a number of the Bajan-Brits when living their early lives in the UK:

> 'I don't know. I think I belong here more than I thought that I belonged in England, actually. When I was in England, I was always considered the outsider. My parents ... my dad always lived in white areas ... then I moved to Leeds, and even (there) my circle of friends was mainly white. I always felt slightly out ... It was always those niggling feelings ... Coming back to a majority black country you feel you fit in'.

> 'Basically, you had to work twice as hard because you were representing the whole black population and people would know everything you did. I used to feel I was under a microscope. When I first came here on vacation, I don't want to be harsh, but I felt normal. I felt like the average person'.

Moreover, such feelings of racial alienation and difference in the UK were further compounded by the fact that elements of a Caribbean gender culture still formed part of this new black skin-white mask identity of the young Bajan-Brits. Thus, attempts at assimilation and egalitarianism were highly problematic. Those 'niggling' things, as pointed out by one of our informants, were based on a social location of disadvantage characterised by elements of Caribbean family ideology and culture. According to Barrow (2001b) Caribbean family, ideology and culture are often characterised by irregular unions, illegitimacy, matrifocality, extensive enduring networks, the value of the mother-child bond, the peripheral role of males and outside children. Moreover, commentators such as Barrow (1996, 2001b), Douglass (1992), Smith (1988) and Roberts and Sinclair (1978) have all argued that social structures in the Caribbean have remained virtually unchanged in the face of economic development and European ideals. It is perhaps not surprising therefore that basic Caribbean family structure has not been dramatically transformed subsequent to living in the UK.

With regards to the gendered role of the Afro-Caribbean male, masculinity is often expressed in terms of outside children, sexual promiscuity and family marginality. It can be argued that such masculine traits present a demarcation from the traditional roles of the white hegemonic male in the English patriarchal system (Dann, 1987; Phillips, 1998). This demarcation often results in a form of Othering. The Afro-Caribbean male identity is constructed of the white hegemonic male's

Other (Fanon 1967; Phillips, 1998a). Such a gendered identity proves to be deviant within the English post-colonial setting.

Moreover, the response to such expressions of deviance seems often to be manifest to the Bajan-Brits in their early socialisation as harassment by the police. Many of our male respondents said that they were constantly harassed by the police in the UK. In several cases, repeated harassment appeared to be a major factor in the final decision to migrate to Barbados. The following two narratives best exemplify the views of our male respondents with respect to the incidence of police harassment:

> 'In England being harassed by the police and walking down the street and not knowing if I was walking past fascists or Nazis and them taking an instant dislike to me. The only time I felt comfortable was in Brixton or somewhere like that. I was stopped before I came here. I took my Dad to the embassy, the Consulate in Tottenham Court Road, and I was outside having a cigarette and a policeman wanted to search my pockets and I just thought forget this bloody ... you know what I mean? I just thought that this stuff was never ending'.

> 'The police would harass us every week. I didn't want my kids growing up in that. It is not their fault that they are black. I never got arrested or nothing. I've only been searched by the police maybe once every two weeks from the age of eighteen to twenty, and they more or less said to us, that it is their policy to search young, black men because it is a good chance that you have something on you'.

Like the gendered role of the male Bajan-Brit, elements of Caribbean female gender structure still remain part of the hybrid identity of the female Bajan-Brit. It is not surprising therefore that most female respondents were single mothers, since matrifocality is an integral feature of Caribbean social structure (Dann, 1987; Barrow, 2001b). However, although culturally accepted within the Caribbean context, matrifocality can be viewed as an aberrant family structure within the English, patriarchal system. As a consequence, many of our single mothers felt discriminated against in England. This discrimination was frequently manifest in a lack of available child care opportunities.

Moreover, several Bajan-Brit single mothers reported that they were unable to rely on the network of kinship support commonly available to West Indian families. For instance, in the Caribbean, a typical role of the extended family, comprising grandparents and close kin, is to undertake much of the childcare for parents who work outside the home. Moreover, since the majority of the parents of single mothers had already returned to Barbados, many of the female informants were bereft of the childcare resource provided by grandparents. It is quite understandable, therefore, that many of the respondents who functioned as matrifocal families felt the need to return to an environment in which such a family structure was fostered.

For example, Mary,[1] a 31 year old single mother who had returned to Barbados one year previously, discussed why she had decided to 'return' to Barbados with her children:

> 'England was really hard and when you are not working and you have children it is very difficult … Childcare is cheap here, plus my parents were here. My mum told me to come down and she and dad would look after them. Plus, their father was no help. He didn't work … We had no money coming in. I was living on social security but I still wanted to have nice things, so I said let me take up mum's offer'.

In another case, an informant explained why she migrated to Barbados. For her, like Mary, return reflected the need to overcome the consequences of being a single mother in England:

> I came over here, because I was a single parent in England, and I didn't want to be in the single parent trap thing, you know, which so many young black women are getting themselves into in the UK.

In a similar fashion, Judy was clear about her need to return to Barbados, viewing her parents as a significant childcare resource:

> Childcare was just too expensive and the kids had grandparents here who were willing to help, you know.

It can be suggested that the motivation of single black mothers to migrate to the security of their extended family is a function of a white patriarchal system, which pathologises single black motherhood on the one hand, and encourages reintegration into a traditional, patriarchal family structure, on the other. Thus, within this particular context, return migration to Barbados may serve to reduce the role of Bajan-Brit single mothers to a more traditional role of daughter, and thereby subject to male authority. For female Bajan-Brits, return migration to the Caribbean is a function of the transnational nature of patriarchy.

However, the decision to 'return' to Barbados by our respondents was based on the recognition that their hybrid identity conferred no advantages within the English patriarchal system. The black-skin white mask hybrid identity of the Bajan-Brits was in itself a liminal existence based on alienation and Othering. By deciding to migrate to Barbados, many of our interviewees sought to 'return' to a place where such 'differences' were the cultural norm, as the following narratives suggest:

> 'I would rather live here with the strifes and trials that you have here than go back to England and have that underlying racism which I find very dangerous'.

'When I stepped off the plane, I said, "thank God I am home"'.

Race and Gender in Barbados: A Theoretical Overview

To explore how Bajan-Brits adjust to and negotiate the racial and gendered structure of Barbados, we employ Said's (1983) *Occidental-Oriental binary framework*. This framework gives rise to a mode of thought based on a European ontological and epistemological binary distinction between the Other, and the European coloniser. This binary distinction is premised on power, domination and a complex hegemony.

The value of Said's discourse lies in the fact that it may be applied to almost any colonial or post-colonial situation where a European construction of the Other exists. In the Barbadian context, what is valued and supported by doctrine, bureaucracy, imagery, vocabulary, scholarship, social and political institutions is a white, English (post)-colonial hegemonic identity. This white English identity is not so surprising for an island which, as already noted, has in past times been described as 'Little England'. Moreover, although having attained political and social enfranchisement for the predominantly black populace, the economic, and by extension, the social and political hegemony of the white planter-class remains virtually unchanged (Beckles, 1996; Lewis, 2001). Beckles (1996: 529) explains that, although there has been some remodelling of the traditional social structure – with more doors now open and glass ceilings raised – the ideological and economic foundations of the nation have remained largely unchanged (see also Potter, Barker, Conway and Klak, 2004). Moreover, Lewis (2001: 163) argues that white domination in Barbados has been so deeply ingrained through social culturalisation, that there is a significant level of acceptance and internalization of the *status quo*.

Although the Oriental discourse of Said (1983) provides an understanding of the social construction of race within post-colonial Barbados, it does not explicitly provide an appreciation of the gender structure with which it intersects. Therefore, this chapter employs the dual frameworks of '*reputation* and *respectability*' (Wilson, 1969, 1973), and '*respect* and *shame*' (Barrow, 2001a) to explain Barbadian gender structure. The perspective of reputation and respectability (Wilson, 1969, 1973) is interpretively centred upon a gendered dichotomy of sexual morality. In the context of the Caribbean, respectability is regarded as being rooted in a Eurocentric culture and is the normative concern of the elite classes and women. Reputation by contrast, is seen as a post-colonial response from those who are unable to attain the prosaic symbols of respectability – and thus characterises the domain of working-class men. Accordingly, working-class West Indian masculinity is defined in terms of reputation – representing 'poor man's riches' – his virility, bragging about sexual conquests, the fathering of many children and boasting of such achievements in public. Other expressive traits that manifest reputation are the use of 'sweet talk', musical ability and displays of erudition in arguments. Wilson's (1969, 1973) conceptual framework has, however, been criticised for its essentialism, as is evident in his classification of reputation as an exclusively male domain. For example, Besson (1989) has argued that his binary gendered analysis serves to obscure and reduce the value of women. However, it can be argued that Wilson's analysis does provide insight into West Indian gender structure, a particular focus here.

Barrow's (2001a) closely linked work on *respect* and *shame* builds upon the earlier emic work of Wilson, (1969, 1973) and provides a further theoretical framework for understanding Caribbean society. Barrow views the complimentary moral codes of respect and shame as integral to understanding Caribbean social relations. She argues that respect and shame refer to personal qualities and reflect sensitivity to public recognition of that quality. If a person knows respect, they also know shame. Thus, one shows respect for oneself, in order to avoid having to feel shame. Barrow's theoretical framework of 'respect and shame', like Wilson's 'reputation and respectability' is strongly gendered. For instance, Barrow (2001a: 204) views a man's self respect as dependent on public expressions of authoritative, self-assertive behaviour. Male shame is based on the inability to gain respect from others. In contrast, a woman's self-respect is reflected in such qualities as modesty, restrain and discretion. This female version of self-respect is very much in keeping with Wilson's (1969) notion of respectability. Moreover, a woman's sexual behaviour is both integral to, and a reflection of, her self-respect. Therefore, her sexual behaviour should indicate qualities of restraint, obedience and faithfulness to her partner. However, Barrow goes beyond Wilson's (1969) thesis and argues that the high incidence of matrifocality, combined with women working outside the home and the increasing use of contraception, have all called into question the clear gender divide of respect. Barrow argues that 'social relations conducted in the idiom of respect are indirect and ambiguous' (2001a:204). Barrow's approach is of significance to the present chapter in so far as it challenges Wilson's tendency to reduce Caribbean gender relations to the binary functionalist constructions of male and female behaviour. Instead, Barrow views the moral code of respect and shame not as mere binary constructions of sex roles, but as complementary, and at times overlapping.

Thus, it is within the context of the dual framework of respectability and reputation (Wilson, 1969, 1973) and shame and reputation (Barrow, 2001a), couched within the Oriental framework of Said (1983), that we interpretively provide a framework for understanding the complexity of Barbadian race and gender norms. Further, these frameworks provide a basis for interpreting how young Bajan-Brit returnees adjust to, and negotiate, local gendered and racialised spaces.

Race and Return: Bajan-Brits as Symbolic Whites

The black skin-white mask hybrid identity, for the most part so alienating in the English setting, appears to have afforded Bajan-Brit returnees an economic and social position of relative advantage and privilege in post-colonial Barbados. Viewed in the context of the Occidental-Oriental framework (Said, 1983), the value of a white English identity in this post-colonial situation can easily be understood. Such an identity, which is often manifested in the use of an English accent, way of dress, work ethic, as well as general patterns of behaviour, is highly discernible within Barbadian culture. Furthermore, these distinctive cultural differences borne by such hybrid individuals of black skin and British origin, have created a new 'inbetween' space within the social and economic milieu of Barbados, (see Potter and Phillips, 2006).

The new liminal space that has thus been etched out, just above the black populace, but below the white-mercantile class, has served, at least to some extent, to challenge the traditional binary constructions of racial identity in Barbados. The 'inbetween' identities of the Bajan-Brits have given rise to a new racial identity, within which the strict boundaries of black and white behaviour are blurred. Thus, race is overshadowed by an overt English identity, which inherently invests returnees with 'symbolic whiteness'. In turn this token whiteness provides a door opener for the social and economic mobility of these Black English individuals in a still predominantly white-dominated hegemonic social order. In addition, this token whiteness is usefully employed by the young English returnees in their negotiations of Barbados' social and economic realms. These basic arguments are elaborated in the account that follows.

The Power of the Black English Accent

There were strong indications that the young British-born returnees employed their English accents as powerful brokering tools to override the extant racial barriers in Barbados. Thus, their accent appeared to provide a unique social and economic privilege within a traditional English hegemonic society. This was freely acknowledged by one of our informants:

> 'There are two ways of going through it in Barbados. If you
> are white or if you've got an accent, you know and you will
> use it to the best of your ability which ever you've got …'.

Several of the young Bajan-Brit returnees articulated the propensity of Barbadians to associate an English accent with symbolic whiteness, and by extension, therefore, power. According to Fanon (1967) this tendency is not surprising in a former colonised setting; it is the articulation of the language of the Mother Country and thereby serves to whiten and to subjugate blackness. For example, one of our informants worked in a college and frequently organised events over the telephone. She related how her token whiteness on the telephone gave rise to surprise when she subsequently appeared in person:

> 'I've had a number of experiences since I have been here in
> terms of like, when I organise events and stuff for college.
> When I get on the phone, especially when I first came here my
> accent was much stronger, and I'd organise things and when I
> would go to collect the cheques or whatever, you could see the

shock on people's faces, because they honestly thought I was white ... and you could tell!'.

Other respondents also articulated similar experiences relating to the association of an English accent with whiteness, focussing in particular, on the advantages that accrue from this in the workplace:

'You get a little more respect in my job because you've got a British accent ...'.

'They are not looking at your colour necessarily now. It's how you speak; the way you were brought up; so you just transgress from one type of barrier to another really'.

In one extreme case, an informant actually reported that:

'... They used to call me a *"pseudo-white person"*'.

The Black English and the Employment Market

The essential privilege that an English accent bestows on the young returnees has been highlighted by means of the preceding excerpts. However, it appeared to be the combination of an English accent and perceptions of a superior English work ethic that frequently served to propel returnees into advantageous professional positions. For example, many Bajan-Brits work in professional jobs, but without the benefit of formal qualifications. For instance, while one returnee held a college certificate in child care, she now manages a major hotel on the West coast. The owner of the hotel is a white Englishman, and it seemed clear that she had been preferred for the job, not least because of her British background.

In a similar fashion, another young returnee had attended an interview for a marketing post at one of the local radio stations. But on hearing her accent, she was offered a job as a radio announcer. At that time she had no broadcasting experience and such a position had never been her career goal. This is what she had to say about how the opportunity presented itself:

'I had applied for jobs before I came here, followed up with phone calls when I arrived here. When I arrived, the Mirror[2] was one of the places that I had applied ... I actually went for a marketing job. When I got to the interview, as soon as I opened my mouth ... they started talking to me about being an announcer on the radio which is something that ... I never ever in a million years ever dreamed of doing. I really didn't want to do it, but opportunity knocked at the door ... You has to go and see what happens. I did an audition a few days later. I think part of it was my accent; they were also impressed by my qualifications and experience too. I think they were charmed by a young English woman ... '.

It is very evident that in seeking and acquiring employment, many of the Bajan-Brit returnees appeared to be highly aware of the common perception held by Barbadians and international employers alike, of a superior 'English way of working'. It was also apparent that they were conscious that this perception was to their advantage in the job market. Such an appreciation was clearly illustrated by one of the interviewees, who emphasised the attraction of her way of talking and her general approach to work:

> 'I was selling advertising. The clients loved me because I was different and yeah … it was "Ella come see me and let me hear you talk". So, I knew that I had something which was an asset to me. It was my work attitude. When everybody dashed off at 5pm, bang on 5pm, 6pm, 7pm, it didn't bother me. I just loved what I was doing, so I quickly got promoted within the company, so therefore there was no need to go back. Professionally it is an advantage because of … I put it down to … this is my personal opinion … colonial days. A lot of Caribbean people still, although it is really changing, because it is a radical concept now, but they still kinda look up to the white person or people that are different to them. So for me, I got treated differently from a professional point of view when I was interacting with people, because I was slightly different and the difference was really the environment which I grew up in. I thought differently. I didn't think Bajan. My work attitude was just getting it done whatever it took. So for me my whole approach was different and the business people tend to like that approach because it made them more financially viable. So professionally, I have done well, and fortunately, my organisation is pro-training, pro-self development. They will give you a year, two years, three years off to do your degree … I am then battling with a personal bias. Number one, professionally it seems that you are getting through easier than the local person and the only reason why I did and many people like me did, is because the work attitude or work approach is different.'

It is noticeable here, that Ella not only mentions that her accent and attitude count in her favour in the workplace, but she also views these advantages as being linked to the colonial era and the historical hegemony of whiteness.

Another defining characteristic that proved an advantage in the employment sector was the mode of dress adopted by the young returnees. In the following excerpt, a young Bajan-Brit tells how his perceived professionalism projected by his appearance helps in the job arena *vis a vis* the attire donned by local workers:

> I used to go to work in a shirt and tie 'cause the guys there wear polo shirts and jeans, so I stood out. Cuff links, the whole shebang – dress hard!

Along much the same lines, another young Bajan-Brit recounts how his friends' discerning appearance and general conduct facilitated entry into managerial positions at relatively early junctures in their chosen careers:

> 'I have three English friends here, and they just came back. They supervise in Price Mart, but for the time that I knew they were living in Barbados they were never working. How did they get a job? Not because of accent, because of the way they conduct themselves. Guy's hair used to be long like an Afro; when he went to the job, all off, clean-cut, shirt and tie. You wouldn't think it is the same person. Only 17 years old, you know, and he supervising 45 year old people that were there in that company for years that could've got the job. Right now there are companies in Barbados that are hungry for young people coming in from overseas 'cause they know that we have the training. All the guys that are foreign in the company they always get through quicker, 'cause Barbadians sit down and watch you like this. That is all they do, they just watch. And, you can always tell where they are, 'cause you could watch them'.

As the above excerpts indicate, young Bajan-Brits recognise the distinct advantages that their English work ethic and other defining characteristics bestow. Thus, it can be concluded that young returnees use these advantages to successfully negotiate the employment sector within the post-colonial spaces of Barbados.

Gender and Return: Adjustment to the Gender System in Barbados

Migration can mean an increase in economic independence and relative autonomy for the migrant (Morokvasic, 1984; Tienda and Booth, 1991; Grieco and Boyd, 2001). Indeed, we have shown that these Bajan-Brit returnees have generally experienced increased economic privilege as a direct result of their perceived Englishness (see Phillips and Potter, 2003). But, on the other hand, they have also entered a highly indigenous gendered system, which serves to change their gender status both in the home and within the wider society. Based on the complementary frameworks of respectability and reputation (Wilson, 1969, 1973), and shame and respect (Barrow, 2001a) as outlined earlier, we now examine how Bajan-Brits, both male and female, negotiate and adjust to their new gender roles, within a post-colonial space.

In general, it would seem that the male Bajan-Brits adjusted quite readily to the reputation system that encouraged black masculinity. However, female returnees appeared to face greater difficulties in adjusting to the diminished independence implicit in the respectability and shame conceptual framework. Hence, much of our focus with regard to issues of gendered adjustment is centred on the difficulties that our female respondents highlighted, particularly with respect to family life, sexism and harassment and friendship patterns. Indeed, our findings reinforce the conclusions of Bueno (1997). It would seem that the return of young female migrants from Western countries to their lesser-developed homelands often results in a strongly gendered problematic. And, the exchange of a Western patriarchal structure for a more traditional one seems all too often to result in some loss of gendered autonomy.

Our female respondents frequently expressed the view that their roles and independence were constrained by the moral code that implies modesty, restraint, obedience and discretion, as a function of the female codes of respectability and shame (Wilson, 1969, 1973; Barrow, 2001a:203). Indeed, many of our respondents felt that their female roles had been circumscribed to the home and family life in Barbados. Furthermore, the general sentiment articulated was that there was a clear distinction between the roles of women and men based upon an inside/outside dichotomy. For instance, women were often encouraged to stay out of the public eye and not to be seen exhibiting 'shameful' behaviour (Barrow, 2001a). In this context, shameful behaviour was viewed as any type of behaviour that went against the local dictates of respectability.

Male respondents by contrast, were actively encouraged to affirm the traits of indigenous masculinity – bragging about sexual prowess, having outside women, and exercising patriarchal authority (Wilson, 1969, 1973). It is not surprising, therefore, that male Bajan-Brit returnees were quite agreeable to their change in circumstance. Hence our male respondents were positive with respect to their adjustment to Barbados, as indicated by the following excerpt:

> 'When it comes to, you know, social gatherings and stuff like that … There is no language barrier. Many times I would be walking down the beach and they would give me a beer and they would say "man come over". They are wonderful at that'.

In contrast, the female Bajan-Brits, who found much of their freedom eroded, were quite vocal about the constraints and conflicts that are generated by confinement to home and family, the restriction of 'outside pursuits', sexism and sexual harassment and friendship patterns.

One informant clearly articulated such restriction in gender norms when she described her Bajan-Brit friend's experience of home life in Barbados. For her friend, as indeed for other female Bajan-Brits, adjustment often translated into relative 'confinement'. In her friend's particular case, adjusting to a traditional female role seemed to be a source of tension between her and her local Barbadian partner:

> 'I think that is another thing as well with my friend Pam, who is married to a Bajan. They (Bajan males) expect you to stay at home and look after the kids and don't have any sort of independence and don't have any sort of life, which is sort of like Veronica as well. She is married to a Bajan and they want to suppress your independence quite a lot, which is why I would never marry a Bajan. Not here anyway. 'Cause I think at the end of the day they have this attitude. Yeah, they will

> marry you but they will do as they please anyway. But you
> are expected to stay at home and look after the kids and run
> the family and whatever else. They are not necessarily geared
> up for career women, a lot of Bajan men. They don't like it!'.

Another female informant recounted how exactly adjusting to the notion of respect and shame has led to a reduction in the roles and activities of her returnee mother, who had been a very active person when she lived in England:

> 'I mean when I think about my father in England. He had a
> few friends but basically he was going to work, and you know,
> doing what he had to do – helping around the house. My
> mother, you know (had) Parents' Teachers Association, and
> Church and you know, West Indian group that they had. It was
> always lively and you know. I guess she had more ... and also
> she was younger when she went up to England, so she had
> friends from training school, while she was training to be a
> nurse and that kind of thing, so it was more that sort of long
> termed friendships, that sort of thing. I wonder what they are
> going to do because my mother has been here (Barbados) on
> two occasions for a six-month period and it is almost like
> cutting off someone's wings. She is so active in England and
> she is into so much and here is like ... '.

Pat's narrative also conveys a similar opinion with regard to the restrictions imposed by the norms of respectability within Barbadian society. Adjustment to respectability had clearly been a difficult process. It had proved so challenging that some actually articulated this in terms of a literal loss:

> 'Yes, definitely, especially people who are not working ... I
> mean you can't do housework everyday. You can't do
> housework every hour and if you are an active person it is
> very, very difficult. You have to stay at home and stare at the
> four walls and just twiddle your thumbs. Because a lot of
> people before they left England even though they were retired,
> they still had a lot to do and it is very difficult to just come
> back and suddenly ... I look at it as a loss whereby you go
> through a lot of phases. When you go through bereavement,
> you go through the anger phase, you go through the sad phase,
> and you know ... that's how I look at it – that's how I view it'.

Moreover, adjusting to respectability had often necessitated a curtailment of what is locally considered as 'shameful' behaviour (Barrow, 2001a). This shameful behaviour is often enacted when the respectable virtues of modesty, restraint and discretion are threatened by the Bajan-Brits' refusal to keep out of the public's eye – the male domain (Barrow, 2001a: 203, 204). It is within this context, that such Western emancipated females found some difficulty in learning the 'dos' and 'don'ts' of restrained respectability. As an instance of this, Jenny recounts her experience of being 'shamed' as a result of smoking in public:

> 'People always in your business! Like when I first came here I used to smoke waiting for the bus and people would pass and say: "oh you shouldn't be smoking". And I would be saying to myself, what does it have to do with you? People are so open and close, whereas in England people just get on with their lives'.

Jenny also describes the dichotomy in female and male spaces and her experiences of this binary gendered construction. For example, female spaces and social activities often serve to promote respectable behaviour and are house-based. In this case, she also experienced 'shame' for venturing into the male domain. Jenny recounts the trials and tribulations of learning about male and female spaces in Barbados:

> 'I find in England you go to pubs, nightclubs, the works, and here, you are looked down a bit like a scorned woman if you party a lot. You know, if you talk to Bajan women they say "no", they don't drink, they don't go partying, don't smoke … they have their own reasons why they think ladies shouldn't go there. But I also notice as well, you only see males in the local rum shops[3] and so on … so I never really felt comfortable to entice women to go partying with me.

A corollary of such gendered adjustment to reputation is the issue of sexism and the associated sexual harassment faced by the young female returnees. The female returnees found themselves to be the object of the local Barbadian male gaze. This male gaze is derived directly from the traditional paternalism of reputation (Wilson, 1969, 1973) and its consequential macho behaviour. Such overt macho behaviour was interpreted by many of the female Bajan-Brit returnees as sexual harassment. For example, Sandra recounts her experiences of being publicly 'harassed' in the streets by local males:

> 'The first thing actually … when I first came here, I found it really difficult, 'cause they would hit on you all the time … I never used to get this attention in England, you know, and I haven't suddenly become a model or anything. So then I used to dress like a tramp, you know – scruffy, thinking that it would deter them. And it didn't. So then, you know, it would kind of embarrass me when they would say "hi beautiful", you know. I am like "hey, leave me alone"'.

Another informant was swift in summing up the situation relating to public harassment, cogently declaring that Barbados 'is still sort of a man's club'.

The workplace also proved to be the battleground for many of the female Bajan-Brit returnees with regards to sexism. For example, in the narrative that follows, Shirley talks about how she grapples with the difficulties of working in an unregulated patriarchal environment. A consequence of this unregulated environment for female Bajan-Brits is being the subject of both subtle and overt sexism. For example, in her narrative, Shirley discusses her feelings of being objectified in the workplace and wider society:

> 'I find it is more difficult being a woman here than in England, definitely. First of all in terms of equality and women being equal to men ... I never really felt that was an issue in England. I would go for interviews quite confidently and do presentations in front of boards, management which were predominantly male – never felt threatened. Here is like double standards ... I would say that sexuality has a lot to do with it because somebody can't see past the physical and not really listen to you or what you are capable of doing. I just find that it is different ... It is like too much emphasis on the woman as a sexual being ... not as much emphasis is put on your capabilities as a woman, your abilities ... it is inequality ... macho bosses coming on to you, depending on how you respond to that, that would have an impact. I am not going to use that to get what I want...and it has hindered me quite a bit'.

For many of the Bajan-Brit returnees, the androcentric nature of Barbadian society was quite real, and was manifest in the absence of laws concerning sexual harassment. However, many female Bajan-Brits' have responded to such sexist challenges by taking a firm stance on this issue and by refusing to accept this aspect of the job culture:

> 'Men here are really chauvinist. You know, they do what you allow them to do ... I don't stand no shit, but then I never worked for a male boss. I think it is about what you allow them to do. Here there are no laws to protect women at work. I find that men are on a totally different level'.

It is clear that the female Bajan-Brits have found themselves contesting the highly pervasive patriarchal system that makes-up the post-colonial spaces of Barbados. It can be argued that such issues experienced in the workplace are the direct expression of a widely accepted assumption that womens' place is in the home. Consequently, there is no available 'protection' for women, legal or otherwise, if they venture into the 'outside' realm of the male domain.

Conclusion

This chapter has sought to contribute to the under-researched discourse on race, gender and return migration. Based upon the demographic profile of young second-generation returnees to Barbados, we find that return migration must be understood from a context of racialised and gendered hybridity. We argue that Bajan-Brit returnees' motivation to migrate can best be understood from the perspective of a white, patriarchal hegemonic society that seeks to Other the hybrid identity of the Afro-Caribbean.

Moreover, utilising the dual interpretative frameworks provided by the polarities *shame/respect* (Barrow, 2001a) and *reputation/respectability* (Wilson, 1969, 1973), interpreted in a post-colonial framework of Orientalism (Said, 1973), we conclude that Bajan-Brit migrants occupy a third ambivalent, although essentially privileged space within extant Barbadian social and economic structures. Theirs appears to be a liminal 'third space' derived by virtue of having been born and raised in England, and yet being of the black race. Accordingly, such returnees are shown to be both advantaged and disadvantaged, transnational and national, and black, but in some senses, symbolically white. Living in Barbados has involved these young, second-generation returnees in countless contestations concerning their extant identities within the post-colonial racialised and gendered spaces of contemporary Barbados.

Acknowledgements

The research described in this chapter was undertaken with the support of funding from The Leverhulme Trust for a project under the title 'The social dynamics of foreign-born and young returning nationals to the Caribbean'.

Notes

1. Names have been changed to guarantee anonymity.
2. The name of the media company has been changed in order to guarantee anonymity.
3. The local rum shop is the embodiment of the male space. It is the place where the male can wax lyrical about his masculine pursuits, and hone his skills of erudition. Therefore, it is a space where females are not welcome.

References

Barrow, C. (1996), *Family in the Caribbean*, Jamaica: Ian Randle Publishers.
Barrow, C. (2001a), 'Reputation and ranking in a Barbadian society', in C. Barrow and R. Reddock (eds), *Caribbean Sociology*, Oxford: James Currey Publishers, 201-213.
Barrow, C. (2001b), 'Men, women and family in the Caribbean', in C. Barrow and R. Reddock (eds), *Caribbean Sociology*, Oxford: James Currey Publishers, 418-426.
Beckles, H. (1996), 'Independence and the social crisis of nationalism in Barbados,' in Beckles, H. and Shepherd, V. (eds), *Caribbean Freedom: Economy and Society from Emancipation to the Present*, Princetown NJ; Marcus Weiners Publishers, 528-539.

Besson, J. (1989), 'Reputation and respectability reconsidered: A new perspective on the Afro-Caribbean peasant women', Paper presented at the Centre of Gender Studies, Hull University.

Bhabha, H. (1994), *The Location of Culture*, London: Routledge.

Boyd, M. (1975), 'The Status of Immigrant women in Canada', *Canadian Review of Sociology and Anthropology*, 12:406-416.

Boyd, M. (1986), 'Immigrant women in Canada', in R. Simmon and C. Brettel (eds), *International Migration: The Female Experience*, New Jersey: Rowan and Allanheld, 128-131.

Bueno, L. (1997), 'Dominican women's experiences of return migration: the life stories of five women', in P. Pessar (ed), *Caribbean Circuits: New Directions in the Study of Caribbean* Migration, New York: Centre for Migration Studies, 14-60.

Chamberlain, M. (1995), 'Family Narratives and Migration Dynamics: Barbadians to Britain', *New West Indian Guide* 69 (3&4): 253-275.

Clarke, E. (1957) (1970), *My Mother who Fathered Me*, London: George Allen and Unwin Ltd.

Dann, G. (1987), *The Barbadian Male: Sexual Attitudes and Practice*, Basingstoke: Macmillan.

Douglass, L. (1992), *The Power of Sentiment: Love, Hierarchy and the Jamaican Family Elite*, Boulder: Westview Press.

Fanon, F. (1967) (1952), *Black Skin, White Masks*, London: Paladin.

Gilroy, P. (1993), *The Black Atlantic: Modernity and Double Consciousness*, London: Verso.

Grasmuck, S. and P. Pessar (1991), *Between Two Islands. Dominican International Migration*, Berkley: University of California Press.

Grieco, E. and M. Boyd (1998), 'Women and Migration: Incorporating Gender into International Migration Theory', Center for the Study of Population, Florida State University, Working Paper 98-139, (http://www.fsu.edu/~popctr/papers/floridastate/1998.html)

Hall, S. (1989), 'New ethnicities', in Morley, R. and Huan-Hsing Chen (eds), *Critical Dialogues in Cultural Studies*, London:Routledge, 441-449.

Henriques, F. (1953), *Family and Colour in Jamaica*, London: Eyre and Spottiswode.

Herskovits, M. (1958), *The Myth of the Negro Past*, Boston: Beacon Press.

Hondagneu-Sotelo, P. (1994), *Gendered Transitions: Mexican Experiences of Immigration*, Berkeley: University of California Press.

Lewis, L. (1998), 'Masculinity and the dance of the dragon', *Feminist Review*, 59:164-185.

Lewis, L. (2001), 'The contestation of race in Barbadian society and the camouflage of conservatism', in Meeks, B. and Lindahl. F. (eds), *New Caribbean Thought: A Reader*, Mona, Kingston: University of the West Indies Press, 144-195.

Lim, L. (1995), 'The Status of Women and International Migration', 29-55 in *International Migration Policies and the Status of Female Migrants*, United Nations Department for Economic and Social Information and Policy Analysis, Population Division, New York: United Nations.

Loomba, A. (1998), *Colonialism/Postcolonialism*, London: Routledge,

Massey, D. (1990), 'The social and economic origins of immigration', *Annals of the American Academy of Political and Social Science*, 510:60-72.

McLeod, J. (2000), *Beginning Postcolonialism*, Manchester: Manchester University Press.

Mercer, K. and I. Julien (1988), 'Race, sexual politics and black masculinity: A dossier', in R. Chapman and J. Rutherford (eds), *Male Order: Unwrapping Masculinities*, London: Lawrence & Wishart, 80-140.

Morokvasic, M. (1984), 'Birds of passage are also women ...', *International Migration Review,* 4:886-907.

Pedrazza, S. (1991), 'Women and migration: the social consequences of gender', *Annual Review of Sociology*, 17: 303-25.

Pessar, P. (1986), 'The role of gender in Dominican settlement in the United States', in J. Nash and H. Safa (eds), *Women and Change in Latin America*, AM: Bergin and Garvey, 273-294.

Pessar, P. (ed) (1997), *Caribbean Circuits: New Directions in the Study of Caribbean Migration*, New York: Centre for Migration Studies.

Pessar, P. and S. Mayler (2001), 'Gender and Transnational Migration', Center for Migration and Development, Princeton University, Working Paper #01-06e, (http://cmd.princeton.edu/Papers_pages/trans_mig.htm).

Phillips, J. (1999), 'Tourist-oriented prostitution in Barbados: the case of the beach boy and the white female tourist', in K. Kempadoo (ed), *Sun, Sex and Gold: Tourism and Sex Work in the* Caribbean, Boulder: Rowan & Littlefield, 183-200.

Phillips, J. and Potter, R.B. (2003), 'Social dynamics of 'foreign-born' and 'young' returning nationals to the Caribbean: a review of the literature', *Reading Geographical Paper*, 197, 21pp.

Potter, R.B. and G. Dann (1987), *Barbados: World Bibliographic Series*, Volume 76, Oxford, Santa Barbara and Denver: Clio Press.

Potter, R.B. (2001), '"Tales of two societies": young return migrants to St. Lucia and Barbados', *Caribbean Geography*, 12, 24-43.

Potter, R.B. (2003), 'Foreign-born and young returning nationals to Barbados: results of a pilot study', *Reading Geographical Paper*, 166, 40pp.

Potter, R.B. and Phillips, J. (2004), 'The rejuvenation of tourism in Barbados: reflections on the Butler model', *Geography,* 89, 240-247.

Potter, R. B. and Phillips, J. (2006) 'Both black and symbolically white: the Bajan-Brit Return Migrants as Post-colonial hybrid', *Ethic and Racial Studies*, (in press).

Potter, R.B., Barker, D., Conway, D. and T. Klak. (2004), *The Contemporary Caribbean*, Prentice-Hall: London and New York.

Pratt, G. and B. Yeoh (2003), 'Transnational (counter) topographies', *Gender, Place and Culture*, 10, 159-166.

Roberts, G. and S. Sinclair (1978), *Women in Jamaica: Patterns of Reproduction and Family*, Millwood: KTO Press.

Ryan, L. (2002), 'Chinese women as transnational migrants: gender and class in global migration narratives', *International Migration*, 40 (2): 93-114.

Said, E. (1978), *Orientalism*, Handsworth: Penguin.

Scott, J. (1986), 'Gender: a useful category of historical analysis', *American Historical Review*, 91: 1053-75.

Simmey, T. (1946), *Welfare and Planning in the West Indies*, London: Oxford University Press.

Smith, R. (1988), *Kinship and Class in the West Indies: A Genealogical study of Jamaica and Guyana*, Cambridge: Cambridge University Press.

Tienda, M. and K. Booth (1991), 'Gender, Migration and Social Change', *International Sociology*, 6: 51-72.

Wilson, P. (1969), 'Reputation and respectability. A suggestion for Caribbean ethnology', *Man*, 4 (1): 70-84.

Wilson, P. (1973), *Crab Antics: The Social Anthropology of English-Speaking Negro Societies of the Caribbean*, New Haven: Yale University Press.

Chapter 5

The Nexus and the Family Tree: Return Migration to Grenada and Transnationalism in Context

Joseph Rodman and Dennis Conway

The transnational habitations of Caribbean migrants contribute to the formation of an increasingly malleable nexus of adaptive relationships between the Caribbean and the wider world. The present situation in many Caribbean societies is one of continuous exchange of people, ideas and capital with North America and Europe; one in which migrants take advantage of their multiple experiences, and potentially alter the social, economic, political, and attitudinal structures of their island homes. Contemporary migrants influence the Caribbean while maintaining a firm footing in far-flung metropolitan centers – Toronto, London, New York City, and Miami/Fort Lauderdale in particular.

In addition to analyzing the larger-scale effects of return migration and transnationalism, this study considers the individual experiences of working-age returnees to Grenada. Many authors have lamented the difficulties of quantifying return migration, often highlighting the lack of satisfactory data on the phenomenon (Ravenstein 1885; Gmelch 1980; Byron 1994). The use of migrant stories addresses this problem by illuminating the varied and complex experiences and relationships inherent in contemporary human mobility (Fog Olwig 1999; Lawson 2000). In their narratives and histories, migrants reveal multifarious transnational networks, which emerge as systems of social change that have serious implications for people and places (McHugh 2000). Migrant narratives are used to reveal individual motivations for migration and return and adjustments at both ends of the migration stream. Furthermore, the narratives highlight the extent to which many migrants operate within a transnational existence, sinuously maintaining identities, jobs and families in two distinct places, and creatively taking advantage of the opportunities they construct for themselves and their families in both locations. This chapter contributes to regional return migration literature with its focus on Grenada, where, unlike most Caribbean islands, little migration research has been conducted (exceptions being Tobias 1975; Pool 1989). More generally, this study examines the decision-making process of return migrants in Grenada, their adjustments upon return, their contribution and potential impact in their homeland and the overseas linkages they maintain upon return.

Transnationalism and return migration are not new to the Caribbean. Ravenstein's (1885) early postulation that each wave of migration has a compensating counter-wave has provided the basis for return migration studies and typifies one aspect of the historical movement of peoples to and from the Caribbean. Return migration from elsewhere in the Caribbean, North America, and Europe has been a common practice in the region since the seventeenth century. Each of the major international migrations of Caribbean people has been complemented by significant counter-migrations (Thomas-Hope 1985). Indeed, it is common to refer to a culture or ideology of migration within the Caribbean social science literature (Rubenstein 1979; Chamberlain 1997; Guarnizo 1997; Fog Olwig 1998; Fog Olwig 1999). As Chamberlain (1997) points out, this culture of migration is often permeated with the expectation of return.

The nature of the Caribbean's culture of migration has changed significantly over the last several decades. Rapid technological improvements in transportation and communication have created a new migration ethos that challenges the notion of return migration as the completion of the migration cycle (Gmelch 1980; Mountz and Wright 1996; McHugh 2000). Recent migration patterns differ considerably from previous experiences. In the context of the Caribbean, Goulbourne (2002: 190) maintains that, 'return is no longer a permanent or static feature but is rather a recurring and dynamic process that includes many returns and re-emigrations within a single lifetime'. This emergent migrant dynamism creates linkages within the Caribbean, and between these islands and the metropoles of North America and Europe. Migrant networks now produce more rapid and more efficient flows of human and financial capital than previous migration streams.

The new patterns of migration in an increasingly mobile world have given rise to multiple definitions and classifications of return migration. However, most important to the study of all types of return migration is the realization that the individual who relocates to a new country continues to maintain ties with the homeland and that these ties influence the possibility of return migration. In the Caribbean context, emigration is treated not as a means of severing relationships with home, friends and family, but as a way of broadening individual opportunities beyond island shores in order to improve an individual's, family's, or community's situation back home (Thomas-Hope 1985). Caribbean people often employ what Carnegie (1982: 11) has termed 'strategic flexibility' in their migration patterns and plans – an approach that allows migrants to quickly adjust to varying situations that arise and create a net of multiple options that permits them to 'hedge against future insecurity'. Carnegie (1982: 12) argues that 'strategic flexibility' is part of a wider ideology of migration in the West Indies that is 'far more pervasive and institutionalized' than it is in North America. The historic institutionalization of migration in the Caribbean combined with the contemporary ease of mobility fosters an expectation of return and, subsequently, generates a multiplicity of spaces for operation, identity and cultural expression.

Contrary to early theories of migration, non-economic factors strongly influence the decision to return (Gmelch 1980; Chamberlain 1997). Migrant's motives for return have included strong family ties in the home country, dissatisfaction with the social status or condition in the receiving country, obligations to relatives, feelings of loyalty or patriotism, the perception of better opportunities in the homeland and nostalgia. Appropriating the long-established 'push-pull' framework, return migration is largely determined by socio-cultural pull-factors in the home society rather than dissatisfaction with conditions in the receiving country (Phillips and Potter 2003).

In an attempt to advance the study of return migration in the Caribbean beyond the description of factors influencing the decision to return, analysis of the impact of migrants in their home country after return emerges as a fresh area of inquiry. While considerable energy has been devoted to assessments of remittances and the investment of returning retirees in the Caribbean, relatively little research has focused on the experiences of younger, working-age returnees. This study does, however. During varied experiences in North America and Europe, young migrants obtain vital education, work skills and understanding. Upon return to their homelands, young returnees carry with them a considerable cache of human, social and financial capital; possessing the potential to effectively convey important ideas, attitudes, skills, and capital. It is the experiences of these younger returning migrants, and their resort to transnational practices and use of transnational networks, that is examined here.

Significance of Return to Grenada

The historical movement of people to and from Grenada can be linked to events on the island that result in 'short bursts of return migration' (Pool 1989: 261). The first effort to build the Panama Canal, starting in 1884, required the labor of tens of thousands of West Indians, including Grenadians. The subsequent abandonment of the project in 1888 can be seen as the first modern event that precipitated the return migration of laborers to their home islands, such as Grenada, in the Caribbean (Marshall 1982; Rubenstein 1979). Pool (1989) cites momentous episodes on the island and in the region, such as the cocoa boom in the late 1920s, Gairy's labor revolt in the early 1950s, the black power revolt in Trinidad and Tobago and the subsequent forming of the New Jewel Movement in Grenada in the early 1970s, as impetuses for return migration to the island. However, in the same study the author acknowledges that migration patterns to and from Grenada have become more consistent since the US invasion in 1983 (Pool 1989). Increased efficiency in international transportation and communication and the establishment of Grenadian social networks abroad have fostered the mobility of Grenadians in and out of their homeland. As Table 5.1 indicates, the movement of Grenadians off their island has markedly increased even in the past few years.

Pool (1989) importantly recognizes the embedded linkages between Grenada and Trinidad and Tobago. He argues that large-scale circulation of Grenadians to Trinidad and Tobago began in the 1910s due to the high wages offered on Trinidad's oilfields. However, upon further review, migration to Trinidad and Tobago is more complex. Valtonen (1996) maintains that the combination of

Table 5.1   The inflow and outflow of Grenadian citizens in their homeland by year

Year	Arrivals				Departures			
	Nationals Abroad	Residents	Total	% of Pop.	Nationals Abroad	Residents	Total	% of Pop.
2001	14,958	47,935	92,593	70.15	12,812	12,347	25,159	28.20
2002	16,258	49,887	66,145	74.14	14,117	43,828	57,945	64.95
2003	14,485	55,719	70,204	78.65	15,940	55,645	71,585	80.20

Source: Government of Grenada 2003; IDB 2004 (figures for 2003 are projected).

under-population in Trinidad and Tobago and the islands' delayed (in comparison to the rest of the West Indies) development of plantations and subsequent labor shortage allowed Trinidad and Tobago to be a destination for migrants from neighboring islands experiencing economic hardships after the decline of the sugar industry in the first half of the twentieth century. The loss of jobs due to the sugar crisis in the Caribbean was not ameliorated through diversification and industrialization in many of the islands. However, Trinidad and Tobago became an attractive destination for other islanders because they had successfully developed their mineral industry in the early part of the twentieth century. Trinidad and Tobago further attracted more migrants as a result of the opening of United States military bases, which created a wave of construction and service jobs. Grenadians took advantage of these opportunities in Trinidad and Tobago throughout the twentieth century – more so than other Caribbean islands largely due to the proximity of the two islands. As a result of these initial migrations from Grenada to Trinidad and Tobago paths and social linkages were established that in turn perpetuated further migration streams. So much so, in fact, that during the 1962

Table 5.2 Out migration in selected Caribbean countries, 1990

Country	Total Population	Total Out-Migrants	Out-Migration Rate
British Virgin Islands	16,105	2,949	26.8
Grenada	**83,838**	**18,687**	**19.1**
Montserrat	10,634	1,958	18.6
St. Kitts & Nevis	40,612	8,309	18.2
St. Vincent	106,482	18,169	15.1
Antigua & Barbuda	59,104	5,620	10.7
Dominica	69,463	7,507	10.1
St. Lucia	133,308	8,483	6.2
US Virgin Islands	101,809	1,524	2.2
Barbados	244,817	4,240	1.9
Guyana	701,654	13,453	1.9
Trinidad & Tobago	1,118,574	8,735	0.8
Jamaica	2,299,675	4,926	0.2
Bahamas	233,228	4,047	0.1

Source: Adapted from Thomas-Hope 2001.

election in Grenada one of the main questions concerned a unitary state between Grenada and Trinidad and Tobago. Although unification never occurred, the presence of Grenadians in Trinidad persists. By 1980 it was estimated that roughly 21,000 Grenadian-born migrants resided in Trinidad and Tobago – this is approximately a quarter of the entire Grenadian population (Valtonen 1996).

Contemporary data concerning international mobility in Grenada indicates that migration and, subsequently, return migration, figure prominently in the present-day lives of Grenadians. The island's out-migration rate is the second highest in the Caribbean (Table 5.2). Grenada's high out-migration rate, combined with migrants' frequent intent to return (Byron 1994; Ahlburg and Brown 1998), suggests that a substantial return migration rate to the island can be expected. A study conducted by the United Nations Economic Commission for Latin America and the Caribbean (UNECLAC 1998) affirms the assumption that significant numbers of migrants are returning to their Caribbean island homes after spending time abroad (Table 5.3).

Table 5.3 Returnees to selected Caribbean countries, 1990

Country	Total Returnees	As Percentage of Population
Antigua & Barbuda	5,937	9.47
British Virgin Islands	1,533	9.52
Grenada	**6,799**	**7.36**
St. Vincent	7,554	7.09
St. Lucia	9,453	6.77

Source: UNECLAC 1998; IDB 2004.

Importantly, the UNECLAC (1998) report also draws attention to the age-distribution of return migrants to Grenada. The return migration literature often focuses interest on the impact of returning retirees (who do in fact constitute a large percentage of returnees) (Gmelch 1992; Thomas-Hope 1992; Byron 1994); however, as Table 5.4 illustrates, a substantial portion of returning migrants to Grenada are young and of working age. The 25-39 year cohort is significantly larger (2,254-33 per cent), than either the more youthful cohort preceding it (716-10 per cent), or the older returnee cohort following it (1,411-21 per cent). These numbers suggest that a significant percentage (over a third) of returnees to Grenada represent, at the very least, a contribution to the domestic work force and, at best, a pool of internationally skilled, educated, and experienced individuals keen to spearhead dynamic, forward-looking national development.

Finally, the UN publication highlights the linkages between Grenada and Trinidad and Tobago (as historically acknowledged by Pool 1989) as well as the increasingly well-trodden paths to the metropoles of the UK, USA and Canada (Table 5.5). The data discussed above, and the experiences of returning Grenadians presented throughout the rest of this article, indicate that return migration is not an ephemeral phenomenon of migration patterns in Grenada, but rather an embedded and significant feature of migration and circulation for the Caribbean region as a whole.

Table 5.4 Age-distribution of returnees to Grenada, 1990

Age	Returnees	Percentage
0-4	29	0.43
5-9	79	1.16
10-14	137	2.02
15-19	250	3.68
20-24	329	4.84
25-29	**667**	**9.81**
30-34	**846**	**12.44**
35-39	**741**	**10.90**
40-44	514	7.56
45-49	404	5.94
50-54	493	7.25
55-59	492	7.24
60-64	533	7.84
65+	1285	18.90
Total	6799	100.00

Source: Adapted from UNECLAC, 1998.

Table 5.5 Distribution of returnees to Grenada by country last lived, 1990

Country	Percentage
Trinidad & Tobago	38.25
Great Britain	22.08
United States	11.64
Canada	6.79
Aruba	3.47
Venezuela	3.35
Barbados	2.82
St. Lucia	1.08
Antigua & Barbuda	0.34
St. Croix	0.25
St. Thomas	0.25
Martinique	0.03
Other	7.84
Not Stated	1.82
Total	100.00

Source: Adapted from UNECLAC, 1998.

Study Methodology

A set of semi-structured, open-ended interviews conducted in the summer of 2003 by one of the authors, is the main empirical basis for this research. The secondary data in Tables 5.1-5.5, provide the more general contextual situation, while our interviews of 17 respondents' recorded life history and migration history data provide 'narratives of experience'. The first-person histories were elicited from a snow-ball sample of returnees, the first wave of whom were identified by local primary informants. The interviews lasted anywhere from forty minutes to two hours in length. Each interview was conducted in a location agreeable to the respondent; often times, their place of employment. Each respondent was asked a common set of general questions, which included biographical information regarding their present age, marital status, educational level and job histories. All respondents were encouraged to be expansive in their answers. In most cases, answers were audio-taped and later transcribed. In a few instances, respondents preferred not to be recorded; in which case, their answers were written down in note form by the interviewing author.

The subject sample consisted of 17 young, working age migrants who had returned to Grenada, from the United Kingdom, Ireland, Canada or the United States. Many of the respondents had lived abroad at multiple times during their lifetime in multiple destinations. Eleven had spent time in the United States, 6 in Canada, 5 in the United Kingdom, and 1 in Ireland. In terms of the sample's gender breakdown, 5 of the respondents were female, and 12 were male. The group's average age was 32 years (at the time of interview). The respondents had spent an average total of 9 years abroad during their lifetime, but their experience characteristically consisted of two sojourns, not one long absence. Each of the respondent's sojourns lasted an average duration of 4.8 years in length; with the longest sojourn being 18 years, and the shortest 'stay away' being 6 months. With reference to their age when they originally left Grenada, 1 left as a child, 13 left as young adults, and 3 had already circulated back and forth during their youth – spending equal amounts of time abroad and in Grenada – before returning as young adults. With reference to the recency of their return, 12 of the respondents had returned to Grenada within the previous 5 years, half of whom had arrived back in Grenada during the preceding 6 months. The longest one of our resident respondents had been in Grenada was 13 years following his time abroad. The average length of residence after return among our sample was a 4 year adaptation period.

The majority of the respondents, regardless of the variation in durations of sojourn(s) abroad, were in general agreement that social re-adaptation to island life came quickly. On the other hand, returnees often cited problems of workplace adaptation upon their return to Grenada, and businessmen and public sector employees were particularly vocal about the perceived inefficiencies they encountered. These background summaries of the snowball sample's demographic and migration experiences then serve as a back drop to the narrative accounts that provide much more in-depth assessments of these return migrant's life-experiences 'back home'.

Intent to Return

Return is a critical element of the migration agenda for the majority of Caribbean migrants (Thomas-Hope 1985; Byron 1994). Realized or only presumed, the migrant's intent to return, or the 'return orientation', must be recognized as not necessarily translating into reality (Thomas-Hope 1985: 159). Byron (1994: 169) points out that 'intentions alone cannot be used as the basis of a model of migration'. However, migrants who mean to return are obviously more likely to do so than those who do not. Surveys of Tongans and Samoans in Australia by Ahlburg and Brown (1998) and of Nevisians in Britain by Byron (1994) demonstrate that the overwhelming majority of migrants, 99 per cent and 71 per cent respectively, intend to return home after varying periods of stay abroad. Furthermore, evidence put forth by De Souza (1998) of the real rates of return for Trinidadians living abroad (between 85-98 per cent for the period 1987-1991) supports the notion that intention to return, while not universally accurate, is more of a reality than a mythical prospect for migrants. Results from interviews conducted in Grenada for this study corroborate the arguments that return figures prominently in the plans of Caribbean migrants.

The following account of a 37 year-old supervisor at Grenada's National Water and Sewerage Authority (NAWASA) highlights several important themes that permeate a migrant's 'return orientation'. Here he recounts his intentions to return to Grenada after spending a little more than three years in Utah where he received a Bachelor's Degree in Environmental Management at Utah Valley State College:

> 'It was not something that I had seriously considered [staying in Utah or the USA]. It was not like I had problems in Utah. Actually, my wife likes it; she really loves it, in Utah. It's different, a lot different than the Caribbean. I didn't have a problem, you know, staying in Utah. But it's not that I wanted to stay in the US ... One [reason] is, I was on study leave. I had a bond with this company [NAWASA] ... I was working here and then I got study leave. So, while I was on study leave they were still paying part of my salary here. I had a commitment to come back here. I still think that I had an obligation to come back'.

> '... Added to that, you know, you always want to come back. You always want to come back home. My parents, my friends, everybody ... I still have my family here. My wife is here, but she is going to go back next month [to Utah]. She's going to go back to finish school; the baby is going to stay here'.

> '... Culturally it's [Utah] a lot different. That's the part that I really mind, but, you know, apart from that I don't mind living in Utah. I feel very comfortable there. They have some very good things about them ... If I don't remember to lock up my car, I don't bother. It's OK. I tell people, you know, I spent three years in Utah and I've seen, you know that bar, club

[The Club] that you put across your steering wheel? I've seen just one'.

'... If I lose my wallet, it's cool. I say 90-95 per cent of the time I'm probably going to get it back. If you stop on the street someone will give you a ride. I don't mind living there'.

'Culturally, you know, I'm used to ... you can go and sit down by the shop and old talk. If you play dominoes, you play cards, you watch cricket ... you can do that. But you can't do that stuff in Utah. You reach a certain age and there are some things that become important to you'.

In addition to underscoring the intent to return to Grenada throughout his stay abroad, this account illustrates the importance of positive factors in the home society as opposed to negative factors in the receiving society for influencing the decision to return. This echoes the argument in the literature that pull-factors are more important than push-factors in the choice to return (Phillips and Potter 2003). Additionally, formal obligations, to employers or families, can sometimes play a significant role in the return of migrants. Transnationalism is touched upon in this relation as the interviewee's wife soon plans on traveling back to Utah, leaving their young child with the father in Grenada. Transnational families are not uncommon in Grenada. While transnationalism, in the context of families and otherwise, will be discussed more thoroughly later, this account does reveal that the intention to return to Grenada often revolves around what migrant parents view as beneficial to their children. Interviewees often linked their desire to return home to the positive upbringing they hope to provide for their children in Grenada. A single mother of one discusses her concerns of raising her daughter abroad compared to the benefits she perceived in Grenada:

'It's hard to live in New York with a baby ... there is no time in New York ... Everyone is very busy. I often worked very late and had to call my sister to baby-sit for me ... I am extremely close to my daughter ... I want her to grow up in Grenada. There is good education and discipline in Grenada. Kids are immature in the US and the UK; they take things for granted'.

The effect of return to Grenada on the children of migrants weighs heavily on migration plans, even when the children or family has yet to be formed. A young, unmarried, returned Grenadian man expands upon the sentiments offered in the previous description while addressing the plans he has for his future family:

'I want them [my children] to understand the Grenadian culture ... I would really like them to understand where their parents – I'm assuming that the lady will be from Grenada also – at least for me, what the culture is, you know ... They need to know where they come from, what the culture of the Caribbean is like, Grenada and the Caribbean as a whole'.

A more circuitous explanation for return to Grenada is offered by a half-Irish, half-Grenadian physician who had spent a combined 23 years living in Ireland and England before deciding to relocate to his place of birth in the Caribbean. Here he justifies his return to Grenada by interestingly linking the 'naturalness' of the island to facilitating a positive upbringing for his children,

> '[Grenada's] a natural place to live in. Whereas England and Ireland, I love Ireland, but England's unnatural. People are friendly in Ireland. I find in Grenada that people are very friendly as well. England I find they are far less friendly, more rigid ... Whereas in Grenada ... you were allowed to breathe more so ... We have pets here. We have five dogs, we have hamsters, we have birds, we have cats, we have a turtle, we have all these things. Children have all that. It's great for the children ... I just think that Grenada is a great place for nature, a great place for living, a great place for bringing up children ... I love on a Sunday morning and, with my youngest daughter, its six o'clock in the morning, I take her to Grand Anse Beach on Sunday morning, we go for 15, 20 minute swim. And I love it. I love to go for a swim in the sea, or go fishing from time to time with my other children as well; they're four of us ... I just think it's more natural, you know'.

The migration culture of the Caribbean is based largely on family relations, particularly on the bond between the parent and child (Fog Olwig 1999). The preceding elucidations provided by Grenadian migrants illuminate, in colorful, if various ways, the importance of the consideration of children in the intent and realization of return.

Beyond kinship motivations and formal obligations, return migration is often part of a more wide-ranging plan by the migrant to gain experience, security and education that they can eventually bring back to their island homes. The two explanations that follow regard return migration as part of a long-term life strategy, less focused on specific career or family commitments. The first account is of a 25 year-old woman who had spent five years in Toronto She is currently trying to re-return to Canada, where the majority of her family resides:

> 'Maybe I go over there ... go back to school, work ... after, maybe at a certain age you can't take to cold anymore ... in your 40s you will come back home and, you know, relax yourself, have a job, have your nice home and everything what you want ... and there go to school to get a good education ... and that's what I want ... work for a while up there, of course ... and after, come back home [to Grenada] to work. At the same time I should be able to have everything I need ... my

security, my house, my car. I'll be able to come back and relax
... while I'm working, of course.

The second story comes from the pastor of a Presbyterian church in Grenada
who has traveled back and forth to the United States, where he acquired his
Bachelor's and Master's degrees, while also spending time in the seminary:

> 'Well it was very easy for me to come back to Grenada
> because I went to the States with a purpose to come back ... to
> get an education, to come back. And I guess ... my experience
> is different in that I went so that I can learn so that I can come
> back so that I can help my denomination grow. There was
> never a thought of remaining [there]'.

This study of return migrant experiences supplements the general trends shown
in other analyses that intention to relocate to island homes figures prominently in
migrant agendas. Indeed, the preceding examples strongly support Chamberlain's
(1997) argument that the intent and expectation of return is embedded in the
culture of migration in the Caribbean. Realisation of return is motivated by a
number of factors, including formal workplace obligations, perceived benefits of
the rearing of children and the situating of return in a 'strategically flexible'
migration schema, all of which have been illuminated in the previous accounts of
returned Grenadians (Carnegie 1982).

Intent to return is of considerable social and economic importance to the home
nation. Migrants who are intending to return are more likely to invest resources in
the homeland. Furthermore, as was stated to be the case in several of the examples
above, migrants are assumed to acquire valuable human capital in the form of
education and work skills, and potentially financial and physical capital as well.
This human and social capital can logically be expected to be invested in the
economic benefit of family, community and nation upon return.

Decision to Return

The investment and return of capital by repatriating migrants to Grenada represents
an economic aspect of the return phenomenon. However, when asked about their
specific motivations for coming back to Grenada returnees very often stated that
non-economic factors determined their decision. When discussing their moves to
Grenada, returnees revealed a strong sense of loyalty and nationalism for their
place of birth. Examining young Trinidadian returnees, De Souza (1998) explains
this attachment to the 'West Indian homeland milieu' by migrant's denotation of
their home island as a place of comfort, security and familiarity, by their failure to
function completely and comfortably in the host country, their lack of roots in the
receiving nation, and their inability to fully relate to foreign cultural norms. The
findings from Grenada point less to an inability to function and adapt to, or even

accept, American and European ways of life, and more to an inherent feeling of loyalty and a desire to contribute to their homeland.

In contrast to Potter's (2003a; 2003b) studies of young returning nationals to St. Lucia and Barbados, returning Grenadians more commonly identified a compulsion to contribute to the Grenadian nation, as opposed to returning for individual or family reasons. Migrant nationalism is expressed in the following account of a 34 year-old Grenadian, who had spent a combined five years in Toronto and New York City obtaining his Bachelor's degree in Geology, before working for 2 years as a materials engineer in the United Arab Emirates, Botswana and Jamaica and subsequently returning to Grenada to work in the public sector:

> '... I always felt that I would come back and live in Grenada. So I wanted to see how Grenada had developed and what was needed ... That was one of my biggest needs – to work with a company that ... work in Grenada that would let me see what was needed; which direction to go ... About one and a half years after [moving back to Grenada] I got an offer to move closer to my field of study, which is materials and testing and so on. I took it up and I'm here, basically ... this is my contribution to Grenada ... I always felt that I really wanted to return to contribute, see what there is to get into. I never felt that I have to come back, but I felt that I wanted to come back'.

A physician pinpoints the appreciation and reward he felt for practicing medicine in Grenada compared to practicing in Britain as a determining factor in his decision to return to the Caribbean:

> '... I thought Grenada needed me. As much as I needed Grenada, Grenada needed me as well. And, you know, in Ireland you're just ... I always fighting for one job. In England everybody's fighting for higher jobs in medicine. And I realized if you died the next day you wouldn't be missed at all. They'd probably be relieved that you died because then another person could get a job. I always felt I wanted, um ... I suppose I felt it my duty to help Grenada. I suppose that's why I'm back here'.

As justification for his return, one respondent simply asked 'Who will take care of my country?'. Even in situations where moving to Grenada after spending considerable time abroad may cause significant stress, feelings of loyalty and nationalism play a large role in migrant's final decision. For instance, after spending the majority of her childhood abroad in Canada and Britain, one female returnee expressed competing emotions in her decision to come back to Grenada, with nationalism finally holding sway:

> 'I knew I did not want to live in someone else's country; this [Grenada] was my home ... [But] people in Grenada are not

> culturally-minded. I didn't, and still don't, identify with the
> culture here. But I am very nationalistic. The pace of life is
> much slower. I have to give myself a pep talk frequently to
> keep myself going and keep myself motivated'.

The overall picture conveyed by the above accounts suggests a less individualistic and materialistic rationalization for return compared to what has been suggested in portions of the literature. Debates on the benefits and detriments of return migrant's impact on national development often revolve around the (in)appropriate investment of resources by the returnee – namely, what has commonly been termed 'conspicuous consumption'. However, the findings from Grenada point to at least a genuine *intent* to contribute to the nation's development and progress. As discussed earlier, intent does not always translate into fruition, but it does constitute a meaningful step in the manifestation of the migrant's desires, and hopes, to help their homeland develop. These narratives suggest that migration and return are not decisions made entirely on the basis of individual and family advancement; rather, small island pride plays a significant role in Grenadian's transnational experiences. Return migrants to Grenada cannot simply be classified as unskilled, failed migrants or inactive retirees; many returnees are in fact enterprising, educated, patriotic, and young. The potential bearing of this returning demographic on the development of Grenada is significant.

Further evidence from Grenada supports the argument of non-economic factors weighing heavily in the decision to return. For young Caribbean men, migration is often part of the maturing process (also see Pool 1989):

> 'I was dying to leave Grenada when I was 16 … I wanted to
> go out and see the world … Become a man so to say … I think
> it's good for children to get away. When they're of age.
> 'Cause I've been away … And I think I've come back a much
> better person, with my eyes open, my mind open to all
> different people of cultures, religions, everything like that.
> You stay in the Caribbean alone you become very close-
> minded, I think'.

Respondents often indicated a highly flexible, spontaneous and impromptu outlook towards their migration decisions about where to go and when. The culture of migration in Grenada and the well-established and broad-based family and social networks abroad allow individuals to impulsively take advantage of life circumstances with transnational mobility an ever-present option. For her first move abroad, a then-17 year-old student explains that she found out just weeks before her departure date that she would be moving to Toronto to be with family:

> 'I had never visited Toronto before moving there, it wasn't
> like that. It was like … I went there the summer after I
> finished high school. My sister just called me and told me I
> was coming up. Yeah, I had no idea I was moving to Toronto
> before that'.

The use of kinship networks facilitates spontaneous movement in the return migration stream as well. The following account comes from an individual, who quit a managerial-level position at a business in Canada at the age of 33 in order to return to Grenada:

> 'I had no job prospects [coming to Grenada]. I said, "Whatever, I'll find a job. I'm talented" ... I came back in '97, in February ... I was deciding either Trinidad or Grenada, 'cause I have family in Trinidad, I have brothers ... And I was born in Trinidad to Grenadian parents. And so I came back ... and I spent 10 days in Trinidad and 10 days in Grenada, but the deciding factor ... I didn't like my family in Trinidad, how I got along with them too well. And I came here [to Grenada] ... So then, I went back [to Canada] and I decided to move, and I moved. I came in February [to Grenada to visit] and I moved back in July, of the same year, 1997'.

All of these biographical vignettes from Grenada highlight the importance of non-economic motivations in the decision to return and, frequently, the spontaneity of these decisions. Much of the impulsiveness inherent in decisions to move to and from Grenada is rooted in transnational networks of friends and family. The examples demonstrate that the family can act as the reason for migration, as was the case for the female student moving to Toronto to be with her family, and also as a mechanism and facilitator for transnational mobility, as was the case for the man who quit his job in Canada and debated his resettlement in Grenada and Trinidad. In addition to demonstrating how nuanced family relations influence migration in Grenada, the findings from these interviews suggest a non-economically focused culture of migration is deeply-embedded. Individuals seem to take strategic advantage of familial and social networks abroad, often in order to fulfill a desire to eventually contribute to Grenada's growth and well-being.

Transnationalism – Family, Home and Entrepreneurialism

The diasporic diffusion of Grenadians to North America and Europe has formed a lattice of transnational socio-cultural and kinship relations that unites family, friends and countrymen across space. Guarnizo (1997) defines transnationalism as 'the web of cultural, social, economic, and political relationships, practices and identities built by migrants across national borders'. Glick Schiller *et al.* (1992) provide a broader definition, referring to transnationalism as 'the process by which immigrants build social fields that link together their country of origin and their country of settlement'. In the case of Grenada, transnationalism is most visibly manifest in new orientations of home, family and work that have resulted from the increasing ease of mobility between the Caribbean and the wider world.

Congruent with Fog Olwig's (1999) findings in Nevis, the cases in Grenada reveal that children of migrants are often raised in homes where at least one parent is absent. Previously cited in this study, the returnee who had spent several years in

Utah casually mentioned in his interview the fact that his wife would soon return to the States to pursue her education, leaving their infant child with the father in Grenada. This account provides a clear example of how the Grenadian population has become diffuse and how migrant culture has saturated life in the region to a point where families have appropriately adapted to the situation and become diffuse themselves. One respondent who works in media relies largely on his personal computer and the internet as his workspace, 'The technical nature of my work allows me to be anywhere... my operation is very flexible'. Indeed, his wife and two children reside in Florida where the respondent visits for 1-2 weeks per month, while much of his work is completed in Grenada, intended for Grenadian consumption. This particular case demonstrates a reversal of home and workplace for those Thomas-Hope (1985) labels as 'perpetual circulators', who typically regard the Caribbean as home and the foreign country as the work-space.

Potter's (2003a; 2003b) studies of young returning nationals to St. Lucia and Barbados indicate that returnees often migrate to the Caribbean with the purpose of starting a business. This proved to be the case among some of the return migrants in Grenada as well, where transnational linkages provide flexible and functional spaces in which entrepreneurial Grenadians are able to operate. Grenadians obtain valuable business education and experience abroad. They are able to relay these skills, capital and experience to the Caribbean and combine them with their more informal, social connections and know-how in order to become successful entrepreneurs in their home island. Such intrepid migrants are able to incorporate their foreign formal experience with their domestic informal experience in order to set up businesses upon their return to Grenada. Here they find they are able to efficiently use their informal relations in Grenada to create businesses and, subsequently, take advantage of their overseas experience and introduce new ideas that foster profitable enterprises.

Creative business opportunities in Grenada, as perceived by returnees through a lens of overseas experience, is captured in the following account:

> 'There are qualified people [in Grenada], but they don't do anything ... You need to know who you are ... People don't have any innovation ... you need to pursue a career and take initiative ... Lots of people are coming back [to Grenada] but they are not doing anything. People lack initiative in business ... and there is a lack of empowerment of people in Grenada ... I started a car business to cater to medical students. I started with one car. I had to have flexibility and I had to experiment. By the second year I had three cars and by the third year I began to diversify'.

The interviewee attributed much of his business success in Grenada to his hard work and to his 9-year experience in Canada, where he enrolled in several business administration courses.

Additionally, through their experiences abroad, returnees are able to occasionally create flexible and strategic transnational enterprises and careers that allow them to keep a foot in Grenada as well as abroad. One respondent recalled how he was able to publish and circulate a low-end magazine that focused on events in Grenada while he was living in south Florida. At the time, the interviewee had enough contact with Grenada, through personal trips, communication with family and friends, and the use of the worldwide web, to make such an endeavor possible. At the time of his interview, this same businessman was in the midst of a 6-week stay in Grenada during which he was attempting to establish a small-scale media and communications outfit. The design was to begin his operation by attracting sponsors in Grenada using the informal, personal connections he had established while working for the state-run TV station during his years of growing up in Grenada, before eventually expanding his company's reach to the wider Caribbean and the USA:

> 'The idea right now is to get it's [his company's] feet wet in the Grenadian culture, Grenadian business culture, establish an identity here. Let people know you're here. Two main partners, we both live overseas … Although from Florida … I'm gonna try to do some stuff from Florida with the Grenada Board of Tourism … The long-term goal, obviously, is to have a multi-media outfit that serves not only Grenada, but the Caribbean and the Grenadian and West Indian markets in North America … There's a lot of mediocrity here. And people settle for it. We wanna show people [in Grenada] that you can do things at a very high standard and still get paid well, and still make a lot of money, which is what we are doing … Just under a month, I mean we managed to set up a company and have it running and we are on-stream to make EC50,000 [about $19,000] in revenue in about a month. That's from a company that started six weeks ago. We had the office for the first two weeks in the car. You know, it's not bad'.

The entrepreneurial energies and aspirations described above, in part, address the call by Gmelch (1980) to look further at the innovative potential of returnees – how migrants' skills and ideas are 'used constructively to the benefit of the home society'. Although he acknowledges the potential impact of returning businessmen, in the final analysis, Gmelch (1980) is skeptical of their influence and claims that the majority of the Barbadian businesses established by returnees turned out to be redundant. However, the experiences from Grenada outlined above indicate a vitality of entrepreneurial spirit that undermines Gmelch's argument. The individuals interviewed for this study who were involved in business demonstrated an eagerness to fill a niche in the Grenadian business market, rather than add a redundant company 'to an already saturated market' for the purpose of individual profit or personal convenience (Gmelch 1980).

Importantly, these returning businessmen to Grenada recognize the potential demonstration effect their enterprises may have on the business community on the island; reinforcing the notion that many returnees have a vested interest in the growth and development of their country. Via overseas experience, returnees are able to see new opportunities for businesses in Grenada. As Gmelch (1987) argues for Barbados, there seems to be more room in Caribbean islands for the introduction of foreign ideas and attitudes in the private sector than in the public sector. Returnees are more readily able to experiment and apply their transnational business experience in the private sector, where growth, development and profit are the motivations for success. Added to this, returnees seem more willing and able to take advantage of the opportunities presented in the private sector than are those who have never left the island.

The diffusion of Grenadian people throughout the wider Atlantic has not so much resulted in the detrimental disruption of migrant orientations of home, family and work, mythically perpetuated by dire circumstances on their island homes that somehow threaten survival. Rather, Grenadians recognize the limitations of the opportunities for personal, family, and even national growth that their homeland offers. This recognition compels many Grenadians to pursue experiences abroad, where they gain critical work skills and education that can eventually be transferred back to Grenada. Through their experiences overseas, migrants establish complex transnational networks of kith and kin, facilitating a fluid exchange of people, capital, and ideas through space and across national borders. The increasingly mobile and transnational nature of Caribbean life often results in the reconfiguration of notions of home(-space), family(-space) and work(-space). As evidenced by the respondent's experiences, the reconfiguration of such spaces relies on established social networks, which transforms transnational migration from a possibly disruptive experience into a manageable, pragmatic and even matter-of-fact life-strategy.

Conclusions

The migration 'narratives' of this study's seventeen respondents, their reasons for moving away and back and their adaptive experiences, form the foundation of this study. These in-depth accounts, volunteered as reflective, and qualitative, 'life-experiences', provide a valuable perspective of the links between sociality and spatiality, and between the macro-level social phenomena and micro-level dynamics of return migration and transnationalism in Grenada. The voices of the migrants reveal the societal processes and meanings through which international mobility is embedded in everyday life. And these complex, distinctively tailored transnational networks – the product of a historically-entrenched culture of migration – *are* very much a part of everyday life in Grenada, and of life experiences of Grenadians at home and abroad; whether they are among the mobile or not. There is little reduction in the scope of human aspiration and degrees of entrepreneurialism across space, as Grenadians extend their field(s) of operation beyond limited island and regional margins. Migrants explain how international movement and transnational enterprises are logical extensions – rather than inconvenient disruptions – of family, home and work. First-person migrant accounts in Grenada highlight the complexities and realities of human agency that are often lost in macro-scale conceptualizations and grand theoretical musings of human migration.

The strategies employed by Grenadian migrants both promote and rely upon a transnational nexus of mobility. This nexus allows Grenadian migrants to realize their intent to return by fostering reconfigured orientations of home, family and work that operate effectively across borders, thereby retaining migrants within multi-local, intraregional and extra-regional socio-cultural fields; and among alternative nodes within intersecting family-based and community-maintained networks of these transnational fields. The findings from this study in Grenada suggest a flexible conceptualization of migration and return. Grenadians treat migration as a practical and viable strategy that encourages individual maturation, family advancement, entrepreneurialism, or sometimes simply acts as an escape for those at a dead end. Indeed, as many authors have acknowledged, migration has become a way of life for many West Indians; it has become an integral part of the culture and, in some cases, it has become mundane (Chamberlain 1997; Guarnizo 1997; Fog Olwig 1998; Fog Olwig 1999).

Importantly, Grenadian returnees often use migration as a means of extending their field of opportunity, operation and experience with a purpose of using that acquired human and financial capital for national growth, progression and benefit. Migration is considered less as a survival decision and more as a strategy for individual, family and national advancement. As families, homes, and careers diffuse through emerging and solidifying transnational lattices, the culture of those at home continues to be saturated with the idea of moving, while the culture of those abroad is infused with the idea of return. Although critics have often been skeptical about the benefits of returning migrants, these Grenadian experiences suggest that contemporary migrants have an acute awareness of how to appropriate evolving transnational webs of kith, kin, capital and culture, not only for their own advancement and life-goals, but for the benefit of their island homes. Small island national pride, is not an incidental aspect of their transnational identity, it is an essential dimension.

References

Ahlburg, D.A. and R.P.C. Brown (1998), 'Migrants' Intentions to Return Home and Capital Transfers: A Study of Tongans and Samoans in Australia', *The Journal of Development Studies*, Dec. 35(2): 125-151.

Byron, M. (1994) *Post-War Caribbean Migration to Britain: The Unfinished Cycle*, Aldershot, England: Avery Publishing.

Carnegie, C.V. (1982), 'Strategic Flexibility in the West Indies: A Social Psychology of Caribbean Migration', *Caribbean Review*, 11 (1): 11-13, 54.

Chamberlain, M. (1997) *Narratives of Exile and Return*, London: The Macmillan Press.

Conway, D. (2000), 'Notions Unbounded: A Critical (Re)Read Suggests that US-Caribbean Circuits Tell the Story Better', *Theoretical and Methodological Issues in Migration Research*, Agozino, B. (ed) Brookfield, VT: Ashgate, 203-226.

De Souza, R.M. (1998), 'The Spell of the Cascadura: West Indian Return Migration as Local Response', *Globalization and Neoliberalism: The Caribbean Context*, T. Klak. Lanham: Rowman & Littlefield Publishing, 227-253.

Fog Olwig, K. (1998), 'Constructing Lives: Migration and Life Stories Among Nevisians', *Caribbean Migration: Globalised Identities*, M. Chamberlain (ed), London: Routledge. 63-80.

Fog Olwig, K. (1999), 'Narratives of the Children Left Behind: Home and Identity in Globalised Caribbean Families', *Journal of Ethnic and Migration Studies*, 25(2): 267-284.

Glick Schiller, N., L. Basch and C. Szanton Blanc (1992), 'Transnationalism: A New Analytical Framework for Understanding Migration', *Annals of the New York Academy of Sciences*, 645: 1-24.

Gmelch, G. (1980), 'Return Migration', *Annual Review of Anthropology*, 9:135-159.

Gmelch, G. (1987), 'Work, Innovation, and Investment: The Impact of Return Migrants in Barbados', *Human Organization*, 46(2): 131-140.

Gmelch, G. (1992), *Double Passage: The Lives of Caribbean Migrants Abroad and Back Home*, Ann Arbor, MI: The University of Michigan Press.

Goulbourne, H. (2002), *Caribbean Transnational Experience*, London: Pluto Press.

Government of Grenada (2003), Grenada immigration statistics obtained through personal communication with member of Grenada's Ministry of Tourism, August, 2003.

Guarnizo, L.E. (1997), 'The Emergence of a Transnational Social Formation of the Mirage of Return Migration among Dominican Transmigrants', *Identities*, 4(2): 281-322.

IDB (2004), US Census Bureau's International Data Base Homepage. Accessed 2 Feb 2004. http://www.census.gov/ipc/www/idbnew.html

Lawson, V.A. (2000), 'Arguments Within Geographies of Movement: The Theoretical Potential of Migrants' Stories', *Progress in Human Geography*, 24(2): 173-189.

Marshall, D.I. (1982), 'The History of Caribbean Migrations', *Caribbean Review*, 11(1): 6-9, 52-53.

McHugh, K.E. (2000), 'Inside, Outside, Upside Down, Backward, Forward, Round and Round: A Case for Ethnographic Studies in Migration', *Progress in Human Geography*, 24(1): 71-89.

Mountz, A. and R. Wright (1996), 'Daily Life in the Transnational Migrant Community of San Agustin, Oaxaca, and Poughkeepsie, New York', *Diaspora*, 5:403-428.

Phillips, J. and R.B. Potter (2003), 'Social Dynamics of "Foreign-Born" and "Young" Returning Nationals to the Caribbean: A Review of the Literature', University of Reading, Department of Geography: Geographical Paper, No. 167.

Pool, G.R. (1989), 'Shifts in Grenadian Migration: An Historical Perspective', *International Migration Review*. Summer 23(2): 238-266.

Potter, R.B. (2003a), '"Foreign-Born" and "Young" Returning Nationals to Barbados: A Pilot Study', University of Reading, Department of Geography: Geographical Paper, No. 166.

Potter, R.B. (2003b), '"Foreign-Born" and "Young" Returning Nationals to St. Lucia: A Pilot Study', University of Reading, Department of Geography: Geographical Paper, No. 168.

Ravenstein, E.G. (1885), 'The Laws of Migration', *Journal of the Statistical Society*, 48(2): 167-219.

Rubenstein, H. (1979), 'The Return Ideology in West Indian Migration', *Papers in Anthropology*, 20(1): 21-38.

Rubenstein, H. (1982), 'The Impact of Remittances in the Rural English Speaking Caribbean: Notes on the Literature', in Stinner et al. (eds), *Return Migration and Remittances: Developing a Caribbean Perspective*, Washington D.C.: Smithsonian Institute, Research Institute on Immigration and Ethnic Studies, Occasional Papers No.3, 237-266.

Rubenstein, H. (1983), 'Remittances and Rural Underdevelopment in the English Speaking Caribbean', *Human Organization,* 42(4): 295-306.

Thomas-Hope, E.M. (1985), 'Return Migration and Its Implications for Caribbean Development', in R.A. Pastor (ed), *Migration and Development in the Caribbean: The Unexpected Connection*, Boulder, Colorado: Westview Press, 157-177.

Thomas-Hope, E.M. (1992), *Explanation in Caribbean Migration*, London: Macmillan.

Thomas-Hope, E. (2001), 'Trends and Patterns of Migration to and from Caribbean Countries', http://www.eclac.cl/celade/proyectos/migracion/ThomasHope.doc.

Tobias, P. (1975), '"How You Gonna Keep 'Em Down in the Tropics, Once They've Dreamed of New York?": Some Aspects of Grenadian Migration', Ph.D. Thesis, Rice University.

UNECLAC (1998), 'A Study of Return Migration to the Organisation of Eastern Caribbean States (OECS) Territories and the British Virgin Islands in the Closing Years of the Twentieth Century: Implications for Social Policy', Published by the United Nations Economic Commission for Latin America and the Caribbean, Subregional Headquarters for the Caribbean, Caribbean Development and Cooperation Committee: Port-of-Spain, Trinidad and Tobago.

Valtonen, K. (1996), 'Bread and Tea: A Study of the Integration of Low-income Immigrants from Other Caribbean Territories into Trinidad', *International Migration Review.* 30(4): 995-1019.

Chapter 6

My Motherland, or
My Mother's Land?
Return Migration and the Experience
of Young British-Trinidadians

Marina Lee-Cunin

'People always tell me that they can't understand why I want
to live here but they really don't understand the issue of
quality of life'.

Introduction

In the 1990s, a small but significant group of young people began to migrate to
Trinidad. Many considered that they were returning 'back home' although they
were not born nor raised there. Further, they held Trinidadian passports which
declared them to be 'citizens by descent' and therefore, were nationals of a country
in which they had never lived. In this chapter, five British-Trinidadians tell their
stories of migration and return. Although these record a distinct and individual
experience for each returnee, nevertheless certain commonalities of the return
process were shared. Perhaps the most significant were their reasons for return
which were concerned with perceptions of gaining a better quality of life in
Trinidad[1] to that which they were presently experiencing in the United Kingdom
(UK).

In general, returnees are aware that they are returning to an island with a
diverse economy and a stable infrastructure. Despite higher levels of
unemployment in the nineties and a growing problem with crime, the successful oil
and gas-based economy ensure that Trinidad and Tobago maintains high levels of
prosperity in comparison with other Caribbean islands that rely heavily on a tourist
economy (World Trade Organisation, 1998). Foreign investment has steadily
increased since the 1970s and many international companies are long established
there bringing with them foreign employees and their families. Therefore,
Trinidadians have been accustomed to the movement of 'foreigners' in and out of
their community, quite often resulting in a less distinctive boundary between the
local and the foreign.

Young returnees in particular do not attract any specific attention. On an official level, the government does not provide any special assistance to them apart from allowing duty and tax concessions for bringing household goods and a vehicle[2] for personal use into the country. Similarly, the Trinidad and Tobago High Commission in London does not offer any advice or support to young returning nationals, nor keep any particular information on the profile of returnees or their levels of re-return to the United Kingdom (UK). Potter (2003a, 2003b) in his recent studies of young, foreign-born returnees noted a similar lack of official information on returnees with respect to the islands of Barbados and St Lucia. A lack of interest is also mirrored at the social level as although the Trinidadian community is aware of this young migrant group, such returnees are perhaps viewed as natural extensions to their parents who also returned to the island on reaching retirement age during the 1980s and early 1990s. Therefore, the returnee experience was an individual one in which the decision to migrate was made individually and carried out individually with support coming from the intimate circle of family and friends.

In order to identify the commonalities and differences in their migratory experiences, the five individuals in this chapter were interviewed at least three times. The initial interview was semi-structured, but in later interviews returnees spoke freely which allowed for a more self-descriptive process as a means of understanding and determining the individual nature of their experience. Particular attention was paid to: 1) the push and pull factors that made them want to leave England in order to live in Trinidad; 2) the issues that they immediately noted were different from the UK; 3) the various adjustments that they perceived they had to make to life in Trinidad; 4) their feelings about where they thought their country of origin was; and e) their general living standards on arriving in Trinidad.

It should be noted that these five individuals were part of a larger (forthcoming) study of young British-Trinidadian returnees. In total, 23 in-depth interviews were conducted during the years 1995-2001. The five individuals in this chapter best represent the cross section of all the returnees interviewed with respect to their different family status, such as single parent, married with children, divorced and/or married to a non-British-Trinidadian spouse, as well as their ethnic background either Afro-Trinidadian and/or Indo-Trinidadian descent.

The returnees also had common characteristics in their overall profile such as: 1) being in the age category of 25-38 years; 2) coming from the same geographical area of London; 3) being university-educated in the UK or gaining professional qualifications there; 4) having all of their work experience in a British environment before returning; 5) gaining their Trinidad nationality in their teens or early twenties; and 6) having two Trinidadian parents who came to the UK in the sixties and seemed to have imbued their children with positive views of the island.

A brief self-description of myself is also in order at this point. I am a London-born and British educated social-anthropologist. My parents are both Trinidadian; my mother is Indo-Trinidadian and my father is of multi-ethnic heritage. My family returned to Trinidad in 1975 where I attended secondary school for four years. My first passport was Trinidadian, I gained my British one much later. I spent five years re-returning to the UK and then returning again to Trinidad. By

1984, my whole family had re-returned to the UK. Towards the end of the 1980s, my parents had a second period of return and by the 1990s, it is sufficient to say that we had established a kind of dual home environment. Similar to the experience of Bueno (1997) in her interviews with Dominican women returnees, I also found that because of my personal experiences, the returnees spoke to me as if I were able to relate to some or all of their experiences, which in some cases, was rightfully so.

This chapter will present the life stories of young British-Trinidadian returnees with a brief introductory profile given prior to each story. The chapter then concludes with a general discussion and analysis of their overall experiences and adjustments to life in Trinidad.

Returnees' Stories

James

James was born in England. He was married to a British-Portuguese woman and described himself as black. His first child was born in Trinidad and the second was born in the UK. Both of his parents were Trinidadian and they returned to Trinidad in the 1980s. James had moved with them and spent three years from aged 15 to 18 attending school in Trinidad; he then returned to the UK for his university education. He had been on several holidays to Trinidad before his first return as an adult in 1997 at the age of 28. He re-returned to England in 2000 with his wife and son and then again returned to Trinidad in 2001 where he currently resides with his family.

> 'When we were young, my Dad used to tell us about black history and Trinidad history and all these old stories about the family. I really enjoyed that. It was a different world to north London. I went to Trinidad to do my 'O' and 'A' levels – my Dad got us Trinidad passports – we already had British ones. That was my first time living in Trinidad. I spent three years there, going to school and stuff. I really enjoyed it. My parents didn't worry about anything with us. We used to go where we wanted with our friends. I was 16, 17 years old and I used to be liming everywhere, but you know, it was safe then. If I was in London at that age, my life would have been so restricted.
>
> I came back to London to go to uni. After graduating I had a lot of different jobs but my aim was always to go back to Trinidad and have my own business. I worked and got married up here and then I went back to Trinidad with my wife. My wife always knew we would be going home to live. She didn't have a problem with that. She had gone down on holiday two times before we made the move. She was more excited than me to go.

White people move to Spain all the time. I mean most black people don't realize that they would actually have a better life back home, especially now. Those English just up and go, no language skills, no job, nothing, but they move because they think, "why shouldn't we, life's going to be better". I think more black people should take that attitude.

My family were completely supportive of the move. My parents were retired back here (Trinidad) and they wanted us to live with them. We stayed with them in the beginning.

I can't really think of anything bad that my parents told us about Trinidad when we were young, it was all positive. Trinidad is different from all the other Caribbean islands because of the races. I think that's why it's more of a friendly, more of an accepting place. I think it's the only place in the world that would accept our kids.

My wife isn't from here. We really didn't have any problems when we came here to live. She passed for a white Trinidadian, you know Spanish or Portuguese one. Her family is actually from Portugal although she's British. Her skin isn't white white, you know. She doesn't view herself as white, more Mediterranean. I got a few looks and comments from black women when I was walking with her in town.

Some people really thought that we lived like the white people in England. If you said you would buy a roti from some place in London, they would say "you do that up there?" like it was something bad. After we lived here for a while though, you make your friends and it gets easier, especially when you tell people that you've been living home for over three years.

I didn't have to work when we first went down. We had money because we sold everything in England. Our house, car, everything. I wanted to start my own business and buy a house. We stayed with my parents and then we rented a flat which was good so we had our independence.

After the first few months, I got a job working in this company doing accounts stuff. The pay wasn't good and after a while I left. It didn't matter too much because we still had savings. I spent my time trying to set up a business but it wasn't that easy. You have to know people to do business and have contacts. Unfortunately, most of my family work in the civil service or for private firms so I didn't have the automatic base of contacts to help me out.

We should have bought a house straight away with the money we came with but we didn't, that was one mistake. Renting gave us independence but it came out of the savings. When we finally decided where we wanted to buy the house and the type of house, it was too expensive for us so we just continued renting.

Our son was born there. My wife hated the experience. That's when you see the real Trinidad, if you know what I mean. The system is slow and the people who work in it gave us the most problems, you know how it is. Once the nurses heard my wife's accent, it was like "who she think she is". One nurse said, "my dear, this is not England you know". I mean not all the nurses and doctors were like that but you could definitely notice a difference in treatment plus you had to pay for everything. It was just straight up unprofessional behaviour but it doesn't surprise me.

It's better for a man in Trinidad than a woman. Trinidad is basically macho even I can see that. If you're foreign, that's a good catch for a lot of women, a passport out of the country. They come around, always doing things for you. You could be a real waste of time or a nobody in England but in Trinidad being a foreign man especially a white man, goes a long way. It means that the woman could leave Trinidad.

Anyway, we came back to London because of low finances, things for babies cost a lot in Trinidad especially medicine and Kwesi (son) was going to start school. That was a bad move because we had to start all over, renting a place and looking for work. Then within the year of coming back to London, we had another boy and that was when we decided to go back home. London has really changed. I didn't like what Kwesi was coming home with from school, all kind of big talk. In Trinidad, he went to a private kindergarten and it was quite advanced. If you pay for education in Trinidad, the children are more advanced than in England. The children have proper discipline and they're polite. I agree with my wife that children stay like children longer in Trinidad than in England.

I love Trinidad and even though we are back here again (UK), it's only to make some more money so we could go back again. I was really surprised at how everything had changed. There seems to be more foreigners, all of these people from Bosnia or wherever. And they look at me as if I'm an immigrant or something. In Trinidad I don't have to deal with any negativeness about being a black man. England supposed to be my home but still I see the black man struggling.

I can definitely see the benefits of living in England though. I mean the social system is good. I got a job when we came back but then I got made redundant so I had to claim, but you know at home, there isn't anything like that.

I see myself as both nationalities. I'm British and I'm Trinidadian. I think people in Trinidad might still consider me as a British person who moved back there but for me, I feel I'm both. I chose to live in Trinidad as an adult and I think that means something. Right now, I feel less British than before. I mean I have a real loyalty to England and I miss a lot of things, but I still have a real feeling inside for Trinidad. I can't explain it. I think if you asked me where I would prefer to die, I would automatically say Trinidad because I feel it's my motherland'.

Vanessa

Vanessa was born in England. She was married to an Englishman and they had two children. She described herself as black. Both of her parents were Trinidadian and they lived in the UK. She had spent one long holiday in Trinidad after graduating in physiotherapy but had not been back since then. She returned in 1997 at age thirty-two with her husband and two children. She re-returned to England in 1998 and is planning to re-return when her children are teenagers.

'I lived in Trinidad for six months after I graduated from college. I knew I could get a job at any time with my profession (physiotherapy) so I thought rather than get straight into the nine to five of London, I would give myself a well-deserved break and go to Trinidad for six months and then come back and start work in January. I got a kind of part-time job helping my cousin in her business and I had a ball with her. I should have stayed but I came back to London and ended up working for the NHS for over ten years. In that time, I got married, had two children and looked old. I think it was always in my mind to live in Trinidad at some point in time.

I had two passports since the 1980s. When I think about my nationality, I'm British and now I'm also Trinidadian. My parents used to tell us when we were younger, you're British, don't let them treat you like you're not. I think that was more about how life in England was for black people when we were young with the racism. But because I also have a Trinidad and Tobago passport, I'm legally a citizen there and my parents are from there so it would be stupid to say that I'm not Trinidadian. I feel I'm both, it's something which I want my children to experience especially because my husband is English, so in a way, the kids are more two nationalities than me!

I was really fed up of life in London. It was just paying bills and work, and the weather was the worst. My husband kept saying why don't we go to Trinidad because we had talked about it from time to time. Then one day I thought why not? I wanted my children to be able to have that feeling of freedom that I saw my cousins' children having when I was there. I wanted to see them playing on the beach and climbing trees and out in the open not stuck in a house in London.

The good thing about moving down here is that you don't need a lot of money to do the move. You give one month's notice, pack up, get your flight and a few hundred quid goes a long way plus you have your family for a place to stay. Once we made the decision, it was a matter of tying up all the loose ends.

My parents were supportive on the whole, although my mother was more negative at first. She said that the health system wasn't good and once the honeymoon was over, we would see how Trinidad really was. She spent a lot of time telling John about how life would be and how people would rip him off in the market or the shops because he was English. "From the time they see your face, price going up".

I think we got even more support for the children here than in England. My son has friends at his kindergarten and he and my other daughter are at their friends a lot. People are always telling us bring the kids and leave them here to play with their children. That's one of the best things about Trinidad. It's a children-friendly society and you don't need to make appointments or have to pick up the kids at a certain time.

I wanted to open my own practice when I came here. We didn't really work for the first two months. But I knew I had to get my business started so I began to put the word about that I did private therapy work. I got a few clients but not really a steady group. It would have been better to have separate premises.

We lived with my gran. We knew once we got settled we would find our own place. The children settled in fairly well although when my son got sick and we went to the doctor, I wasn't really happy. The doctor was nice but everything just seemed really lacking in facilities. Also we had to pay the doctor and when you're used to the NHS, to have to pay for health care is a bit of a shock and medicine was quite expensive.

I had some friends from the time I was there before and we got together again. I made a few friends with parents at the kids' nursery. One time when we came to collect my son from his friend's birthday party, the parents invited us in and now we're friends. I know sometimes it might be because we're from England and that might be a reason to make friends with us more. But I think people are really friendly in Trinidad.

I noticed that people, especially women tended to be more friendly to John. It was because he was English, white. I didn't get the same treatment. Some women were out of order because they would be openly flirting with him. He's not like that though. One of my friends said that some marriages don't last in Trinidad, I mean with a white guy because the man has too much local temptation down here. The women can be so forward.

The race thing, well, people were always on about it. Black people, Indian people, Syrian people, whatever. I know my husband is white and I'm black. We don't have a problem with it. People, especially family, would make comments on the children and who looked like the father or mother. Basically one of my sons is lighter in complexion so they would say "he looks like his father, he real handsome" and the other one looks like me. Sometimes that annoyed me because that's where Trinidad can appear really small-minded to think about things like skin colour. Once I said my kids are black and my cousin said "you shouldn't say that, how they black,

they fair" because their father is English. So I said in England,
my kids are black. Also people still use the word "Negro"
down here and after a while, I just let that go. The older black
people use it all the time. I don't say nothing about that
anymore because I would just get "this isn't England" or
something like that.

I would say most of my friends were black but that wasn't
out of anything in particular. I mean we met Indian people and
other people as well but it was a matter of who you met where.
Most of the people we met were through family and we ended
up socialising with those people on the weekends. I had Asian
friends in England so some of the stuff I would hear about
Indian people, it was just prejudice. But the thing is Asians
have got a closed culture and don't like to mix in marriage and
they do have a thing about skin colour too, so although I didn't
see it in a direct way, I know that Indian people would have
been more likely to be racist against black people, not all but
just it wouldn't surprise me, you know what I mean? I heard
people saying in Trinidad that Indians object to their daughter
marrying a black man.

After six months, we had to be more realistic because we
weren't in full-time employment and our finances were really
low what with the children's schooling and general living
expenses and paying for the doctor when the children got ill. I
decided that I would apply to work at the hospital but when I
thought about it I thought if I'm going to do that then I might
as well go back to England. John wanted to stay longer in
Trinidad and give it more of a chance. I felt like we were not
achieving what we came for and we should go back and make
some more money and then come back again. I don't regret
going down though, not one bit.

Now that I came back to England, I appreciate it more. I
think I have two motherlands, Trinidad and England. I didn't
feel that as strongly before but I do now. I think it's important
to recognize England because that's where I spent most of my
life'.

Denise

Denise was born in Trinidad and came to England with her family at two years old.
She described herself as a black, single professional woman with a degree in
marketing and a variety of work experience in British private and public
institutions. She made regular trips to Trinidad before returning at aged thrity-
seven in 1997. Her mother had returned to Trinidad as a retiree, her father had
passed away. Denise still lives in Trinidad.

'I was born in Trinidad and came to England at two years old.
I came to Trinidad once with my family when I was 11 and the
next time I came back was in my mid-20s for a holiday. After

that time, I came down almost every year for six years until I was in my mid-30s. Then I came here permanently to live.

The main reasons for my decision to leave England were career prospects, better quality of life and perhaps wanting to have a life-long relationship as well as starting a family. I had gained a lot of work experience in a variety of positions but I really felt as if I wasn't reaching my full potential at work. The only way was through self-employment and so I took the plunge and did it. At first, I had too much work, then after a several months, there was a big lull and I had little or no work and I began to build up debt.

My mother retired in Trinidad and said that I should come down and open my business there. I never took it seriously but one weekend, I thought about it and it made sense. It was only a matter of sorting out my debts which in fact I couldn't because I didn't have the money! I gave my landlord notice and a ticket and that was that.

I didn't go down with a lot of money. I think I had a couple hundred pounds which was enough to live without work for a month or more but I decided to find work straight away, accumulate some cash and then set up my business. I don't know about other people but I found it fairly easy to get a job. I splashed my CV everywhere and in two weeks I was employed. The pay wasn't that good but I think simply getting myself into that framework of work was the most positive thing for me to do. I think my self-employed experience in London was really appreciated by my boss as he knew I could work independently and responsibly. I was out of the house, I was able to meet people, make some money, I met so many people through work, good and bad I might add, but good friends are a lifeline to a single woman.

At work, some women can be really bitchy especially if the men pay you attention, which they do because of the accent and being new of course. But really, any new female in the office attracts the usual male attention. They [the men] want to help you out even if you don't need it, especially when you don't need it! They want to tell you things you already know and introduce you to the Trini way of doing things. In some conversations, the line between sexual harassment and friendliness is really blurred! I don't mean serious sexual harassment, I never experienced that but just the jokes and comments they make, you couldn't really get away with some of that stuff in England within the work environment. I just laugh. But when some of the women see that, it's like looking for trouble because they get a serious attitude with you and give you one line conversation. I like to have good relations with my colleagues but here, terms like professionalism and good working relations and even the definition of colleagues have their own cultural meaning. But I've got some great life long women friends here.

I was quite shocked when I came here because I forgot how much racism there is, it can be real bad. I always loved how in Trinidad people were from everywhere and in England, I've always had friends from everywhere, it's London, right? But I listened to some people talking about Indian people and one was talking about coolies and I just felt quite sick. These people are our age! Sometimes I try to explain how racism affects different groups of people of colour in England but they don't get it. And then my Indian friend here said that in her house, her family can't stand black people and using the phrase "dem nigga people" is normal for some of her relatives. I was so shocked because they were very friendly to me when I visited her home. The politics here also encourages people to stay prejudiced.

I lived with my mother at first but I knew I would need my own space soon. After a year and a half, I was able to move out. My apartment is small and it's expensive but I need to be in a safe area being on my own. The crime situation has definitely got worse since I've been here. My social night life is more limited than before. However, crime is bad in England as well, just because Trinidad is small we feel it more intensely.

Now, there are a lot of single women, divorced or perhaps in a relationship but not married. I used to think that it would be the worse thing to be 40 and not be married or have children down here. I knew how in small societies, people think something is wrong with you. It's too much pressure. People see me as a career woman which I don't always like because it's a stereotype. I am in a stable relationship now but to tell you something, once or twice I said I was seeing someone to avoid having to say I was single. But a friend of mine told me that someone at work said she never saw me with any man so I must be with a married man. I laughed plenty but it says something that if a woman is in her late 30s and not married she must be "the other woman".

In my relationship, although Tony (partner) has travelled a lot, he's a real Trini man with his little macho behaviour and t'ing. I had so many conversations with him about women and men and sexism but he can't see it. I think that's one of the problems of having a relationship with a Trinidadian man. They're not aware of women's issues or gender issues like some men in England. Men still view women very traditionally even if you work in a professional position, they still expect you to cook or do women's type work at home. But I have to say some of the women are just as bad. If a married man is cheating, some women will blame the wife for not looking after her man well by carrying on with her career or something. I've heard intelligent women say things like that.

I have always felt and will continue to feel that Trinidad is more my motherland than England. Of course, Africa is my motherland but in an immediate way, Trinidad is. It's just something you feel inside or something you just know. I'm very happy here. I'm settled, I have no intention of going back to England. It's so freeing not to be a minority anymore. I know that sometimes people behave really stupid down here, small-minded but I just remember what it was like in England and that makes me feel better. I have the beach here, I go down the islands, I go clubbing and to nice restaurants. In England you have to be wealthy to have this lifestyle, here it's free. People always tell me that they can't understand why I want to live here but they really don't understand the issue of quality of life. That's the over-riding thing. Quality of life'.

Annette

Annette was born in England. She described herself as an Afro-Trinidadian divorced woman with one son. She graduated in business and accountancy. Her Trinidadian parents had returned to Trinidad and she had visited them on several holidays during her education at school and university. She returned to Trinidad at aged twenty-seven in 1998 and still lives there.

'My parents are Trinidadian and they live here. I came quite a lot to Trinidad when my parents came back to live and I was doing my 'A' levels and at university. I had three month holidays so I felt like I had two homes. I got my Trinidad passport when I doing 'A' levels.

I came to Trinidad for two reasons, my career and my child. England was okay but being a single parent was horrible there. My son's father wasn't very supportive, he would turn up when he wanted and didn't financially support him so it wasn't a good way for my son to be brought up. I thought about what would happen when we came here because Daniel wouldn't really see his father but that was his father's choice, I mean if he was a good father, I wouldn't have come back home to live.

I definitely have more support for my child here than in England. I have family here to help out and it's just easier. I hated living alone in that flat and when I needed anything, it was very stressful. Here, my son has aunts, uncles, grandparents. I can actually work full-time and my son is in school doing well.

I did temping after university and then I got married and everything changed. I stayed on doing temp stuff and then I got pregnant. My relationship wasn't going well and I was separated by the time Daniel was born, living in a small flat on benefits with no good support as such. I didn't know what to do so I talked with my Mum and told her the situation honestly. She said just pack up and come to Trinidad.

I worked really hard for my degree and here I have a good job where I'm using my skills. I got the job through contact.

Everything here is contact. It's how things work here but I'm benefiting from that so I don't want to criticise things. The pay isn't good but I'm living with my parents and so I don't have to pay rent.

My office is alright. There are a few stupid women there, older ones. They probably think because I'm young and from England that I don't know anything. They make comments like "you have to get used to how Trinidadians are" and sometimes tease me about my accent. I don't have much to do with them and my mother said, that they must think I'm acting hoity toity. I'm not. I find their conversations boring. They're too small-minded. The men here are too silly sometimes. I'm not really interested in having a relationship as yet. When you're a mother, you have different priorities. In England, I didn't really get the opportunity to progress myself because of my personal situation. In Trinidad, I had a chance to organise my life better. I don't mind going back to England but I want to make sure that I have the same life as I have here. I don't know if that could really happen.

I pay this lady to take Daniel to and from school and my mother is home to look after him until I reach home from work. Having a child is expensive in Trinidad because you have to pay for healthcare but I think my son has benefited so much from being with his grandparents. I'm not too keen on the education system here. They really push the children. Daniel had homework and had to do tests and he wasn't even five! I thought that was ridiculous, but it's the system here. One or two people at work said how they heard the children were bad in English schools and the teachers used to get beat up. I tried to explain that was some schools not all.

You know I don't like when someone at work hears something about England and then they come and ask me about it. Then I have to defend England. I don't have to but I feel I have to because they don't understand what life is like there. If I don't defend it then I will hear "that's why I like my Trinidad, places like England have bad things" or something like that. I feel like they are directing those comments to me personally. It really upsets me sometimes because I feel like an outsider. But I talked with my Mum. She said it's all because of jealousy, if they got a chance to live there, they would go running. She's right.

I have some nice, intelligent friends here and I go out with them every now and then. Mostly though on weekends, I prefer to spend time with Daniel, visiting cousins or friends around the area. It's nice where we live because we can walk to people's houses.

The race thing is just stupid. I have friends who are black, another who is part Chinese, my other friend is Indian. I don't care about that race thing. I'm aware of my African heritage and my son knows who he is. When I hear prejudiced people speaking, I just don't listen. Like at work, sometimes this woman will say something like "black people can't run nothing". She's black by the way. Or she will say something about "Indian people this or that" or make comments about skin colour like "if she catch too much sun, it don't suit her" or "that man really black, he blue black". I know people are prejudiced but I never got any prejudiced treatment down here.

Sometimes I talk with Savi (East Indian friend) about race stuff. Her family would probably disown her if she married a black man. She had a black boyfriend and they didn't even know. That happens with Asian girls in England and black guys. If I married again I would prefer him to be black, but to be honest, it just depends doesn't it? There's this guy, a white Trini, he's one of our clients, nice guy and he's also asking me when I get some free time we could go for a drink. I'm not interested in a relationship as yet but what I'm saying is that the race thing can be irrelevant sometimes.

I think in Trinidad there are really two groups of people, the intelligent ones and the small minded ones. The intelligent people usually travel abroad or studied away or something like that. You can tell from their conversation, but they're in the minority. The majority of people get their information from cable television which is all American.

I haven't gone back to England in five years and I don't really miss it. I would like Daniel to have some of his education up there but that wouldn't be a sole reason for me moving back there. I don't feel less British or more Trinidadian. All I know is here is home for now. Most Trinidadians don't have a problem with me saying I'm from England but they do if I say I'm from Trinidad. I think maybe your motherland is sometimes a bit of where other people think you are from'.

Sophie

Sophie was born in England and spent eight years of her teenage life in Trinidad. She returned to England at aged 18 and gained qualifications in nursery education. After marriage and one child, she moved to Trinidad at aged 30 in 1996. Her father had also returned to live in Trinidad. Her second child was born in Trinidad in 1997. She re-returned to England in 1998 with her two children after her marriage breakdown. She described herself as Indo-Trinidadian.

'My ex-husband was the same as me, British-Trinidadian. We were born in London and both lived in Trinidad when we were teenagers for a few years and then came back to London. We met in London in our mid-20s, got married and had a daughter born up in London. My parents are Trinidadian, his mother is Trinidadian and his father is English.

His mother moved back to Trinidad just before us. My Dad was also in Trinidad at that time. I wanted to move for a better life, for my daughter and her health, and a better climate. We both had qualifications and felt we could get jobs there easily. We used to always say we were Trinidadian when we were younger. We never really said we were British although we were.

I think it took us six months from when we decided to move to actually being in Trinidad. We had no resistance only support for us moving. We saved for the move. We organised renting out our house in London. My mum paid for my daughter's flight, his Dad paid for the dog, we just had our flights to pay for. We took everything with us, the car, washing machine, freezer, everything. My advice to anyone moving now, is don't take those things with you like a microwave and fridge because they will blow up down there.

In the beginning, things were easy and convenient because we planned it that way. We lived with my ex-mother in law. She didn't take any money from us at that time for rent or anything. We only made that arrangement when I started working. We didn't work for three months as we had saved enough beforehand to live like that. You have to plan, plan, plan when you are going to do something like this.

My daughter went to kindergarten there, more than one. We weren't aware of everything that was going on; when we found out we moved her. In one place, the teacher put sellotape over her mouth to stop her speaking. She was two and a half! Then she didn't want to use the bathroom and we found out it was because of a bad experience at school with the disabled toilet assistant who wasn't trained to care for children. She learnt quite a lot though. She learnt phonetics at two and a half. The school would give homework to three-year-old children. She definitely learned more than she would have in England but there were big differences in how children were cared for.

When you have a job you could live anywhere. For me, it was very easy to get a job. My ex also found a job later but I worked first. I handed out my CV to schools on Friday and by Monday I was offered two jobs at two schools. For one school, it was a case of being in the right place at the right time. A teacher had gone off sick that Friday with no replacement and my CV appeared on the Principal's desk that same day. Actually that school was considered the most prestigious school in the country, private, catholic school. I know that I got the job because of who I was. I was from England, I had British qualifications. I was also a Catholic and the principal was from Ireland so all of that. The school organised everything for me from getting a bank account at this branch that was highly selective in whose accounts they serviced, to getting all my tax and working papers sorted. The name of the school opened all doors for me. Everything was positive. The salary was very low but I made extra salary by giving lessons after school. It was the network that you had being a teacher of that school that was compensation for the pay.

I had plenty of friends. I met them mainly through teaching at the school. I also had friends through my ex-husband's family. I went out a lot and had a good social life.

There was definitely racism down there. I thought one black teacher took it out on some white children. She would make certain comments where you could just tell. Then there were some people who didn't like black people at all. Some people used to look if I was with my mother in law. The black and Indian thing.

My driving test experience was a real laugh. The driving teacher told me I had to give a bribe to the examiner beforehand. I gave 200 dollars and the teacher came back and said it wasn't enough. I then decided I wouldn't pay anything because it was ridiculous. I think the man wanted more money because I was from England. My teacher told me if I wasn't going to pay I would have to take my chances. Well, I knew from the time I saw the man's face that I wasn't going to pass the test. I never even drove out of the compound. It was a joke. The test lasted two minutes. He told me I failed because I didn't follow the correct procedure for looking in the mirror and all that. My uncle said that I should have shopped around for my bribe. Only in Trinidad! Everyone knew that when it came to driving tests and license, you had to pass something. I don't know if it's still like that now.

Everything changed when we split up. He was having an affair. He met her at his work. I heard she used to bring food for him. I moved from his mother's house to my father's. It was difficult travelling to work from my parents' home and I was pregnant so I changed job to teaching at the university's primary school. That was another top school. Having taught at the first school, I had no trouble in getting a next one.

Thomas was born in Trinidad. The childbirth experience was a do-it-yourself one. I went private but really, I could have had him at home. There was no pain relief, no special care as such. In hindsight, it was a better experience for the baby to be born completely natural. I found the antenatal care of the doctors really good but some of the nurses were not professional at all when I was in labour. I think it's to do with when they hear your accent or a race thing, I think it's both because the nurses were black and me being Indian plus when they heard the accent.

I never had any intention of coming back to England until the situation got bad. I was involved in divorce and custody of the children down there. I experienced the Trinidad court system and all I can say is things weren't good at all. The lawyers, the decisions, the attitude of courts, it was Trini style. The police were okay though when we had problems with him (ex-husband). They were quite sympathetic. I eventually left Trinidad when my son was ten months old and daughter was just under five. It was the best decision for the welfare of the children. I don't regret it. At least, up here, there is support for single parents and divorce doesn't have so much stigma. In Trinidad, it's still a stigma to be a divorced woman.

I consider myself both nationalities. I'm British and Trinidadian but I consider my motherland to be Trinidad, I don't know why I just do. I feel it is'.

Discussion

Common themes recur throughout these returnees' narratives. With respect to the 'push' factors of migration, it was notable that the returnees did not point out specific negative factors influencing their decision to leave England. Issues of racism and gender discrimination, for example, were not given as direct reasons for migration yet in the case of James and Annette, these issues were indirectly relevant. James' story illustrated that life for him, as a black man was considerably more positive in Trinidad than in England. This could be evidenced by his recent comments on his return to England in 2000, where he felt that members from the newly arrived Eastern European communities were condescendingly viewing him as a black immigrant, lesser in status than themselves rather than viewing him as a British man. He maintained that as a black man, the feeling of always 'struggling', as well as still being perceived as an outsider in the land of his birth was something that he did not experience in Trinidad. He further stated that Trinidad was the only country where he felt that his ethnically-mixed children would be accepted. Denise also mentioned that living in Trinidad 'freed' her from the experience of always being considered an ethnic minority.

Annette's story of her life in London illustrated the gendered experience of single parenthood in Britain. She experienced poverty when she was dependent on the social security system for assistance as well as social isolation while living alone with her son. One of her main reasons for migrating was the lack of financial and emotional support from her ex-husband with respect to their son. She considered that if she had received more support from her ex, she would not have left England.

Other push factors focussed on the returnees' general reasons of wanting a change of environment from their employment settings, their urbanised dwellings and/or the climate. With respect to the returnees who had children, Vanessa stated that she wanted her children to experience a more open, physical environment, while James, prior to his second return, said that he preferred for his children to be educated in Trinidad as he thought that children were more educationally advanced, were polite and disciplined, and experienced a longer childhood.

Finally, it should be noted that four of the five returnees stated that living in Trinidad was an option that they had always considered. Indeed all of the returnees had obtained their Trinidadian passports in their teenage years or early 20s and were aware of the fact that they could exercise their right of living on the island at any time. Perhaps these two issues had some influence on why many did not name specific negative push factors for wishing to migrate.

As other research has suggested (Guarnizo, 1997; Gmelch, 1992; Potter, 2003a), pull factors appeared to be more influential in the case of return migration to the Caribbean. This was also the case with these five returnees. They all had a strong family network in Trinidad on which to rely and four of the five returnees stated that their parents encouraged them to move to Trinidad and live with them. This meant that they had an immediate home in which to reside as well as parental support. Therefore, the phrase 'going back home' could be said to not only refer to what Anwar (1979) described as 'the myth of return' where migrants held onto the dream of eventually returning to their homeland but in the case of these young returnees, 'going back home' took on a subsequent meaning as a reference to returning to live with their parents in the same household as they did when they were children.

Having parents who were themselves returnees, suggested that the pull factor was not simply a matter of having an existing and supportive family network in Trinidad but that the young returnees were being pulled by following the same path as their returnee parents; a path that appeared to be tried and tested with seemingly successful results. Further, returnee parents were able to act as mediators for these young returnees, generally guiding them from their life in London to their new one in Trinidad, as well as being able to interpret and translate the various socio-cultural differences and dynamics of the UK and Trinidad, which the young returnees might encounter when adjusting to their new lives. A simple example is illustrated when Annette stated that at times she felt as if she was being placed in a position of defending the UK against Trinidad by her colleagues. Her mother's response to Annette was that it was most likely a case of jealousy, as those individuals did not have the opportunity to live abroad as she had.

The young returnees' feelings concerning their identity and nationality also played a significant role in being pulled towards living in Trinidad. None of the five returnees stated that they had considered themselves solely British only Sophie stated that in her younger life she had considered herself to be Trinidadian rather than British. All of the returnees felt that they had dual nationality. In this sense, dual nationality was being interpreted in the broad sense of dual heritage rather than the legal sense. However, it is interesting that when asked which country they considered to be their motherland, their answers were less cohesive as a group.

Sagiv (1998) noted that the potency of the word 'motherland' was such that it immediately evoked feelings of longing and commitment and its emotional force drew upon primordial and mythic sources. Therefore, James having stated that he had two nationalities, then said that he felt Trinidad was his motherland because it was where he wished to die. Denise and Sophie also stated that Trinidad was their considered motherland for reasons that were hard to articulate, referring to it as a feeling inside, while Annette noted that at times, your motherland was in part defined by how others viewed your heritage. Vanessa stated that on her return to England, she now felt that she had two motherlands as it was important to recognise England as the place of her birth and where she grew up.

The other large pull factor mentioned concerned the quality of life. To these returnees, Trinidad represented the ability for them to gain a better quality of life in terms of having quality leisure time, having career opportunities particularly with respect to self-employment, and more enjoyable weather. James noted that the English community migrated to Spain for similar reasons. Vanessa, Annette and Sophie stated that they wanted their children to enjoy better health, be able to play outdoors more, as well as live among their extended family.

In terms of career and employment, James, Denise and Vanessa specifically wanted to start their own business in Trinidad which they perceived as something they could do much more easily than in the UK. Annette noted that because of childcare support in Trinidad, she would be able to work full-time and have a career unlike in London. Sophie felt that because she and her husband had qualifications that they would be able to secure positions fairly easily. Further, although all of the returnees were initially going to live with their families, they expected at some stage to live independently, thereby further enhancing their quality of life.

All of the returnees had either spent long holidays in Trinidad or had lived there for several years during their teenage lives. Also, with the exception of Vanessa who had not been to Trinidad for about ten years, the other four returnees had made recent visits prior to migrating, although none of them specifically visited Trinidad on holiday as part of their decision-making process for migration. In fact, it would seem that none of the returnees took more than six months from their decision to migrate to their actual arrival in Trinidad and the following factors seemed to be influential in this process:

1. Returnees had already obtained their Trinidadian passports so in a legal capacity they were entitled to migrate at any time they chose.
2. Returnees did not have to organise accommodation in Trinidad as they had already arranged to initially reside with their parents or grandparents.
3. Returnees only had to give the statutory one month's notice in their respective employment positions in the UK, with the exception of Denise who was already self-employed.
4. Returnees had few assets which were difficult to liquidate. James had a house and car to sell whereas Sophie rented her house via an agent and the other returnees were in rented accommodation so only needed to give one month's notice to their landlords.
5. Returnees had an extremely favourable UK sterling-Trinidad dollar exchange rate[3] which meant that the initial set up and living costs in Trinidad were very affordable to them; it also meant that all of them did not have to seek employment or start their businesses in the first few months of arrival if they so chose.
6. Returnees had examples of successful return experiences as the majority of them had returnee parents who had also successfully returned.

Four of the five returnees stated that they had complete support for their move to Trinidad; only Vanessa mentioned that her mother was not initially supportive of her move as she was concerned that her daughter's English husband might be perceived by Trinidadians as an affluent but naïve tourist who could easily be exploited.

Once in Trinidad, the returnees spoke of overall positive experiences in their first few months. Four of the five returnees chose not to work for at least two months so that period of time could be viewed as a kind of 'honeymoon' period of arrival. However, this is not to say that the returnees did not have some overall reactions relating to adjustment of life in Trinidad. Although none used the phrase 'culture shock', it was apparent that the one major area of culture shock that they experienced was in relation to issues of race and ethnicity.

All of the returnees expressed surprise at what they saw as high levels of racism existing in Trinidadian society and the polarisation of the Afro-Trinidadian and Indo-Trinidadian communities. In Trinidad, ethnicity permeates all of the society's social, cultural, political and economic institutions and practises. (Yelvington, 1993:1) Politics, in particular is organised along ethnic lines and in turn, ethnicity itself has been said to be politically-derived utilising colonial stereotypes in the struggle to control state resources (Yelvington, 1993:11-12). The various tensions that exist between the Afro-Trinidadian and Indo-Trinidadian communities have been well documented in studies on Trinidadian ethnicity from within and outside of the country (Eriksen, 1992; Brereton, 1981; Dabydeen and Samaroo, 1996).

It is also worth making a brief note of the political environment that the returnees entered when they arrived in Trinidad during the 1990s. In 1987, the People's National Movement (PNM), the ruling party for 30 years and largely viewed as the Afro-Trinidadian party, was defeated by the multi-ethnic National Alliance for Reconstruction (NAR). However, the NAR also splintered along ethnic lines and one result was the formation of a new East Indian-based party called the United National Congress (UNC) formed by Basdeo Panday. In 1991, the PNM was re-elected to power but in 1996, the UNC won the national elections and Trinidad and Tobago had its first East Indian Prime Minister. Therefore, young returnees were migrating to a society in the midst of distinctive political-ethnic change; a country which was governed by an Indo-Trinidadian party but whose national symbols of the steelpan, the calypso and carnival were predominantly considered to be African-derived. Further there were demographic changes in that during the nineties, the Indo-Trinidadian community surpassed the Afro-Trinidadian community as the largest ethnic group.[4] Therefore, it was foreseeable that discussions of race and ethnicity would have occurred among returnees and their family and friends, as part of the everyday conversations about what was happening in the country and even as part of educating the new returnees about the society in which they had chosen to live.

Regarding the returnees' own views of Trinidadian ethnicity, they seemed directly or indirectly appreciative of the multi-ethnic society of Trinidad. Denise noted that despite her parents telling her that she was British, so that she would be aware of her rights as a British citizen, she always loved that people were from 'everywhere' in Trinidad. James, who often framed his return experience in reference to the larger context of his identity as a black man, stated that he thought that Trinidad was the only country where his mixed-race children would be accepted and that in his opinion, the black community was still experiencing unequal treatment with limited opportunities in Britain.

James and Denise both had white British spouses while Sophie's husband at that time was of mixed parentage. Being involved in ethnically-mixed marriages appeared to have some influence on why some returnees were particularly sensitive to racial and ethnic tensions. James was aware that a black man with a white wife was frowned upon by some female members of the Afro-Trinidadian community. It is suggested that black women may perceive relationships between black men and white women as an affront to the black female identity (Norment, 1999). There exists a body of literature that refers to the specific historical, socio-economic and political dynamics and social consequences concerning interracial relationships (Hodes, 1999; Alibhai-Brown, 2001).

On the contrary, Vanessa noted that her white spouse attracted much positive attention from people in general, and especially from Trinidadian women. Although it did not cause any rift in her relationship, she was made aware by her friends that some interracial relationships did not survive in Trinidad, as the foreign husband often succumbed to 'local temptation'. This attention did not only result from an underlying colour-class system where skin colour is linked to social status and respectability, but also from James' observation that foreign men were sometimes viewed by local women as a 'passport out of the country'. This phenomenon could be placed within the broader context of female migration, where marriage to foreign men is linked to issues of women experiencing economic discrimination and social exclusion in their home country, as well as perceiving that their role in the domestic family model is limited.

It was interesting that all the returnees stated that they either refused to participate in conversations about race or tried to explain to Trinidadians the dynamics of race and racism as they applied to both Asian and black communities in the UK. Further, some tried to understand the tensions by using their British experience, for example, Vanessa stated that she had Asian friends in England and therefore felt that some of the attitudes expressed in Trinidad were due to lack of personal knowledge of another culture. However, she also thought that the Indian community in both countries shared the same characteristics of being likely to hold racist views against the black community and being averse to any intimate relationships between Asian women and black men. As Lal (2004) concluded, 'almost everywhere where Indians and blacks form part of the population, there is the perception that Indians are not merely apprehensive of them, but likely to observe a caste-like discrimination against blacks'. Likewise, Annette noted that her Indian friend had to keep her relationship with her black boyfriend secret from her family which she thought was similar to the situation in Britain regarding Asian and black relationships.

The use of the racist terms 'coolie' and 'nigger' to describe Indo and Afro-Trinidadians was particularly shocking to returnees. Denise stated that the attitudes behind those terms were something that she expected from the older generation but not from hers. Equally, Vanessa was surprised to discover that use of the word 'Negro' to refer to African-Trinidadians was still commonplace among Trinidadians of all ethnicities. Such terminology in the UK has long moved to the category of extinct, colonial words by both the black and white communities there, whereas in Trinidad, such words are still in use and the historical, socio-economic and political meaning behind them are either downplayed, not understood, or have been somewhat desensitised.

The references to skin colour made by Annette's work colleagues however are telling of the complex and deep-rooted colonial concepts of race and colour that still exist in Trinidad. A person's 'colour' although one visible sign of what could be loosely termed as racial ancestry, has also been described as a scale with a continuous range of values which included skin colour, hair and facial characteristics defined by the endpoints of 'black' and 'white' (Segal, 1993:87). Vanessa's comment that her children were defined as black in England employed the political context of that word, whereas her relative's response that the children were not black but fair illustrated this concept of colour as a pattern of 'upward identification and downward distancing' (Segal, 1997:91). These examples illustrate the multi-faceted nature of Trinidadian-specific concepts of race, ethnicity and colour to which the returnees had to adjust.

Even friendships were mentioned within the context of ethnicity. Due to the working environment, Annette, Denise and Sophie were more able to meet people of different ethnic backgrounds and therefore, form friendships. However, Vanessa stated that her friendships were primarily made among the black community but her comprehensive explanation of why this occurred was common in the new arrival experience. Relatives tended to introduce their friends to the returnees and such individuals were more likely to be of the same ethnic group.

Women's identity and the macho nature of Trinidadian society were two of the gender-specific issues mentioned by some returnees in relation to adjustment. Despite the status of women in Trinidad comparing favourably in many respects with that of other middle-income developing countries, on a social level, there were still many gender-based constraints. Women's identities still appeared to be framed within the limited boundaries of marriage and children and social attitudes reflected the primacy of the male role. James had little hesitation in stating that life in Trinidad was better for a man than for a woman and that Trinidad was essentially a macho society; Denise suggested that women were still expected to carry out the traditional female roles in the house. Women may even collude with men to reinforce a particular gender identity, as Denise pointed out when she bemoaned the situation of women blaming the wife for a husband's unfaithfulness suggesting that the wife should not have put her career before her husband.

Denise also spoke of the pressures that she perceived to exist for single women in a small society, even going so far as to create a non-existent relationship in order to avoid difficult questions as to why she was single. However, with the rising rate of divorce[5] in Trinidad, single women as a social group could be said to be increasing in numbers, notwithstanding the rate of re-marriage. She further observed that being a single women in her late thirties appeared to imply that she was involved with a married man as purposely choosing to be a single was an option that was not considered, while Sophie concluded that despite the high rate of divorce, she felt it was still a stigma to be a divorced woman in Trinidad.

Although referring specifically to the foreign man's experience, James further said that Trinidadian women would offer to do things for the man, while Vanessa noted that Trinidadian women were very 'forward' in their approaches to men. In fact, the above comments could neatly exemplify some of the perceived causes regarding the breakdown of Sophie's marriage, as it would appear that her ex-husband succumbed to 'local temptation' through a Trinidadian woman who brought meals for him at work; and even though Sophie did not feel blamed for his infidelity, her career was certainly a significant factor as a teacher at one of the most prestigious schools in the country.

The work environment also raised issues of adjustment of which some were gender-specific. Denise and Annette had to adjust to a new culture of work where men made jokes and comments to their female colleagues, that might have been considered sexual harassment in the British workplace. Securing friendships at work was also complicated for these two returnees. Denise commented that any new female employee would have attracted male attention in the office, but also as a newcomer to the country this added another dimension to the attraction. While male attention was centred on introducing her to Trinidadian customs at work, it also served to create negative feelings from other women employees who would not befriend her when they thought she was being too responsive to the men. Female returnees in Potter's (2003a, 2003b) studies also referred to similar issues of sexual competition. Annette on the other hand, purposely distanced herself from her female colleagues because of the age difference and what she saw as their 'small minded' conversations.

Another work-related adjustment issue was having previous 'contacts'. Having established contacts was seen as a necessary factor by returnees in any employment venture in Trinidad. However, for James and Vanessa who wanted to set up a business, they did not appear to have been fully prepared for what self-employment entailed, as none mentioned that they previously researched their intended markets or made any business enquiries prior to arriving in Trinidad. Gmelch (1980) argued that such a lack of preparation on the part of returnees contributed to the reverse culture shock that some experienced. Contacts were also instrumental in Annette's case for securing employment and for Sophie, having taught at a top private school, she easily gained further employment at another good school based on her first employer's reputation.

Annette, Denise and Sophie also secured employment fairly quickly and it would seem that their status as returnees and their British qualifications and experiences were key to the process. All three agreed that their salaries were low, although that did not seem to be detrimental to their living standards.

For the returnees with children, there were some specific issues of adjustment in relation to the health and education systems. Over a period of time of living in Trinidad, they found that the various costs involving children were quite high. When their children fell ill, the reality of having to pay for health services was something specifically different from their experience of the free health service in Britain. Returnees complained that both the cost of doctors and medicine were expensive. Further, James' wife and Sophie both had children born in Trinidad and described an unfriendly environment during childbirth which they attributed directly to their status as returnees. They believed that their British accents and a difference in ethnicity to the attending Afro-Trinidadian nurses largely contributed to the negative atmosphere. Further, their demands for professional care standards were assumed by the staff to be made within the context of unfavourably comparing Trinidad with the UK.

With respect to education matters, first it must be noted that Trinidadian kindergartens and nursery schools are all privately run. Returnees placed their children in these facilities fairly soon after arrival, in part to assist the children in the settling in process. However such schools were not cheap and both James and Vanessa on their return to the UK cited the cost of their children's education as one of the expenses which they could not afford to maintain, given that they were not securely employed while in Trinidad. Concerning the actual education that their children received, Annette thought that children in Trinidad were pushed to learn from too young an age; while Sophie related some negative experiences involving the care of her child, but felt overall that her daughter learned more in Trinidad than she would have at the same age in England. James was more positive about the system when comparing it to the UK, noting that in addition to advanced learning, children in Trinidad were polite and disciplined and able to experience a longer period of childhood. This was one of the main reasons for his eventual second return to Trinidad.

The small-minded attitudes of some people were mentioned by returnees largely in the context of attitudes towards race and ethnicity, such as references to particular ethnic groups based on racist stereotypes and comments about skin colour, as well as biased gender labelling concerning single women as pointed out by Denise. Annette spoke of small-minded attitudes in relation to her feeling like an outsider. At work, she was teased about her accent and felt defensive of the UK and her life. Disassociating herself from social interaction with these people brought her some isolation and she concluded that there were only two groups of people on the island, the small minded ones who had not travelled abroad and the intelligent ones who had or were educated abroad.

Predictably, the returnees' reactions to overall issues of difference found them resorting to accepting things the way they were. With respect to the attitudes they encountered regarding race, racism and gender bias, their adjustment resulted in the adoption of a fairly pragmatic approach of distancing themselves from conversations of the issue at hand. Sophie's driving test experience while pointing to a form of corruption not expected in the UK, was simply laughed off as something that she thought could only happen in Trinidad. However, in her interaction with the legal system via her divorce and custody of her children, she was much more apprehensive of the differences between the two countries and the experience proved to be so distressing that it was one of the major factors in her decision to return to the UK.

There were few adjustment issues with respect to friendships. Women faced some difficulties making friends with Trinidadian women in part due to sexual competition. Further, one returnee questioned whether friendship boosted individuals' social standing as she felt that some people expressly made friends with her and her husband simply because they were from England. However on the whole, returnees agreed that they had made some good friends in Trinidad, noting that in general, Trinidadians are a friendly people.

Conclusion

Despite encountering various issues of adjustment, the five returnees including the two who had re-returned to the UK, spoke favourably of their experience of return. They spoke of enjoying a better quality of life with particular reference to their leisure time and enjoyed their ability to access their extended family network. In particular, returnees who had children felt that Trinidad was simply a more child-friendly society than the UK.

It was interesting to note that in their descriptions of difficult situations, the returnees generally tended to make comparisons with the UK not as a means of criticising the country but more as a way of trying to balance their understanding of what they were experiencing as returnees. They were mindful of their outsider position even though they did not refer to it directly. This was exemplified in their adoption of a position of distance when faced with complicated local issues.

Finally, it is suggested that given their experiences, these five individuals challenged the settler-sojourner model of migration in a number of ways. Firstly, their status as returnees or returning nationals was confined in a sense to a legal category as most of them inherited their right to Trinidadian nationality through descent. They had not spent most of their lives in Trinidad and in most cases had not been born there. Secondly, unlike traditional returnees, some of these young returnees re-returned to the UK with the hope of a second return to Trinidad in the future and one returnee simply moved back and forth between the countries. However, the experience of exclusion in the UK together with the returnees' apparent inheritance of their parents' own nostalgia for returning to their homeland, had created a deep passion among young returnees for Trinidad being their motherland and in this sense, they represented the emotions of the more traditional model of a returnee. It is hoped therefore, that further research conducted into the experience of young returnees to Trinidad will enhance our understanding of contemporary migration patterns of the wider Caribbean region and its diasporic communities, as well as affirm the new developing transnational identities that constitute the Trinidadian community.

Notes

1. Trinidad is part of the twin island republic of Trinidad and Tobago. This chapter concentrates on the experience of returnees to the more ethnically-diverse island of Trinidad which has a specifically different history to that of Tobago.
2. It should be noted that requirements for allowing vehicles into the country under the returning national banner have been restricted over the years.
3. From 1997-2002, the exchange rate on average was 1GB Sterling to 10TT Dollars.

4. In 1998, the ethnic breakdown of Trinidad and Tobago was 40.3 per cent Indo-Trinidadian, 39.5 per cent Afro-Trinidadian, 18.4 per cent Mixed Origin and 1.8 per cent White/Chinese/Other.
5. In 1996, 1500 divorces were granted, but by 2004, the figure stood at 5000 (Maharaj, 2004).

References

Alibhai-Brown, Y. (2001), *Mixed Feelings: The complex lives of mixed-race Britons*, UK: The Women's Press.

Anwar, M. (1979), *The Myth of Return: Pakistanis in Britain*, London: Heinemann.

Brereton, B. (1981), *The History of Modern Trinidad 1783-1962*, London: Heinemann.

Bueno, L. (1997), 'Dominican women's experience of return migration: The life stories of five women', in Pessar, P. (ed) (1997), *Caribbean Circuits New Directions in the Study of Caribbean Migration*, New York: Centre for Migration Studies of New York.

Central Intelligence Agency (1998), *The World Factbook* http://www.cia.gov/cia/publications/factbook/geos/td.html downloaded 01/11/1999.

Dabydeen, D. and Samaroo, B. (eds) (1996), *Across the Dark Waters: Ethnicity and Indian identity in the Caribbean*, London: Heinemann Caribbean.

Eriksen, T.H. (1992), *Us and Them in Modern Societies: Ethnicity and Nationalism in Mauritius, Trinidad and Beyond*, Oslo: Norwegian University Press, 1992.

Gimelch, G. (1980), 'Return migration', *Annual Review of Anthropology*; 9: 135-59.

Gimelch, G. (1992), *Double passage: the lives of Caribbean migrants abroad and back home*, Ann Arbor: University of Michigan Press.

Guanizo, L.E. (1997), 'Going Home: Gender, Class and Household Transformation among Dominican Return Migrants', in Pessar, P. (ed) (1997), *Caribbean Circuits New Directions in the Study of Caribbean Migration*, New York: Centre for Migration Studies of New York.

Hodes, M. (1999), *White Women, Black Men: Illicit Sex in the Nineteenth-century South*, USA: Yale University Press.

Lal, V. (2004), 'Labour and Longing', *India-Seminar.com* www.India-seminar.com/2004/538/538%20vinay%201al.htm downloaded 05/05/2004.

Maharaj, D. (2004), '5,000 Divorces a year in T&T', *Trinidad and Tobago Express* 19/04/2004.

Norment, L. (1999), 'Black Women, White Men, Black Men, White Women – interracial relations', *Ebony* November 1999.

Potter, R.B. (2003a), *Foreign-born and young returning nationals to Barbados: results of a pilot study*, Reading: University of Reading: Reading Geographical Paper No. 166.

Potter, R.B. (2003b), *Foreign-born and young returning nationals to St Lucia: results of a pilot study*, Reading: Reading University: Reading Geographical Paper No. 168.

Sagiv, A. (1998), 'Zionism and the Myth of Motherland', *Azure* Autumn 5759 www.azure.org.il/5-articles3.html downloaded 06/04/2003.

Segal, D. (1993), 'Race and Colour in pre-independence Trinidad and Tobago', in Yelvington, K. (ed) (1993), *Trinidad Ethnicity*, Tennessee: University of Tennessee Press.

World Trade Organisation (1998), 'Trade Policy Reviews: First Press Release, Secretariat and Government Summaries: Trinidad and Tobago: November 1998', Genega, Switzerland: 05/11/1998 www.wto.org/english/tratop_e/tpr_e/tp87_e.htm downloaded 11/10/2002.

Yelvington, K. (ed) (1993), *Trinidad Ethnicity*, Tennessee: University of Tennessee Press.

Chapter 7

No Place Like Home: Returnee R&R (Retention and Rejection) in the Caribbean Homeland

Roger-Mark De Souza

'I remain here because I've seen no better place than home'.
– Lennox Poujade, Trinidadian returnee[1]

Caribbean migrants, even after decades abroad, identify strongly with their homelands, retaining what observers call an 'ideology of return' (Philpott, 1973, Gmelch, 1992). These migrants rarely wish to rupture ties with the island homeland, and often express an intention to return – on retirement, when things get better, when fortunes are made, or when moving objectives are realized (Potter et al, 2004). For those who do come back home permanently, return migration is considered the natural completion of the migration cycle, often expressed as a return to the land of their birth because there is 'no place like home'.

What does 'home' mean for returned migrants? How is the concept of 'home' tied to national loyalties? How do returnees perceive normative spatial units of analysis like nation and identity in times of easy transnational flows of information, people, and goods? This chapter deconstructs the notion of home in the context of Caribbean return migration for a group of returnees from Trinidad and Tobago (T&T). Through an exploration of the social, physical, and personal dimensions of 'home' it illustrates how these dimensions condition returnees' experiences in the homeland – either by reinforcing the decision to return or by rejecting it and encouraging the migrant to leave again.

Deconstructing Notions of Home and Return

Historically, home has been viewed as a static concept denoting a distinction between a sending and receiving nation. Postcolonial theorists, however, have dismantled the idea of a nation fixed to any one territorial space. Transnational discourse posits home as part of a system, often articulated in the Caribbean context of migration as a livelihood strategy, with socio-cultural dimensions. This transnational approach stresses the adaptability of Caribbean peoples whereby they maximize options available to them. Observers have referred to this adaptive approach as 'strategic flexibility' with migrants having two homes, involving identity shifts between the two worlds (Ho, 1991; Connell and King, 1999).

Others stress migrants' perceptions as the driving force behind the conceptualization of home. The underlying belief is that home is a voluntary notion, fashioned by desire and individual will. Durschmidt claims that 'what is considered to be home largely derives from the person's ability to generate a special relationship to a place, less from the physical setting' (cited in Morakhovski, 2001, 3). The saliency of perception is illustrated in the variance in responses migrants give when asked to identify their home. Western (1992), for example, notes that Caribbean migrants identify different locales as home depending on who is asking the question.

Mobility decision-making is also influenced by individual voluntarism and perception. The individual, as the ultimate decision-maker, bases the decision to move on a personal image of return that is projected onto a perceived environment in the future (Thomas-Hope, 1992). Economic, historical, structural or political factors impinge on the decision-making process in so far as they are incorporated into the individual's perception of migration.

Image formation about home is emphasized in the inclusive and exclusionary ways the concept has been applied, particularly in political claims about civic incorporation, citizenship and security. In T&T, such discourse historically emphasized nationalism in local politics. The country's first Prime Minister, Dr. Eric Williams, stated 'There can be no Mother India ... no Mother Africa ... no Mother England ... A nation, like an individual, can only have one Mother. The only Mother we recognize is Mother Trinidad and Tobago' (cited in Millett, 1993: 85).

More recently, host countries have used home and citizenship in a post September 11 world to rally support for national security. In 2002, for example, England's Home Secretary articulated a political philosophy on security and immigration, outlining why the left's response to the rise of the European far right should include ensuring that immigrants can speak English at home (Blunkett, 2002). A recent incarnation of this discourse found itself in the US government's labeling anti-terrorist measures as 'homeland security' concerns. Brown suggests that US dominant culture has been using the rhetoric of home for many years to appease its fears of outsiders invading 'our home, taking our jobs, committing crimes and going on welfare' (Brown, 1998: 6).

A different version of this political discourse raises the question of home and the nation in the context of 'deterrriorialized nation states' in which expatriates are considered part of the original homeland (Basch et al, 1994). Politicians use this approach to engage migrant enclaves overseas in homeland politics. Grenada provides an extreme example with its political leaders claiming a total adult population of 90,000: 30,000 living on the island and the other 60,000 Grenadians living overseas but still calling Grenada home (Basch et al, 1994).

The (re)construction of physical spaces has been used as a unifying concept in urban contexts. Brown (1998) describes how Puerto Rican youth in New York reconstruct 'home' through gang-related territorial practices, particularly when they feel excluded and disenfranchised from traditional conceptions of citizenship. Expatriates' construction or maintenance of a house in the homeland carries similar connotations. Migrants typically use such houses to signal success abroad or an intention to return (Bash et al, 1994; De Souza, 1998; Gmelch, 1992; Richardson, 1983).

The physical limitations of small homeland territories, such as many Caribbean islands, may also have a unifying aspect. It has been noted that while Caribbean islands' environmental limits, small territorial size, and vulnerabilities have spurred emigration (Potter et al. 2004; Richardson, 1983), socio-geographic and cultural associations tie migrants to their island homelands (De Souza, 1998). Connell and King observe that 'The national boundedness of islands and the small scale of their societies create a distinct sense of belonging' (1991, 17).

The recognition that home has many dimensions – as a locale, as a perception, as a political tool and as a means of identification – reinforces the fluidity of the concept itself. Levitt (2001) notes that although many scholars have abandoned the notion of fixed identities for one that acknowledges their malleability, much research still assumes that individuals have a 'master' overarching identity that is fundamentally rooted in a single place. At the same time, transnational community members develop varied, and sometimes conflicting, identities. Rouse (1991) recognized this fluidity among Mexican Americans whom he considered 'bifocal' because their multiple roles enabled them to view the world simultaneously through different types of lens.

Such bifocality is often reflected in the habit of comparing 'home' to 'non-home'. This is what Clifford calls a 'history of locations, and a location of histories' (cited in Morakhovski 2001, 2). This contrast leads to what has been called imaginings, memories and visions of homes and futures. Assessments of 'home' and 'non-home' often revolve around questions of personal security and happiness. Sherrington (2004) notes that home is the place nostalgically associated with comfort, kin, and an abundance of food and drink available without the need for cash. It is also the place where money is in short supply, work is hard, and there are fewer goods and services available. The migrant's host society, the 'not home', is the place that offers services and goods, but living there means needing lots of cash, being surrounded by non-kin, being subjected to crime and eating impure food. Western (1992) also noticed migrants' feelings of ambivalence towards places. He notes that Barbadian Londoners see home as 'an admission of uncertainty and ambiguity … There is London and there is Barbados, and the one isn't a constituent of the larger other, but rather a contender with it' (1992, 257).

These evaluations of home and non-home affect individual mobility decisions. Migrants decide to move by comparing the perceived locale and environment to the 'real or objective' environment (Thomas-Hope, 1992: 31). For the migrant contemplating return, there are distinct 'real' environments at three different stages of the return process. The first, at the time of departure, is the original country. Here, the overwhelming influence on the intention to return home is the combined image of return that is collectively and individually held. At the second stage, while the migrant is still in the host society, 'messages' received while abroad transform the original image of the homeland. In the third stage, in the homeland after return, the returnee builds on expectations of return, memories of life abroad, and the experience of reintegration to assess mobility options. Overall, messages that reinforce these images of home and return come from information, circumstances, occurrences, lifecourse transitions and visits to the particular locale. These messages either confer new meaning or reinforce old perceptions about the validity of home images and return options.

Investigating Return and Home

Conceptually, home has at least four connotations: a locality, a perception, a fluidity, and a comparative lens. Physically, return assumes perhaps as many forms, including seasonal circulation, long-term return, return visitation, and transnational movement (De Souza, 1998). Even though it is said that numerous Caribbean migrants hope to return home indefinitely, it is not clear how many actually do so (Western, 1992). Recent estimates, however, put the return flow in the Eastern Caribbean at between seven and ten percent of the region's population (UNECLAC, 1998). For the purpose of this analysis, I characterize return as the re-entry of a citizen who left with the intention of spending at least a year abroad, and upon returning intended to spend at least a year in the homeland. The United Nations has suggested one year as an appropriate time frame for defining a migrant, and research suggests that 'overstayers', those who stay beyond the time allocated on their visas, do indeed leave the host country within one year (Conway, 1994).

For this analysis, information on return was garnered from T&T's Central Statistical Office, contemporary newspaper reports, and a survey and interviews with 100 return migrants. Informants were contacted through a non-probability snowball sampling technique based on friendship and family networks. The perceptual dimensions of home were revealed in the questionnaires and interviews. Three general areas were examined: the details of the respondent's departure from T&T, the circumstances surrounding return and the experience of adaptation and readjustment to T&T.

A preliminary profile of the returnees interviewed suggests a reasonable mixture of respondents: at the time of return, 61 per cent of the sample was not married; 54 per cent were 30 years and younger; the two largest ethnic groups were represented: 49 per cent were Afro-Trinidadian and 18 percent Indo-Trinidadian; 60 per cent were female and 5 per cent were blue-collar workers. The host countries from which migrants returned were primarily Canada, the United States and the United Kingdom. Other countries where migrants had lived, or returned from, included France, Holland, Spain, Guatemala, Mexico, Venezuela, St. Martin,

Jamaica, and Barbados. Most migrants in the sample returned to T&T over the course of three decades – from the 1970s to the 1990s.

In the process of carrying out this work, it became clear that perceptions of home – reflected in issues of identity, security, and identification (i.e. recognition from others) – were important motivations for return and for deciding to stay, or not to stay, in the homeland. Despite disappointments and difficulties with the homeland, the majority of returnees in the sample were happy to be back. All things considered, 69 per cent were happy to be in T&T, 21 per cent were unhappy, and ten percent were either indifferent/resigned or ambivalent. Those who were unhappy were contemplating re-emigration; those who were happy have either accepted the difficulties of being back or have developed strategies to cope with such difficulties; while those who are ambivalent represented potential re-emigrants.

Perceptions of Home: Personal, Social and Spatial Dimensions

What exactly is the returnee's understanding of home? According to Webster's II University dictionary (1988), home denotes a valued place of origin that affords security and happiness. These two components were identified by respondents who consider T&T home. They derive security and happiness from the fact that their instinctive sense of belonging cannot be denied. At one level, this sense of belonging comes from a direct identification with the community of their birth.

When away from Trinidad, migrants' personal identification with T&T as home was reinforced by an inability to fully relate to the host country. This detachment is a common experience among expatriates. Breyten Breytenbach describes it as such:

> 'To be away from your natural environment is to be deprived
> of ever again functioning completely and fitting in
> instinctively. No other surroundings can replace the shared
> and unquestioned and thereby indigenous feeling of belonging
> made up of smells, sounds, gestures, and natural mimicry'
> (Breytenbach 1993: 70).

Several returnees were unable to personally identify with the host country. Sally Walcott spent five years studying in Guelph, Canada, where she felt 'at home', but never 'connected'. This feeling of connection represents an important component of the vision of home. Charlene Lee Son, a 25-year-old who spent seven years in Canada, reflects the fluidity of the notion of home when she compares her two homes in T&T Canada:

> 'Canada is home in the sense that I got to know myself better
> there. Trinidad is home in the sense that it was where I was
> brought up, learnt important values of life, discipline and

> tolerance, and it was where I had my long-standing friendship-
> based relations'.

Margaret Clark remarks that home for her is a question of belonging, 'As soon as I finished my studies I returned to Trinidad. I was sad to leave Canada, but I was happy to go home. I felt victorious that *I was back in my territory. I was where I belonged*.² Most of the returnees lamented that they did not identify personally with the host society and were happy to feel this connection upon their return.

Beyond this idea of a personal identification with the homeland is a social dimension wherein there is always a place in T&T for the expatriate. This definition of home extends beyond Robert Frost's (1915) well-known response of '(Home is) the place where, when you have to go there, (t)hey have to take you in'. Respondents stress others' recognition of their place in society. Abroad, respondents often felt that they were relegated to the status of an immigrant which made them feel that they didn't fully belong.

For migrants, acceptance is reinforced by interpersonal relations. Joanna Pascall describes what she missed about T&T while she worked for one year in France:

> 'People were nice to me, but not nice to each other. There was
> always a problem with neighbors, you had to lower the music
> as the neighbors would complain, even when they were not
> there you got into the habit of lowering the music. You were
> choked by people around, even when they were not there'.

Indira Rampersaud was tired of calling British friends to ensure that it was okay to visit: 'In England, you couldn't visit people just like that. You always had to call first. You have to be invited. When we would call people, they would tell us, "We're having lunch now".' This is a cliché among expatriates. Kandis Mohammed directly compares this with the sense of belonging she feels in T&T, 'In Canada I was always a number. In Trinidad people always know someone related to you. My brother is a musician, and my dad's father was a government minister. People know you on a one to one basis'.

The positive associations with home are reinforced by memories of an inability to physically recreate 'home' while abroad. Larry Joseph laments that apartment life in Washington, D.C. never made him feel at home. He states:

> 'I didn't like living in an apartment ... The hallways were dark
> and gloomy ... In the States everyone was in an apartment,
> whereas in Trinidad everyone was in a house. You could walk
> outside and touch lawn. I loved coming home for a visit, you
> (would) step out of the plane and sunshine (would) hit you'.

While the security that migrants feel at home is personal in nature, many also spoke of feeling alienated by the host country political discourse. When confronted with situations of racial discrimination, denials of citizenship rights, and policies of segregation, returnees turned to ethnic neighborhoods as a way of reconstructing physical spaces that brought a home-like recognition. Davindra Ramlalsingh talks about his involvement in such a neighborhood when he was in Winnipeg, Canada:

> 'I got involved in Folkarama which at the time was an exclusive West Indian club. There were about fifteen to twenty of us in the cricket club. There were also football and culture clubs, and national associations. In 1969-1970, we brought out Winnipeg Steel Orchestra ... Then there was every migrant subset for one week with Caripeg, it was just like Caribana. *It made people recognize that we were there*'.

In summary, home for this group of returnees refers to three dimensions: first, personal – a place where respondents feel they belong; second, social – a place where they have recognized status; and third, spatial – a place where they can establish a domicile. These three dimensions reinforce returnee expectations of return and provide insight into ways that returnees experience the move back home.

Returnee Retention: Is 'Being Home' Enough?

> 'I wouldn't leave again, meh navel string bury (is buried) here'. – Bridget Mooteram, Trinidadian returnee

Bridget stresses her attachment to home as the main motivation for her remaining in T&T. She refers to an old tradition of burying the umbilical cord at the foot of a tree to inculcate a sense of home and belonging in the child. Like Bridget, those returnees who are happy to be back want to remain because their return has provided security in personal, societal and spatial dimensions of their lives. Homeland attachments reinforce this feeling of 'being home', and despite any difficulties returnees face, they ultimately believe that they are better off in their homeland than in a foreign land.

An important force keeping returnees in T&T is the memory of hardships abroad. Irma Spicer is aware of the difficulties of life abroad for Trinidadians. She explains:

> 'Life is certainly harder in Canada. We used to have these big limes (get togethers), wonderful BBQs, and I would talk to my dad's friends. These are people who have washed dishes, cleaned windows, and worked as taxi drivers and waiters. They were so down and depressed'.

Roxanne Charles agrees that life abroad is difficult, and suggests that it is no easier in T&T, but location-specific assets make it better in the homeland. She gives examples, 'I lived in Canada for thirteen years. In Canada you work your tail off. Here you work just as hard, but you have family, friends, good weather, the beach and you do not need as much money to enjoy yourself'.

Sutton and Basch have indicated that part of the reluctance of Caribbean peoples to be assimilated into US society is an unwillingness to be Americanized into low status black America which is often categorized by low incomes and high school drop-out and unemployment rates (Sutton, 1987:21). As one Caribbean professional asserted, 'We saw their second-class status and didn't want any of it' (Quoted in Basch, 1987: 170). The majority of the returnees in the sample experienced upward (professional and social) mobility on return to T&T and a strong part of their desire to remain is the knowledge that if they were to re-emigrate they would be returning to a lower status.

These feelings of belonging and recognition create the sense of security that returnees sought in coming home. Rita Pierre is comforted by the fact that she can rely on the support of others in T&T. She elaborates, 'T&T is home in the sense that it is where you feel comfortable. I know that if something happens to me that someone will always help. If my car breaks down, someone will always stop to help'. Random assistance given to a stranded motorist is a common anecdote among migrants and is used to emphasize community-based acceptance. Richard Johnston, a 44 year-old executive who lived in Canada for seven years, agrees:

> 'It is easier to struggle here than abroad. Here you will never go hungry. If you can't find anything else there will always be ground provision.[3] In Canada that's not the case. (In T&T) you could go by a partner (friend) and get a meal. *You know that you belong here*'.

Richard associates belonging, support from others, and security with being in the homeland.

Memories also remind migrants of an inability to fully reconstruct home while abroad. The creation of expatriate Caribbean communities provides support and recognition among kin. This interaction is marked by organizations and festivities that cater specifically to the immigrant population. Kasinitz remarks that, '... enterprises and institutions (...) form the locus where certain aspects of a group's culture may be selected and reified so as to become the symbols of an ethnic identity within the host nation and where sentiment, symbol, and geography may become intertwined ... In a city where people are used to thinking of ethnicity and geography as connected, the founding of neighborhoods may also help to create boundaries that others recognize ... Thus churches, clubs, meeting halls, and bakeries that cater to a population in one area serve to make the ethnic group real in the minds of both members and outsiders' (1992: 39).

These Caribbean communities provide support and acceptance – a feeling of home – for new arrivals. At the same time, these communities may isolate long-term migrants from the local population. It is partly because these neighborhoods were away from everyone and everything else that Hayden Marcelle does not envision re-emigrating to Canada. He explains:

> 'Canadians associated us with certain behavioral patterns and, in fact, the vast majority was aloof. We were polarized in certain areas, and you had to drive for hours to get there. We knew very few Canadians. Christmas was dreadful. It was too cold, no one came to visit, if you had a party your friends had to come from afar. *It was nothing like home*'.

In Hayden's opinion these neighborhoods did not lead to assimilation, but to isolation and the creation of barriers. In an interesting twist, Irma Spicer supports this contention by claiming that her Trinidadian friends criticized her for interacting with non-Caribbean people:

> 'On campus there were a lot of Trinis (Trinidadians), I joined the Caribbean Students' Association, and had a group of West Indian friends. But a lot of people go abroad and "discover" that they are black, you have to understand the psyche, some people go so far as to change their names. I didn't have to prove myself, I knew who I was. I never didn't fit in and I had a lot of friends: West Indian, Canadian and international. Some Trinis would drop remarks saying that I could fit in because I was not as dark as they were ... I also used to beat pan (play the steel-pan). It was a way for me to express myself and use my talent – it also kept me connected. I met a lot of Trinis there. I also got some of my Canadian friends to beat pan. It was funny, because I had white friends some Trinis would tell me that I was giving up my nationality ...'.

Irma's experience reflects how race, identity, and social interaction may reinforce feelings of isolation in the host society, even within Caribbean communities.

These home/non-home associations are reinforced by the unhappiness of friends and relatives who live abroad. The periodic visits of these friends and relatives recall the hardships of life in the non-home and contribute to positive image formation regarding home. Yet these imaginings are based on partial realities because these visitors usually return for short periods during festive times and are thus exposed to an incomplete image of T&T. Kandis Mohammed talks about her siblings who live abroad:

> 'My brother comes down every year for Carnival. He has a totally Trinidadian accent. When he has to leave he is full of tears. And my sister has to drive one hour to go to work and she only has a half an hour for lunch. Not me! I will only leave Trinidad if there is a military government. And I will only stay away for as long as the situation prevailed. I will always come back'.

Richard Chen Wing similarly remembers his close friends and co-workers who have left:

> 'Despite T&T's social and economic difficulties, the peace,
> friendliness of the people and the weather are far better than
> in other countries ... A lot of my friends have emigrated. That
> was a question that I had to ask myself; "Should I (re-)
> emigrate or not?" In 1986 there was another wave of
> emigration (among my friends). It was a turning point in my
> life. Out of ten close friends in Trinidad, nine left to live
> abroad. I am glad not to have left ... As a foreigner applying
> for jobs, it's common knowledge that I would have to work
> twice as hard. I had status here. I had a chance of moving up
> the ladder. It is easier here. Up there it is more difficult for
> my friends who left good jobs here. They are often told that
> they are overqualified for the job. When you go abroad you
> run a greater risk of not making it'.

The difficulties that Richard's friends experience enabled him to weigh the advantages and disadvantages of re-emigration, ultimately reinforcing the positive image of home, and confirming his decision to stay. Richard and Kandis' comments provide some evidence of the challenges that qualified returnees may face in combining the personal and professional expectations of being home.

While some returnees remain because they consider T&T home, another group was able to make a home for themselves abroad. Even though this group felt a personal identification with T&T, their ability to reconstruct a home for themselves overseas demonstrates the role of individual will in home formation. Among this group was Rhonda Williams. Rhonda felt at home in Canada, but she deliberately did not get involved in activities that were Caribbean while abroad. She explains why:

> 'I didn't go to Toronto to recreate T&T in Toronto. I didn't
> go there to make West Indian friends. I made a lot of
> Canadian friends ... I had a strong sense of being Trinidadian
> so I was able to reject all the "West Indianness" when I was
> up there. I didn't feel the need to assert my
> "Trinidadianness". I didn't need to have a "Rambo Trini
> accent" to prove a point'.

Nigel Khan had a slightly different perception. He constructed a home for himself in Canada, but feels a certain ambiguity about where he feels at home:

> 'When I was there, Canada was never home. Now I feel that
> Canada is more home. I feel safer there. Even though my
> parents are here, I have a closer knit with friends and family
> in Canada. ... When I was up there T&T was home. Now I am
> becoming prejudiced, I am looking for bad things here, such
> as the crime. There is still a lot of racial prejudice in Canada;
> it is the worst thing up there, though it is not as bad as the
> United States. I didn't like the unfriendliness of the people,

> they tend to remain aloof. But in Canada I became part of the
> fabric of the society. I love baseball. I was part of Toronto. I
> had roots there. I had a nice clique, we would have nice limes
> (get togethers)'.

This kind of ambiguity about home may give rise to a transnational lifestyle. Sutton situates transnationalism in a socio-cultural framework which she describes as 'a bi-directional flow of peoples, ideas, practices, and ideologies between the Caribbean region and (the host society)' (1987:20). Basch extends this definition to include the political realm, and suggests that 'West Indians (are) becoming more directly involved in both the political life of the home countries and the US – with the involvements at the two ends representing not opposing poles, but rather a single field of action comprised of a diverse but yet unitary set of interests for the immigrants' (1987: 162).

Kasinitz (1992) suggests that this phenomenon has been further facilitated by liberal citizenship laws. All Anglophone Caribbean citizens maintain full citizenship rights even if they become citizens of another country, and in some cases such rights are extended to their foreign-born children and grandchildren. Yet T&T citizens were the last in the region to receive this privilege. Until 1992, Trinidadians were legally allowed to only hold dual status (i.e. Trinidadian citizenship-foreign residency). This meant that officially the allegiance was always with T&T and that transnationalism, at least at the official level, had not accelerated the assimilation of Trinidadians abroad. Of the 61 respondents who possessed dual status in the sample, only 18 per cent had dual citizenship. Basch (1987) indicates that this is not uncommon and generally Caribbean people are reluctant to become citizens of their host country.

Returnee Rejection: No Reason to Feel 'At Home' Here

> 'There is no reason for me to be "home". I don't feel as if I
> owe T&T anything. I am bitter about being back. Here it is
> still a matter of who you know. People are still getting jobs
> based on old laurels. For me home is where you could pay
> bills, pay rent. I am ready to leave at any time'.
> – Shirley Duke, Trinidadian returnee

Several individuals, like Shirley, are contemplating re-emigration because they have been disappointed with their move back home. For those who are thinking of leaving, thoughts of re-emigration are the result of expectations of home not being met.

One key component of being comfortable on return is being able to identify with T&T. The situation of a young married couple, Robert and Ann-Marie Lee Heung, is interesting as it reveals the conflict that some families face in deciding to remain in T&T. On the one hand Robert feels very Trinidadian, while Ann-Marie, a third generation Chinese-Trinidadian, feels trapped and unhappy. She explains:

> 'I like Canada. As a child I would spend every summer there. For me Canada has been home for a long time. In Trinidad we grew up like true Chinese. I never ate doubles or went down south.[4] At home we ate only Chinese food. My grandparents were from China. My parents were more Trinidadian ... I went to a private school where we called each other by our Chinese names, but stopped when we got teased about it ... Now most of my family is in Canada. This time here in Trinidad is not permanent. We have no house here ... Conversations with people here are so limited and unstimulating. All they talk about is other people's business or the movies. Trinidad is not permanent for me. I remember going for my income tax clearance (to be able to travel), and being questioned about why I was traveling so often. I was really upset. I do not want to be stuck here forever. I am not sinking with the ship. Especially with all this coup thing and all that'.[5]

Even though Ann-Marie was born in T&T and emigrated at the age of 22, she does not personally identify with the land of her birth. Throughout her childhood as her attachment to T&T was being restricted, her identification with Canada was constantly being fomented through annual summer visits. She eventually emigrated with no intention of returning, but marriage changed the circumstances. Her deeply-felt lack of identification with T&T is reaffirmed on her return as a tied-spouse. She experiences value conflict and alienation, and would like to re-emigrate, but her husband has a strong affinity to staying. Uncertainty prevails in terms of a clear dissonance in each one's future migration plans, their marital stability, or both.

Another common problem that encourages returnees to leave is the resentment, or lack of recognition, of foreign qualifications and experience. Laurence Bain explains that his sister re-emigrated to England after a year back in Trinidad:

> 'My sister came back to live and couldn't make it. She got a job in the hospital. (In England), she was a well-off nurse, a district nurse. Here, she had a difficult time because of the system of hiring and because they took her on a temporary basis. She felt that they wouldn't recognize her experience, especially with her qualifications'.

Randy Noel, a 50-year-old marketing consultant, had comparable experiences:

> 'This is the second time that I've returned to Trinidad. When I was separated from my first wife I came down alone, and then I essentially returned to Canada as she was having

problems with my first son and I needed to go there to help out. When I returned (the second time) in 1990 I found it difficult to get a job. Many of the businesses are family owned and run, and very often there is no entree for the qualified ... The techniques that I've tried to implement here have been met with a certain degree of resistance. I have been told that those techniques are good for Canada'.

David Joseph, a hotel management consultant, also experienced problems similar to Randy Noel's when he tried to implement innovative ideas on his return. He came back in 1987 after 25 years in Canada where he worked in hotel management. He explains:

'In Tobago there were a lot of political interests running (X) hotel. I was watched very closely, as if under suspicion. Soon after my arrival I had a run-in with the General Manager who told me that my approach was good for Canada, but would not work in Tobago. I was made to leave'.

In the year following my initial interviews, 3 per cent of my sample had left primarily due to an inability to secure suitable employment or because of dissatisfaction with work. Harry Rampersaud, an opera singer, is one of them. He claims, 'People don't want us. "X" (A talent show on local television) often features locals who have gone abroad and done well. But she, (the organizer of the show), has her people and she didn't want us. I don't think people care'. A local newspaper ran an article on Harry's departure. The article notes that he was leaving because of 'disenchantment with the way things have gone for him' and quotes Harry as saying, 'It is with some sadness that I am saying good-bye to my homeland, T&T, ... but I am a singer, and not yet a "has been", so while I still have the gift I must try and use it to the best of my ability'. The article concludes that he plans to settle in Miami, Florida or Paris, France where opportunities are available to him that he cannot find at home.[6]

Harry's case is not the norm, however. Even though further education, training, and research are other areas of professional development where returnees experience difficulty, it is not the strongest driving force for re-emigration. Bremner (1983) suggests that a barrier to the re-entry of professionals is remuneration, but concludes that even if enough monetary incentives were offered expatriate scientists would still not return. Abroad they enjoy the satisfaction of being in a society where their work is understood and considered important. He suggests that they are accorded a dignity that will be denied in their communities of origin. This lack of emphasis on economic push/pull factors is supported in the sample. Among those who are unhappy in T&T, only two individuals are contemplating re-emigration because of economic reasons. This is not unusual. Those who have economic concerns are unlikely to return in the first place.[7] For those who voluntarily return, the recognition that 'home' represents a haven and a satisfactory life-style is more personal than professional.

One of the major disappointments of return is the threat to personal security that is posed by increasing levels of criminality. Indira Rampersaud, Harry the opera singer's wife, explained the curtailing effect of crime on their dream return after 36 years in England:

> 'We love the fruits here, but you can't even enjoy the fruits from your own back yard. People come into your yard and steal your fruits and you can't even chase them away. With so much burglar proofing you almost feel as if you're in a jail ... In England, when you go on holidays you let the police know so that they can keep an eye on your house. In Trinidad, the police are the last people that you let know because they would go into your house themselves'.

One year after our initial interview, I re-interviewed Indira two days prior to their re-departure from T&T. She indicated that while their case was complicated by Harry's inability to find students to teach and her inability to find a job, Harry was mugged in downtown Port-of-Spain a few weeks earlier. This experience cemented their decision to leave.

Another returnee, 43-year-old Nigel Khan, a sales manager of a telecommunications firm, was able to find employment in T&T, yet is making plans to re-emigrate to Canada because he finds the level of crime daunting. He intimates:

> 'At the moment when we left (in 1986) I was not committed to leaving. I had a good job in Trinidad, but was forced to go as our papers had come through. I went to land (to enter the host country to secure legal status), and come back to Trinidad.[8] When I came back I discovered that someone had broken into our house and stole some of our valuables. That encouraged me to just pack up and leave ... I decided to return for two reasons: firstly, my marriage broke up and secondly, I got an excellent job offer in Trinidad ... It meant that I would be taken care of financially and deep down inside I wanted to come back ... A few months after my return, I was robbed of my car at gunpoint. It's ironical that crime that pushed me to leave the first time and once again I am preparing to leave for the same reason'.

Insecurity also reflects itself in a lack of opportunities for children. Roxanne Charles, who returned when her marriage dissolved, thinks that she will only leave again if the education of her children necessitated a change. This is an option because of her dual citizen status:

'In Trinidad I spend a lot of money on fees to send the
children to (private) school. My eldest daughter was eight
when we came to Trinidad. She couldn't pick-up ... We have
since discovered that she is learning disabled and I pay
$1,800 TT ($424US at the exchange rate then) a term to send
her to school. If T&T cannot cater to the educational needs of
my children I will leave again'.

In a similar case, Colin Warner explains that he would only leave if his family's
welfare was threatened:

'We will leave if there was an ongoing situation like the
coup, a situation where there is a total breakdown of
amenities. We know that sometime in the future a situation
like the coup will happen again and many more lives will be
lost. But we plan to stay here for a while. We are looking into
buying property, and we just bought a second car'.

The notion of home – particularly failed expectations of personal security and
societal recognition that should come from being home – are significant
determinants in migrants' desire to leave the homeland again. At the same time, a
number of other factors, some fortuitous, influence the desire and ability to leave.

Non-Home Motivations for Leaving or Staying: What's Home Got to Do With It?

'I love Trinidad and Tobago. There will always be that
connection, but it's not like an umbilical cord, necessary for
the sustenance of life'.
– Irma Spicer, Trinidadian returnee

While expectations of 'being home' may weigh heavily in the decision to return
home, other factors encouraged migrants to leave again or to stay in T&T. In this
sample, these include lifecourse transitions, particularly those in the early stages of
household and career formation, kinship obligations back home, and an
unwillingness to move again.

Several young returnees suggested that an increasingly interdependent world
and the ease of travel impinge on their decision to leave again. Twenty-six year-old
Denise Pouchet explains:

'Home will always be home, but apart from family I see no
guarantee that I will not emigrate again, at least temporarily.
This is not due to any dissatisfaction with my country (after
all certain aspects of any country in today's world would be
cause for dissatisfaction), but because in my present field

(and any other I am likely to get into), travel isn't only enticing, it's a distinct possibility. People of my generation and of future ones are not citizens of any one country, but of the world'.

For many of the young returnees, T&T is where they would prefer to live, but they do not see themselves as committed to settling on the islands.

Another important consideration in the decision to leave is the migrant's willingness to return in the first place. Carla Ramroop, a 24-year-old British-trained dentist, explains why reintegration is more difficult for her:

'I didn't want to come back. My parents insisted that I come back as they didn't approve of my boyfriend. Even though I went to England when I was nineteen and only stayed for five years, it was there that I really grew up. I am not really settled here. I am in transition, and I am here for as long as it lasts'.

Carla makes a clear association between a willingness to return, belonging, and security. In fact, she overturns one of the popular returnee clichés, when she notes:

'People in Trinidad are crude. The other day my car shut down around three in the afternoon. Men in passing cars yelled out abuses and I got a lot of cat calls. In England the Automobile Association would have come. I sat down in the car with the windows rolled up. Policemen passed, they didn't even stop once. Abroad they are more receptive, more polite. I am not settling down in Trinidad, my heart is in England'.

In the sample, the migrants who returned involuntarily, similar to Carla, are thinking of leaving again.

The migrants who didn't want to stay on the island didn't have a sense of T&T as home – and, paradoxically, experienced homelessness on their return. At a personal level, they didn't have a strong sense of a Trinidadian identity. Socially, they felt that the society was unwilling to accept their status as a Trinidadian with foreign experience and expertise or they didn't have a feeling of security that their vision of home should provide. Finally, others were relatively young and in the early stages of family or career formation and not ready to settle in one place. Those who didn't return to T&T of their own volition were looking for an opportunity to leave again.

Others want to leave, but remain in T&T because of familial ties. Nicola Ragoonath is on the verge of re-emigrating, but is kept on the homefront by the presence of her elderly mother who refuses to leave:

'I am contemplating leaving for New York. I have two sisters there. If only my mother would come up with me, I would go

up. I would go with the intention of returning. I could retire
nicely here if I came back. At the moment I have no house
and land here as it is too expensive and the mortgage is too
long. The less that I have the easier it will be for me to leave.
Knowing what I know now, I wouldn't advise anyone to
return'.

Nicola has tried to convince her mother to emigrate, but her mother refuses to
and Nicola wants to be close to her mother. Cynthia Tiwari also wanted to remain
to be close to her father who was left alone after her mother's death, but she will
soon be joining her fiancé in Barbados. She is willing to leave as she will never be
too far from her father. Janet Beckles, a single 43-year-old mother, was not tied by
caring for an older parent, but by ensuring her son's well-being. However, as her
son grew old enough to adjust, she acted upon her desire to move:

'I decided to leave when (my son) was four years old.
Trinidad has nothing to offer young people. It is an ideal
country for people under ten and over 70. The lack of order,
inefficient utilities, unreliable communication and transport
networks; any improvement has been cosmetic. In Canada
there is a law for everything. In Trinidad there is so much
lawlessness, with regard to crime, traffic, regulations ... things
are not ordered'.

Once her son was old enough, Janet left Trinidad again for Canada. Marisa
Namsoo, on the other hand, decided against re-emigrating as she felt family support
in case of hardship would make up for any economic benefits derived from leaving
again:

'We thought of going away to live, but my husband is well
established here, and we thought that it was better to be a big
fish in a little pond. And we know that if we were to ever
ketch arse (have a difficult time) at least we have family
support'.

Another group of returnees gave different reasons for staying. Some cited the
difficulty of uprooting oneself again. Returning was already a flurry of activity, and
once resettled in T&T, they cannot imagine having to start the process over once
again. Brian Le Blanc and his wife remain because relocating requires a significant
financial outlay that they simply do not have. Furthermore, at home they have risen
in status by having lived abroad, and they are materially better off, having brought
back all their possessions with them. Brian explains:

'Our lifestyle in the States was much better. We were richer
and we had more value for our money ... We lived well. We
had a boat, two cars and we were renting with the option of

> buying. We felt at home. If we could go back financially, we
> would. We sold the cars and the boat. We brought back all
> our household items. To remigrate now is harder, we need
> substantial savings, we need more money'.

Comparing the home to the non-home, Calvin Goddard believes that readjusting once again to the host society will be too difficult. He elaborates on re-entry difficulties in the host country:

> 'The three toughest things for people to adapt to when they
> go abroad are:
> 1. Adjustment to family situations up there. With the terrible
>   exchange rate, it is not that easy to compete with siblings
>   who remained abroad.
> 2. (In T&T) academics are not highly valued. You could have
>   attained a middle-high (level) position here over the years,
>   and when you go up there you start at the bottom because
>   you don't have the qualifications.
> 3. The coldness is a deterrent. It is not necessarily the weather,
>   but more a question of being out in the cold. A lack of
>   social acceptance could be an excuse, a rationalization not
>   to leave again'.

Another respondent, Richard Noel, talked about the difficultly of having two homes and the necessity to establish residency (roots) in one place:

> 'In Canada you had a kind of home feeling, but you need to
> put down your roots somewhere, so we no longer have two
> homes. Trinidad is our first choice. To leave again is a very
> hard decision. Here I now have my own job, I have a home.
> To start over at this age (44 years) is not easy. I suppose if it
> were a question of security. If things got so bad that my
> daughter's security was threatened, I might leave again'.

As Richard suggests, age is an important factor that deters re-emigration. Not only is it difficult to start over at a later age, but for some, like Barbara Roach, this means that they have no means of support abroad, 'I would go back at anytime, but I don't have the means to live there. I wouldn't get any pension up there'. These examples illustrate once again that re-migration might paradoxically be harder for some qualified, educated Trinidadians who have been enjoyed professional success in the Caribbean.

Assessing 'Home' as a Determinant in Return Decision Making

This chapter has offered some insight into Caribbean migrants' perspectives on home and the role those perspectives play in mobility decisions. The migrants who lived abroad for several years and returned home were motivated by the allure of 'no place like home'. Examining this motivation provides a context for a critical deconstruction of notions such as transnationalism, home, and return and offers some conclusions about the relationship of home to return decision making.

Home is multidimensional, at once individual and collective, static and fluid. At an individual level, home represents an attachment to the land of birth, a societal acceptance of one's identity as a Caribbean person, and a desire to physically reconstruct the cognitive idea of home. The physical dimensions of home – a house or domicile – is the culmination of the home-return association. It embodies an intention to return and to stay, success abroad, shelter and security, and an ability to construct a personal space according to one's preferences, providing recognition and a place in the home community.

Home for some of the returnees in this sample was a static concept. It clearly meant their Caribbean homeland. Even though the transnational literature challenges normative conceptualizations of home, the concept of home as a static entity still has some validity for these returnees. For others, particularly those who were younger or who had a weak attachment to T&T, home was a fluid concept. This fluidity sometimes gives rise to ambiguity and uncertainty reflected in differing expectations about home. Generally, all the returnees hoped for happiness, recognition, and security. When these expectations were not fulfilled, some returnees left again. Others hedged their bets. They came back, but retained their dual citizenship. They didn't want to forego the option of re-emigration should the return fail or they wanted to pass on the benefits of dual citizenship to their children.

Mobility decisions based on expectations of home are tied to collective image formation that is sustained by direct experiences and memories. The respondents in this sample – first-generation migrants – identified with the Caribbean homeland in an immediate way. Their memories of T&T are vivid and direct, much more than it probably would be for their children. These individuals recalled memories, and received new messages, about home and return, which they subsequently transformed into a meaningful image of security and happiness. This image was filtered through a system of values, reference groups, aspirations and goals, and helped cement the decision to move back home.

Home is also a means of identity and affirmation. Despite global connections, particularly ease of travel and free information flows, living in a foreign land created a distance between the land where many of the migrants lived and the home where they acquired their identity. When these migrants were refused an identity and an identification in the host nation, the homeland became a strong focal point for positive image formation and symbolic associations of home, security, and belonging. In a sense, migrants sought to overcome a sense of homelessness during their time overseas and wanted to reaffirm their Caribbean identity. At these times, home served as a means of self affirmation. For some of these migrants, the return home was a conscious resistance to second-class citizenship abroad, to a transnational lifestyle and to having two homes.

Home is a powerful contextual filter in mobility decision making. Home is used to draw comparisons between home/non-home, security/rootlessness and real/perceived environments. Returnees measure this notion of home – their sense of identity, the recognition of their rightful place in society, and a feeling of security – against the potential benefits and pitfalls of life abroad.

Overall, home can serve as an incentive and a disincentive to remain. Those migrants who chose to stay in T&T are willing to do so for a variety of reasons. These reasons include memories and examples of the difficulties of life abroad, kinship ties, and the difficulty of recreating 'home' abroad. Overall, homeland attachments and 'being home' are key motivators for keeping these returnees in their homeland. Others chose to remain in the homeland, even though they are disappointed with some aspects of the return. They believe that the benefits of being home far outweigh the economic benefits of being in a non-home society where they will never instinctively fit in. Some migrants are happy to shuttle back and forth, maintaining two homes, keeping 'one foot in, one foot out' as one returnee stated.

Some returnees are kept in the homeland by other forces – such as kinship obligations – and feel 'homeless' in the Caribbean. They often want to leave again. For them, the failure of the return represents the failure to find a home again in the homeland. Others willingly leave to reconstruct a primary home for themselves abroad, maintaining an attachment to the land of their birth, but functioning in a new home where they have been able to create a niche for themselves. Many of these migrants live and function in Caribbean neighborhoods where they enjoy the benefits of a Caribbean milieu that offers the advantages of life overseas.

Using the construct of home to examine return certainly does not provide the complete story on return motivations, but it gives a perspective on the role of home and homeland attachments in mobility decision making and in identity formation in periods of circulation and transnationalism. Not only do homeland associations offer R&R – rest and relaxation – expressed as security and happiness by these migrants, they also serve as a driving force for returnee retention and rejection. Ultimately, it is the migrants themselves who highlight these associations of home with identity, recognition, and security, suggesting that for them a physical return to the Caribbean homeland best exemplifies the belief that there's 'no place like home'.

Notes

1. All names used for respondents are fictitious in order to preserve their identity.
2. All italics in migrants' statements denote my emphasis.
3. Ground provision refers to root crops such as eddoes, dasheen, yam and potatoes.
4. Ann Marie implies that she has had little contact with the Indo-Trinidadian population of Trinidad: *doubles*, a popular delicacy of East Indian origin, is a small bread split in two pieces with curried *channa* (chick-peas) in between; *South* refers to the southern

part of the island where many Indo-Trinidadians live. (In reality, there is a larger Indo-Trinidadian population in Central Trinidad.)

5. In reference to the failed attempted coup d'etat of 1990.
6. In order to maintain Harry's anonymity I do not give the reference for the article.
7. See Daniel (1983) for a discussion of economic concerns as a practical barrier to the re-entry of Trinidadian expatriates.
8. This strategy of entering the host country to secure legal status and subsequently returning to T&T was common among those who had dual citizenship or residency.

References

Bash, Linda G. (1987), 'The Politics of Caribbeanization: Vincentians and Grenadians in New York', in Constance R. Sutton and Elsa M. Chaney (eds), *Caribbean Life in New York City: Sociocultural Dimensions*, New York: Center for Migration Studies, 160–181.

Bash, Linda G., Glick Schiller, Nina, and Cristina Szanton Blanc (1994), *Nations Unbound: Transnational Projects, Postcolonial Predicaments, and Deterritorialized Nation States*, Langhorne, Pennsylvania: Gordon and Breach.

Blunkett, David (2002), 'What does citizenship mean today?', *Observer*, 15 September, 2002, Accessed on line at http://www.blacklondon.org.uk/news/2002/20020915a.htm on September 1, 2004.

Bremner, Theodore (1983), 'The Caribbean Expatriate: Barriers to Returning, Perspectives of the Natural Scientist', in *Caribbean Immigration to the United States*, Roy S. Bryce-Laporte and Delores M. Mortimer (eds), Washington, D.C.: RIIES Occasional Papers, No.1, 149-157.

Breytenbach, Breyten (1993), *Return to Paradise*, Orlando, Florida: Harcourt Brace and Company.

Brown, Monica. (1998), 'Neither Here Nor There: Nuyorican Literature, Home, and the "American" National Symbolic', JSRI Working Paper #42, The Julian Samora Research Institute, Michigan State University, East Lansing, Michigan.

Conway, Dennis (1994), 'The Complexity of Caribbean Migration', *Caribbean Affairs*, 7(4): 96-119.

Connell, John Connell and King, Russell (1999), 'Island Migration in a Changing World', in *Small Worlds, Global Lives*, Russell King and John Connell (eds), Pinter, New York, 1-26.

Daniel, Edwin (1983), 'Perspectives on the Total Utilization of Manpower and the Caribbean Expatriate: Barriers to Returning', in *Caribbean Immigration to the United States*, Roy S. Bryce-Laporte and Delores M. Mortimer (eds), Washington, D.C.: RIIES Occasional Papers, No.1, 158-168.

De Souza, Roger-Mark (1998), 'The Spell of the Cascadura: West Indian Return Migration', in Thomas Klak (ed), *Globalization and Neoliberalism The Caribbean Context*, Maryland: Rowman and Littlefield, 227-253.

Frost, Robert (1915), 'The Death of the Hired Man', in *North of Boston*, Henry Holt and Company, New York.

Gmelch, George (1992), *Double Passage: The Lives of Caribbean Migrants Abroad and Back Home*, Michigan: The University of Michigan Press.

Harbison, Sarah (1981), 'Family Structure and Family Strategy in Migration Decision

Making', in *Migration Decision Making: Multidisciplinary Approaches to Microlevel Studies in Developed and Developing Countries*, Gordon F. De Jong and Robert W. Gardner (eds), New York: Pergamon Press Inc.

Ho, Christine (1991), *Salt-water trinnies: Afro-Trinidadian immigrant networks and non-assimilation in Los Angeles*, New York: AMS Press, Inc.

Kasinitz, Philip (1992), *Caribbean New York: Black Immigrants and the Politics of Race*, New York: Cornell University Press.

Levitt, Peggy (2001), *The Transnational Villagers*, University of California Press.

Mendes, John (1986), *Cote ci, cote la: T&T Dictionary*, Port-of-Spain, Trinidad.

Millett, Trevor (1993), *The Chinese in Trinidad*, Port-of-Spain: Inprint Caribbean Limited.

Morakhovski, Dimitri (2001), 'Travel practices and perpetual change: "home" in the era of movement', Paper presented at TASA (The Australian Sociological Association) 2001 Conference, The University of Sydney, 13-15 December 2001.

Philpott, Stuart (1973), *West Indian Migration: The Montserrat Case*, Athlone Press, London.

Philpott, Stuart (1999), 'The Breath of "The Beast": Migration, Volcanic Disaster, Place and Identity in Montserrat', in *Small Worlds, Global Lives*, Russell King and John Connell (eds), Pinter, New York, 137-159.

Potter, Robert, Barker, David, Conway, Dennis and Klak, Thomas (2004), *The Contemporary Caribbean*, Pearson Prentice Hall, Essex, England.

Richardson, Bonham C. (1983), *Caribbean Migrants – Environment and Human Survival on St. Kitts and Nevis*, Knoxville: The University of Tennessee Press, 1983.

Rouse, Roger (1991), 'Mexican Migration and the Social Space of Postmodernism', *Diaspora*, 1:8-23.

Sherrington, Richard (2004), 'The discomforts of "home": influences and imaginings of rural pasts and urban futures in Dar es Salaam, Tanzania', Paper presented at the Association of Social Anthropologists of the UK and Commonwealth, 2004, Durham, United Kingdom.

Sutton, Constance R. (1987), 'The Caribbeanization of New York City and the Emergence of a Transnational Socio-cultural System', in Constance R. Sutton and Elsa M. Chaney (eds), *Caribbean Life in New York City: Sociocultural Dimensions*, New York: Center for Migration Studies.

Thomas-Hope, Elizabeth (1992), *Explanation in Caribbean Migration – Perception of the Image: Jamaica, Barbados, St. Vincent*, Warwick University Caribbean Studies, London: The Macmillan Press Ltd.

UNECLAC (1998), 'A Study of Return Migration to the Organization of Eastern Caribbean States (OECS) Territories and the British Virgin Islands in the Closing Years of the Twentieth Century', United Nations Economic Commission for Latin America and the Caribbean, Port of Spain, T&T.

Western, John (1992), *A Passage to England*, Minneapolis: University of Minnesota Press.

Chapter 8

Return Migration to Trinidad and Tobago: Motives, Consequences and the Prospect of Re-Migration

Godfrey St. Bernard

With respect to the Caribbean Sub-Region, there has been a plethora of studies of migratory processes – examining the process and its developmental consequences – particularly in the context of Jamaica, the Lesser Antillean islands of the Eastern Caribbean, and the Hispanic Caribbean (Maunder, 1955; Roberts and Mills, 1958; Tidrick, 1966; Ebanks et al, 1979; Kritz, 1981; Roberts, 1981; Smith, 1981; Segal; 1987; Richardson, 1983; Thomas-Hope, 1992). Notably, Boodhoo and Baksh (1981) undertook a study sampling three different migrant groups – potential Guyanese emigrants, Guyanese immigrants who reside permanently in England and returnees – to understand their migratory processes and what these consequential losses – or 'brain drain' – of the country's educated could mean to the national development of Guyana. One of the present chapter's intentions is to stimulate a similar interest in migratory processes and their developmental consequences in Trinidad and Tobago, during the post-independence era and in particular, the 1990s.

Prior to the 1960s, emigration from Trinidad and Tobago was not as profound as in a number of neighbouring Caribbean islands and was offset by immigration that resulted in net gains in the resident population during that period. In the late 1960s, however, emigration emerged as a public concern that became manifest in the context of the 'brain drain' phenomenon. This stimulated public interest and resulted in an early study that sought to examine the implications of the 'brain drain' on prospects for national development in Trinidad and Tobago (Rampersad and Pujadas, 1970). Since then, and particularly during the 1990s, there have been no systematic studies of migratory processes in Trinidad and Tobago. In other Caribbean islands that were net losers of population (with emigration exceeding immigration), researchers (Girling, 1974, Ebanks et al, 1979, Cooper, 1985; Anderson, 1985) channeled their energies towards studying emigration and its impact on domestic spheres particularly with regard to labor market, public welfare services and the provision of health services.

Since the 1980s, there has been a proliferation of studies of in-migration and in particular, some have focused upon return migration in Caribbean societies (Rubenstein, 1982; Thomas-Hope, 1985, 1999; Gmelch, 1987, 1992; Muschkin,

1993; Byron, 1994, 2000; Plaza, 2002). These authors have focused primarily on return migration to countries such as St. Vincent, Jamaica, Barbados, Puerto Rico and St. Kitts and Nevis. With the exception of Puerto Rico, because of its special Commonwealth relationship with the United States, all other Caribbean cases have been characterized by steady flows of emigrants mainly to the United Kingdom and to a lesser extent, the United States and Canada during the decades of the 1950s and 1960s. In many cases, these studies examined returnees' desire to return and their actual experiences upon return to the Caribbean, with the target populations often being retirees or elderly returnees.

Return migration is a critical development issue that should interest policy-makers and other allied stakeholders in Trinidad and Tobago. Yet, prior to this collection, only one article by De Souza (1998) had specifically investigated return migration in Trinidad and Tobago. Together with the two earlier Chapters by Lee-Cunin and De Souza, our three chapters constitute a definite addition to the formal literature examining return migration in Trinidad and Tobago. In terms of its general context, this chapter details the differential experiences of returnees who lived in Trinidad and Tobago in the late 1990s. It focuses upon their motivations to return, the consequences of their return, their thoughts regarding the exercise of choice to return to the host country, and their consideration of re-migration, *in the majority of cases* to their recently-departed, metropolitan country. The findings in this study are also examined in order to assess the extent to which they are consistent with or depart from two sets of findings – the more qualitative findings on return to Trinidad and Tobago articulated by De Souza (1998) and observations on return migration experiences in other Anglophone settings.

Literature Review – Return Migration in the Caribbean

Despite the number of studies examining migratory processes in Caribbean societies, return migration has rarely been the main focus. Instead, most research has focused on emigration and its consequences, such migration being primarily due to pressures associated with over-population, and in particular, labour shortages that had intensified over time. In the post-World War II era, such problems had intensified further as several countries in the region had become net losers of population due to migration. This reinforced the importance of studying emigration, if only to examine its effects on the development status of countries. Insofar as the process of emigration was gaining momentum, studies targeting the impact of remittances and the phenomenon of return migration were inevitable. According to Stinner and de Albuquerque (1982: xxxix): 'Migrants to the metropole and to other societies within the region do return and this return conveys important demographic, socio-economic and political implications for the original sending society and the migrants themselves'.

While recognizing the changing profile of return migrants in the Caribbean, Byron (2000) notes that substantial proportions are still likely to be elderly returnees especially from Britain where they had migrated prior to the 1962 Commonwealth Immigration Act. She alluded to the fact that there were significant regional inflows of return migrants from Britain during the 1970s and 1980s, due to some extent to the return of the Post-War wave of migrants who sought to await their guaranteed pensions before retirement. Hypothesizing a regional trend, Byron (2000) noted that return migration from the United States and Canada is likely to become much more commonplace beyond 2000. She also noted that there is likely to be an influx of returnees of a younger age as contractual labour – agricultural farm workers, in the main – recruited from a number of Caribbean islands to work in the United States and Canada, complete their contractual obligations. The impact of globalization and the proliferation of industries in sectors such as tourism, retail sales and hi-tech, constitute a second set of factors that she identified as being instrumental in prompting the return of younger returnees to their Caribbean homelands. A third factor was the deportation of young Caribbean nationals who violated visa restrictions or had criminal records (Byron 2000). Increasingly, Caribbean nationals have been migrating to the North Atlantic for short sojourns during which they can earn sufficient income to fund major domestic projects such as housing (Conway 2002).

Based upon fieldwork conducted during the mid-1960s, Philpott (1973) examined West Indian migratory processes with a view to discerning their implications within institutional spheres in small domestic settings. His study focused specifically on the island of Montserrat where migration to Britain was heavily concentrated in the 20-45 age group. In his view, such emigration was consistent with a 'migrant ideology' that was based upon a high likelihood of return – a prospect being stimulated by domestic commitments in the form of remittances to the island. Philpott also documented cases in which Montserratians embraced 'migration ideologies' and returned to the island to become petty entrepreneurs investing in 'rum shops', transportation and agriculture. While such decisions enhanced the social status of the returnees within domestic spheres, the outcomes were short-lived insofar as the Montserratian economy was incapable of guaranteeing them a sustainable livelihood. The end result was the eventual re-migration of several returnees.

With reference to the international migration patterns of Trinidadians and Tobagonians, De Souza (1998) noted that long-term emigration mainly involving unskilled individuals persisted until the 1970s. However, from about the mid-1970s onward, he acknowledged that there was a change in the occupational profile of migrants to the extent that emigrants were increasingly drawn from among highly educated professionals including doctors, nurses, teachers and scientists. Not surprising, De Souza alluded to the notion that recent cohorts of returnees have often been professionals who possess attributes deemed to be critical to prompting national and regional development initiatives. De Souza's definition of a returnee, like this present study, uses a one year absence as a definitional criterion. In the context of this research, De Souza's work is also important because of its qualitative approach to discerning the key motivational factors influencing a return to Trinidad and Tobago. According to De Souza, these factors include the homeland milieu, family reunification, return obligations, and professional fulfillments.

The return of second-generation Caribbean migrants is an additional consideration in the emerging body of literature on return migration among younger persons of Caribbean heritage (Potter 2001, 2003a, 2003b; Potter and Phillips 2002). For example, Plaza (2002) examines the adjustment experiences of a very small sample of second-generation British Caribbean returnees to Barbados and Jamaica. He notes that factors associated with race, skin colour and gender are principal motivators in the decision to return to the Caribbean. According to Plaza, this group of returnees generally secures professionals jobs, but at the same time experiences great difficulty in interactions with their local peers while making the transition to a professional life in the Caribbean.

Altogether, the earlier studies suggest that return migration has several dimensions which deserve examination: namely, selectivity and direction, motivation for return, consequences and the exercise of choice by considering re-migration. In terms of selectivity and direction, the age and functional classification of returnees are critical factors that can be evaluated and assessed in the context of emergent patterns in contemporary Trinidad and Tobago. Motivations for return are contingent upon personal aspirations, expectations and activities at personal or group levels, bearing in mind existing institutional structures whether material and/or non-material in sending and host environments. In accordance with earlier observations by Davison (1967) and Patterson (1968), it should not be surprising that Stinner and de Albuquerque (1982) have been able to extract five principal domains to capture the broad spectrum of motivations. The five include; socio-economic maladaptation, life course transitions, expiration, termination or violation of contractual arrangements, homeland linkage and societal socio-economic situations. The consequential experiences of return migrants and choices associated with re-migration are somewhat related, insofar as the latter is often contingent on the former; as observed by Philpott (1973) and Plaza (2002).

Materials and Methods

The data from the 2000 Population and Housing Census for Trinidad and Tobago are still being processed and when made available, they should provide a means for gauging the magnitude of return migration during the 1990-2000 intercensal period. As a precautionary measure, ECLAC (1998) has already noted that the census in Trinidad and Tobago elicited data on respondents' place of residence at one, five and ten years ago and suggested that such a 'point-in-time' approach does not capture all movements that occurred during the 1990s decade. This is because some nationals may have returned and re-migrated or died during the 1990-2000 period, thereby eliminating them from inclusion in any enumeration process. We can, however, utilize available travel statistics to gauge the magnitude of return migration during the 1990-1998 period. Accordingly, Table 8.1 shows that at least 240 immigrants may have entered Trinidad and Tobago on a permanent basis, with another 6,270 entering on a temporary or 'quasi-permanent' basis. Since there is nothing to distinguish between immigrants who have been citizens of Trinidad and

Tobago from those who have been aliens, the number of return migrants during 1990-1998 should not be in excess of 6,270 and could in reality be substantially less. Furthermore, this estimate is based on the assumption that the official international travel statistics are reliable. These are far from inconsequential considerations, since they have implications for evaluating decisions surrounding the selection of the sample that is used in this chapter.

Table 8.1   Immigration statistics – immigrants with potentially permanent intentions

Year	Student Residents	Permanent Immigrants	Potentially Permanent Intentions	Temporary/ Quasi- Permanent Immigrants	Total Immigrants
	$c1$	$c2$	$c3=c1+c2$	$c4$	$c5=c3+c4$
1990	22	-	22	327	349
1991	53	-	53	652	705
1992	24	3	27	421	448
1993	25	-	25	635	660
1994	9	2	11	676	687
1995	49	11	60	481	541
1996	21	-	21	663	684
1997	11	5	16	1268	1284
1998	-	5	5	1147	1152
1990-1998	214	26	240	6270	6510

Source: Annual Statistical Digest, 1998/1999 – Central Statistical Office.

In 1998, the Ministry of Social and Community Development, under the auspices of the Government of the Republic of Trinidad and Tobago, undertook a Survey of Return Migration in Trinidad and Tobago. For definitional purposes, return migrants were identified as nationals of Trinidad and Tobago who had lived abroad continuously for at least a year and returned within the last twenty years, with an intention to remain in Trinidad and Tobago. The main objectives of the survey were to 1) analyze current trends in external migration in Trinidad and Tobago, 2) assess the reasons for return migration and 3) determine the impact of return migration on the economy as a whole. It should be noted that the survey was constrained by a limited budget and by the absence of a spatially explicit stratified random sampling frame (see Berry and Baker 1968), from which one could draw a nationally representative sample of return migrants.

During the fourth quarter of 1997 and the first quarter of 1998, the Continuous Sample Survey of Population (CSSP) conducted household inquiries to determine whether any occupants had been abroad for at least one year and returned home to stay. The sample was selected using a two stage sampling design. At the first stage of selection, 480 enumeration districts[1] were selected randomly. At the second stage, systematic sampling was used to select clusters of twenty households based upon a random start and a sampling interval that was determined by the number of households in enumeration districts. Altogether, 9,600 households were selected from the main geographic domains of the country.[2] Analysis of the completed survey revealed that 4 per cent of all households had at least one return migrant and each of these households had an average of 1.1 to 1.25 return migrants. The anticipated number of respondent households with at least one returnee to be chosen for sampling was expected to be between 422 and 480. However, after a first test phase had been completed of the national sample of 9,600 households, 294 return migrants were targeted for enumeration with responses being obtained from only 251; the remainder either refusing to co-operate, or being represented as a 'no contact'.

Characteristics of Returnees – Sample Survey

In the context of the CSSP sample survey, return migration is defined as the re-entry of a migrant who left his/her country of birth to go abroad for some period, usually a minimum of one year and upon his/her return has intentions to remain permanently. Irrespective of the time frame defining permanency, it is clear that return migration is contingent upon emigration at some time in the past. Thus, the decision to return is likely to have been influenced by a variety of motivational factors, and is thus likely to spawn intended and unintended consequences for returnees. Depending upon the likelihood of such varied consequences and on the influence of other contextual and social factors, returnees will be faced with making important livelihood choices including a consideration to re-migrate. This chapter, therefore, examines these motivational factors, consequences and choices in the context of variations in key socio-demographic attributes such as gender, ethnicity, current age, education, certification and current economic activity.

Percentage distributions of the 251 returnees in the sample are shown in Table 8.2, categorized into a number of characteristics: gender, age cohorts, ethnic identity, highest educational attainment and educational qualifications. There was a greater proportion of female returnees (59 per cent) than male returnees (40 per cent) and just over one half (51 per cent) of the returnees were aged 20-44 years while just over one quarter (26 per cent) were aged 45-64 years. There was a

Table 8.2 Socio-demographic characteristics of the returnees

Socio-demographic characteristics of returnees (n = 251)	Percentage of returnees
Gender	
Male	39.8
Female	59.4
Age-cohorts	
Less than 20 years	10.0
20-44 years	51.4
45-64 years	26.3
65 years and over	12.4
Ethnic Group	
Indian	35.5
African	35.5
Mixed	25.1
Chinese	1.2
Caucasian	2.0
Highest Education Level	
None	1.2
Primary	30.7
Secondary	43.8
Tertiary	22.7
Certification achievements	
None	38.6
School leaving	4.8
O Level/ CXC/ Senior Cambridge	17.1
A Level/HSC	2.4
Diploma/Certificate	16.3
First Degree	12.4
Post-graduate Degree	7.2

Source: These estimates were generated using data from the Survey of Return Migrants, Government of Trinidad and Tobago, 1998.

larger proportion aged 65 years and over (12 per cent) than under 20 years (10 per cent). Traditionally, Trinidad and Tobago has an ethnically diverse population. During the 1990s, the respective proportions have been as follows: persons of African descent (40 per cent), persons of East Indian descent (40 per cent), persons of Mixed descent (18 per cent) with the remaining groups – those of European, Chinese, Portuguese, Syrian, Lebanese and Amerindian descent each accounting for less than 1 per cent of the country's total population. Accordingly, Table 8.2

also shows that the ethnic breakdown of returnees in the sample is consistent with the overall population patterns, except that the persons of East Indian and African origins are slightly under-represented, while those of 'Mixed-race' origin are slightly over-represented in the sample of returnees. The majority of returnees (almost 76 per cent) did not attain education above the secondary level, the greatest share claiming to have attained secondary education (44 per cent). In contrast, 23 per cent attained tertiary level[3] education. According to their educational qualifications, higher proportions of returnees had completed tertiary level education (38 per cent), 'O' level passes (17 per cent) or primary school leaving certification (5 per cent).

Table 8.3 Emigration experiences of the returnees

Emigration experiences of returnees (n = 251)	Percentage of returnees
Host country of first entry	
Canada	30.7
United Kingdom	15.5
United States	39.4
English-speaking Caribbean country	6.8
Central/South American country	4.0
European country (excluding the United Kingdom)	1.2
Middle Eastern country	1.6
Main activity in host country – first entry	
Work	33.5
Study	26.5
Work and study	21.5
Home duties	7.2
Nothing	8.0
Other	2.4
Length of stay in host country – first entry	
Less than 1 year	1.2
1 year	15.9
2 years	16.3
3 years	15.5
4 years	8.4
5-9 years	22.8
10-19 years	10.4
20 years and over	9.6
Exposure only to host country – first exposure	
Yes	92.4
No	6.8

Source: These estimates were generated using data from the Survey of Return Migrants, Government of Trinidad and Tobago, 1998.

Table 8.3 captures the migration experiences of the returnees in the sample. Almost 86 per cent of the returnees had emigrated to metropolitan North America and England, with proportions as follows: the United States (39 per cent), Canada (31 per cent) and England (16 per cent). These observations are consistent with others' observations that North America is the primary choice for Trinidadian emigrants, with England in third place, especially in the aftermath of the 1962 Commonwealth Immigration Act. A further 11 per cent of returnees had emigrated to regional neighbors; either another English-speaking Caribbean country (7 per cent) or to Central and South America (4 per cent). The Middle East was a region where some returnees (1.6 per cent) had migrated, perhaps as a result of shared cultural systems such as Islam and the exploitation of petroleum resources. The vast majority of returnees (nearly 82 per cent) engaged primarily in work-related activities, studies or both during their sojourn overseas. Table 8.3 also reveals that 23 per cent of the returnees spent 5-9 years in the country where they had emigrated from Trinidad and Tobago, 20 percent had spent at least 10 years abroad, and 10 per cent had stayed away for more than 20 years. This suggests that a sizeable proportion of returnees (53 per cent) have spent five years or more living abroad. In contrast, fewer (32 per cent) had relatively short sojourns between 1 to 2 years in length. Interestingly, the vast majority of returnees (92 per cent) spent their time in the host country to which they had emigrated, with only a few moving on to settle elsewhere, before making the decision to return home.

For returnees aged 15 years and over, Table 8.4 depicts variations in their educational level, the main reasons for migrating to the primary host country and length of stay in the host country. Variations are further categorized by returnees' age-cohorts. According to this sample, which excludes returning dependent children, the majority of returnees had not continued their education beyond the secondary level. However, among returnees aged 65 years and over, the majority had no more than primary level education. Age reverses the educational profile, with younger returnees being much more educated and older returnees being correspondingly less educated. As much as 58 per cent of the returnees cited the pursuit of work-related activities or studying as their main reason for emigrating; though a larger proportion had cited work-related reasons (30 per cent) than studying (27 per cent). Work-related activities and educational study featured predominantly as the main reasons for migration for more than half of the returnees aged 20-44 years (68 per cent) and just about half among their counterparts aged 45-64 years (50 per cent). In comparison to the other age groups, the proportion of returnees citing health as a principal reason for emigration was highest (6 per cent) among those aged 45-64 years, very possibly because of their need to seek medical treatment abroad. Although as many as 30 per cent of the returnees aged 65 years and over cited the pursuit of work-related activities as instrumental in their migration, the majority among this elderly cohort (37 per cent) indicated they had emigrated to join relatives abroad. With respect to returnees aged 15 years and over, the largest share (almost a quarter) spent 5-9 years in their primary host country. While a similar proportion was evident among returnees aged 20-44 years, the situation was different among returnees aged 45-64 years and 65 years and over; with the greatest proportions (24.2 per cent and 40 per cent respectively) spending 10-19 years and at least 20 years respectively. Clearly, emigration from Trinidad and Tobago has been and still is a common option for the young, regardless of the departure decade, while return is not so readily determined by age selectivity. The elderly are retiring and returning, but so are middle-aged cohorts, in appreciable proportions.

Table 8.4   Percentage of returnees by educational level, main reason for emigrating to primary host country and length of stay according to age – returnees 15 years and over

	15 years and over	15-19 years	20-44 years	45-64 years	65 years and over
Total (n)	**233**	**8**	**129**	**66**	**30**
Educational level					
No more than Primary	28.7	-	14.7	39.4	66.7
Secondary	46.4	25.0	56.6	36.4	16.7
Tertiary	24.5	75.0	27.9	27.3	16.7
Main reason for emigrating to primary host country					
To study	27.5	12.5	34.9	24.2	6.7
Family leaving	10.7	50.0	10.9	7.6	6.7
Join relatives	17.2	12.5	13.2	16.7	36.7
Retired	0.4	-	-	1.5	-
Seek employment	24.5	25.0	27.9	19.7	20.0
To take up job	6.0	-	5.4	6.1	10.0
Get married	-	-	-	-	-
Health	2.6	-	0.8	6.1	3.3
Vacation	6.0	-	3.1	7.6	16.7
Other	3.9	-	2.3	9.1	-
Length of stay in host country – first entry					
Less than 2 years	14.2	25.0	13.2	15.2	13.3
2 years	16.3	25.0	20.9	10.6	6.7
3 years	16.3	12.5	20.2	9.1	16.7
4 years	8.2	12.5	11.6	4.5	-
5-9 years	24.0	25.0	27.9	22.7	10.0
10-19 years	11.2	-	4.5	24.2	13.3
20 + years	9.9	-	1.2	13.6	40.0

Return Motivations and Consequences

Motivation to Return

Returnees' motivations are examined from the standpoint of principal stimuli that prompted their decision to return home; namely, the encouragement from friends and relatives, inducements and arrangements that are likely to guarantee their return sooner or later and the systematic accumulation of physical and human capital, whether with particularistic or universalistic intentions. These four main

Table 8.5   Percentage of returnees 15 years and over with specific motivational reasons for return

	15 years and over	15-19 years	20-44 years	45-64 years	65 years and over
Total (n)	233	8	129	66	30
MOTIVATIONAL FACTOR #1: Main reason for return home					
Studies completed	18.0	-	24.8	12.1	6.7
Family leaving	10.7	50.0	9.3	12.1	3.3
Joinr	26.6	50.0	20.2	33.3	33.3
Retired	5.2	-	-	7.6	23.3
Seek employment	2.1	-	3.9	-	-
To take up job	3.4	-	1.6	7.6	6.7
Visa expired	6.4	-	9.3	4.5	-
Deported	4.3	-	7.0	1.5	-
Health	8.6	-	8.5	7.6	13.3
Other	13.3	-	14.0	12.1	16.7
MOTIVATIONAL FACTOR #2: Encouragement to return home from friends and relatives					
Encouragement	11.2	25.0	8.5	16.7	6.7
No encouragement	87.1	75.0	89.9	81.8	90.0
MOTIVATIONAL FACTOR #3: Nature of preparation for home return					
No preparations	60.1	87.5	61.2	51.5	66.7
Sent money	10.7	12.5	10.1	9.1	16.7
Bought property	5.6	-	3.1	7.6	13.1
Applied for job	6.4	-	5.4	12.1	-
Completed studies/training	16.3	-	19.4	18.2	3.3
MOTIVATIONAL FACTOR #4: Personal resources brought by returnee					
Household effects	41.2	25.0	32.6	48.5	66.7
Money to set up business	4.7	-	4.7	7.6	-
Machinery and equipment	2.6	-	2.3	4.5	-
Skills for country's development	24.9	-	29.5	27.3	6.7

motivational factors are presented in Table 8.5 to allow generalizations to be made about the varying motives which have influenced decision-making of returnees aged 15 years and over. Variations are further categorized by returnees' age-cohorts. The results in Table 8.5 reveal that the principal reason cited for returning home was to join relatives, with 27 per cent of returnees giving such a response. While a desire to join relatives was cited as a main reason among returnees aged 45-64 years and 65 years and over, completion of educational studies was cited by those aged 20-44 years. Among returnees aged 65 years and over, noteworthy proportions indicated that their main reason for returning home was due either to retirement (23 per cent) or health (13 per cent). In general, relatively few returnees indicated that they were encouraged by friends and relatives to return home, this being the case for each of the four age cohorts. Thus, it is clear that the decision to return home, in many cases, was devoid of much positive inducement from friends and relatives and more likely was a product of the returnees' own volition.

For the majority of returnees, no preparatory inducements were made to facilitate their return; this being the case irrespective of their age-cohort. For those with any/some family-or kin- inducement, the majority indicated they had completed studies/training in order to enhance their prospects of participation in gainful pursuits upon their return. This was observed to be the case for returnees in general and those aged 20-44 years and 45-64 years in particular. In contrast, returnees aged 65 years and over were more likely to have sent money or purchased property; capital transactions that were likely to have strengthened their resolve to return home. In terms of the systematic accumulation of physical and human capital, returnees were more likely to bring household effects with them, than to bring money to set up a business; or import machinery and equipment to establish a business; or have aspirations to use their innate skills/training to foster national development initiatives. And, these findings are consistent across the four age cohorts; with the acquisition of household effects from abroad being most pronounced among elderly returnees aged 65 years and over. On the other hand, among returnees aged 20-44 years and 45-64 years, more than a quarter (30 per cent and 27 per cent respectively) indicated that they wished to use their innate skills and expertise to facilitate national development in Trinidad and Tobago.

The Consequences of Return

Returnees were also asked about their adaptation experiences and the consequences of their return home. The range of possible consequences (in Table 8.6) includes various outcomes and divergent experiences in the context of returnees' social and economic pursuits. This therefore covers emergent problems, resettlement initiatives and perceptions pertaining to a host of stimuli with respect to public service delivery, neighbours' response to their return and their self-assessment of the success of the actual decision to return. The latter is classified as either positive or negative and is assessed according to specific socio-demographic characteristics of returnees and to some extent, their experiences and perceptions since returning.

Table 8.6 Percentage of returnees 15 years and over by outcomes of return

Consequences for returnees on their return home return migrants aged 15 years and over (n = 233)	Percentage of returnees
Current economic activity	
Student	1.3
House keeping	18.5
Self-employed	9.0
Employer	1.7
Employee	39.9
Seeking work	6.9
Want work and available	7.4
Setting up business	0.4
Do not want work	14.2
Main problems faced by returnees on return	
None (148)	1.7
Employment (32)	97.0
High cost of living (13)	0.4
High customs duties (6)	98.3
Government service (6)	16.7
Source of assistance/help with settling on return	
Government – Yes	1.7
Government – No	97.0
Private organization – Yes	0.4
Private organization – No	98.3
Family and friends – Yes	16.7
Family and friends – No	82.0
Perception of neighbourhood's response to return	
Favourable	88.8
Unfavourable	6.0
Involvement with community groups and organizations	
Yes	18.5
No	79.8
Returnee's perception of his/her decision to return	
Positive	62.3
Negative	35.2

Source: These estimates were generated using data from the Survey of Return Migrants, Government of Trinidad and Tobago, 1998.

According to the tabulations in Table 8.6, approximately half (51 per cent) of the returnees aged 15 years and over indicated that they were employed at the time of the survey. As much as 14 percent indicated that they either sought work or wanted work and was available while another 14 per cent did not want any work. The majority (64 percent) of returnees faced no major problems upon their return with much smaller proportions claiming to have had problems with employment (14 per cent), high cost of living (6 per cent) and high customs duties (3 per cent).With regard to resettling in Trinidad and Tobago, returnees received virtually no assistance from the government and private organizations and just 17 per cent claimed that they obtained assistance from family and friends. A relatively small proportion (19 per cent) of returnees indicated they got involved in community groups and organizations. Generally, it is worth noting that a relatively large number of returnees (89 per cent) indicated that they received a favourable reception from their neighbours on returning to Trinidad and Tobago.

Table 8.7   Percentage of returnees 15 years and over by their perception of public services

Perceptions of public services return migrants aged 15 years and over (n = 233)	Percentage
Electricity Supply	
Satisfied	64.4
Less than satisfied	33.1
Don't know	0.9
Water provision	
Satisfied	39.4
Less than satisfied	58.4
Don't know	0.4
Transportation infrastructure	
Satisfied	58.8
Less than satisfied	36.9
Don't know	2.6
Health Care Services	
Satisfied	26.2
Less than satisfied	59.2
Don't know	12.9
Education System	
Satisfied	54.0
Less than satisfied	36.9
Don't know	6.9

Source: These estimates were generated using data from the Survey of Return Migrants, Government of Trinidad and Tobago, 1998.

Returnees' perceptions of the quality of public service delivery were assessed for critical domains such as electricity, water, transportation, police, health care and education. These perceptions are described as either yielding satisfaction or dissatisfaction in Table 8.7. The returnees expressed greater levels of satisfaction than dissatisfaction with electricity services (64 per cent satisfied) and to a somewhat lesser extent, with transportation services (58 per cent satisfied) and education services (54 per cent satisfied). In contrast, they expressed greater levels of dissatisfaction than satisfaction with water provision (58 per cent dissatisfied) and health care (59 per cent dissatisfied).

According to the results of the survey (see Tables 8.6, 8.8 and 8.9), a greater proportion of returnees expressed positive rather than negative feelings about their decision to return home (62 per cent as opposed to 35 per cent). Table 8.8 shows that such perceptions persist irrespective of returnees' gender, ethnic origin and highest education level. They also persist in every age group for persons 20 years

Table 8.8   Returnee's perception of his/her decision to return home according to socio-demographic characteristics of respondents

	Percentage distributions	
Perception of decision to return home	Positive	Negative
All returnees aged 15 years and over (n=233)	62.3	35.2
Sex		
Male (n = 96)	60.4	37.5
Female (n = 137)	63.5	33.6
Age		
Less than 20 years (n = 8)	25.0	75.0
20-44 years (n = 129)	58.9	38.0
45-64 years (n = 66)	62.1	36.4
65 years and over (n = 30)	86.7	10.3
Ethnic group		
Indian (n = 86)	55.8	43.0
African (n = 85)	62.4	32.9
Mixed (n = 54)	72.2	27.8
Chinese (n = 3)	66.7	33.3
Caucasian (n = 5)	60.0	20.0
Highest education level		
No more than Primary (n = 67)	68.7	28.4
Secondary (n = 108)	53.7	44.4
Tertiary (n = 57)	71.9	26.3
Length of stay in host country – first entry		
Less than 2 years (n = 33)	81.8	18.2
2 years (n = 38)	60.5	34.2
3 years (n = 38)	47.4	50.0
4 years (n = 19)	31.6	63.2
5-9 years (n = 56)	64.3	32.1
10-19 years (n = 26)	69.2	30.8
20 years and over (n = 23)	73.9	26.1

Source: These estimates were generated using data from the Survey of Return Migrants, Government of Trinidad and Tobago, 1998b.

or older and among returnees who spent at least 5 years overseas. Table 8.9 also shows that a greater proportion of returnees expressed positive rather than negative feelings about their return decision irrespective of their level of satisfaction with all of the critical public service domains except transportation. According to Table 8.10, the feelings about the return decision were no different for those returnees who obtained assistance with settling from friends and those who had no assistance – greater proportions expressed positive rather than negative feelings. This was also the case among returnees who were employees, employers, self employed, engaged in home duties and not interested in obtaining work. A noteworthy observation relates to the overwhelmingly greater proportion of returnees expressing positive rather than negative feelings (74 per cent as opposed to 26 per cent) among those who had been involved in community groups and other organizations despite their relatively small level of participation in such activities.

Table 8.9   Returnee's perception of his/her decision to return home according to perceptions of public services

	Percentage distributions	
Perception of decision to return home	**Positive**	**Negative**
All returnees aged 15 years and over (n=233)	62.3	35.2
Electricity service		
Satisfied (n = 150)	65.3	33.3
Less than satisfied (n = 77)	59.7	40.3
Water provision		
Satisfied (n = 92)	64.1	34.8
Less than satisfied (n = 136)	63.2	36.0
Transportation infrastructure		
Satisfied (n = 137)	72.3	27.0
Less than satisfied (n = 86)	48.8	50.0
Police		
Satisfied (n = 104)	67.3	30.8
Less than satisfied (n = 99)	59.6	40.4
Health care services		
Satisfied (n = 61)	68.9	29.5
Less than satisfied (n = 138)	60.1	39.1
Educational system		
Satisfied (n = 126)	70.6	28.6
Less than satisfied (n = 86)	51.2	47.7

Source: These estimates were generated using data from the Survey of Return Migrants, Government of Trinidad and Tobago, 1998.

Less favourable perceptions with negative feelings outweighing positive feelings were evident among returnees who were seeking work, those who wanted work and were available and the extremely small proportion that had experienced unfavourable responses within their neighbourhood.

Table 8.10 Returnee's perception of his/her decision to return home according to experiences on return home

	Percentage distribution	
Perception of decision to return home	**Positive**	**Negative**
All returnees aged 15 years and over (n=233)	62.3	35.2
Current economic activity		
Student (n = 3)	66.7	33.3
House keeping (n = 43)	62.8	37.2
Self-employed (n = 21)	52.4	42.9
Employer (n = 4)	75.0	25.0
Employee (n = 93)	68.8	30.1
Seeking work (n = 16)	31.3	68.8
Want work and available (n = 17)	47.1	47.1
Setting up business (n = 1)	100.0	-
Do not want work (n = 33)	72.7	24.2
Source of assistance/help with settling on return		
Family and friends – Yes (n = 39)	56.4	43.6
Family and friends – No (n = 191)	64.4	34.0
Perception of neighbourhood's response to return		
Favourable (n = 207)	66.2	33.3
Unfavourable (n = 14)	35.7	64.3
Involvement with community groups and organizations	74.4	25.6
Yes (n = 43)	60.2	38.2
No (n = 186)		

Source: These estimates were generated using data from the Survey of Return Migrants, Government of Trinidad and Tobago, 1998.

Returnee's Exercise of Choice

Having returned, some returnees may have had experiences that precipitated some consideration of re-migration to the metropolitan country they left. In other instances, there may have actually been no such considerations despite having been exposed to unpleasant experiences. To this end, returnees' commitment to fulfilling their return decision is reinforced when they do not entertain prospects of re-migration. The survey data indicates that a greater proportion of returnees (55 per cent as opposed to 42 per cent) have harboured thoughts of re-migration to the host country, rather than resettling indefinitely in their homeland (Tables 8.11, 8.12 and

8.13). Table 8.11 shows that such a pattern is evident irrespective of returnees' gender. At the same time, it is worth noting that returnees aged 65 years or older, those with tertiary level education and those who have spent at least 20 years abroad are among the least likely to consider returning to host countries. In particular, a relatively greater proportion (73 per cent) of returnees aged 65 years or older indicated that they had never given re-migration a thought; an outcome that is consistent with a similar observation among returnees who spent at least 20 years overseas before returning home (70 per cent).

Table 8.11   Choices of returnees on return home according to socio-demographic characteristics of respondents

	Percentage distribution	
Consider returning to host country	Yes	No
All returnees aged 15 years and over (n=233)	55.4	42.1
Gender		
Male (n = 96)	57.3	40.6
Female (n = 137)	54.0	43.1
Age-cohorts		
Less than 20 years (n = 8)	75.0	25.0
20-44 years (n = 129)	60.5	36.4
45-64 years (n = 66)	57.6	40.9
65 years and over (n = 30)	23.3	73.3
Ethnic Group		
Indian (n = 86)	59.3	39.5
African (n = 85)	54.1	41.2
Mixed (n = 54)	55.6	44.4
Chinese (n = 3)	-	100.0
Caucasian (n = 5)	40.0	40.0
Highest Education Level		
No more than Primary (n = 67)	53.7	43.3
Secondary (n = 108)	60.2	38.0
Tertiary (n = 57)	49.1	49.1
Length of Stay in Host Country – First Entry		
Less than 2 years (n = 33)	54.5	45.5
2 years (n = 38)	68.4	26.3
3 years (n = 38)	55.3	42.1
4 years (n = 19)	63.2	31.6
5-9 years (n = 56)	53.6	42.9
10-19 years (n = 26)	57.7	42.3
20 years and over (n = 23)	30.4	69.6

Source: These estimates were generated using data from the Survey of Return Migrants, Government of Trinidad and Tobago, 1998.

Table 8.12   Choices of returnees on return home according to perceptions of public services

	Percentage distribution	
Consider returning to host country	Yes	No
All returnees aged 15 years and over (n=233)	55.4	42.1
Electricity services		
Satisfied (n = 150)	53.3	45.3
Less than Satisfied (n = 77)	57.4	37.7
Water provision		
Satisfied (n = 92)	54.3	44.6
Less than satisfied (n = 136)	57.4	41.9
Transportation infrastructure		
Satisfied (n = 137)	51.1	48.2
Less than satisfied (n = 86)	65.1	33.7
Police		
Satisfied (n = 104)	49.0	49.0
Less than satisfied (n = 99)	63.6	36.4
Health care services		
Satisfied (n = 61)	55.7	42.6
Less than satisfied (n = 138)	58.0	41.3
Educational system		
Satisfied (n = 126)	51.6	47.6
Less than satisfied (n = 86)	67.4	31.4

Source: These estimates were generated using data from the Survey of Return Migrants, Government of Trinidad and Tobago, 1998.

With regard to public services such as electricity, water, transportation, health care and education, Table 8.12 reveals that it did not matter whether returnees were satisfied or less than satisfied with the quality of the delivery. In all cases, there appeared to be a greater proportion of returnees harbouring thoughts of re-migration though this tendency was more pronounced among those who were less than satisfied. This was observed to be especially true in the case of service delivery in areas such as education, transportation and police services.

Data in Table 8.13 demonstrate links between returnees' experiences since returning home and whether or not, they have harboured thoughts of re-migration. Despite a general pattern suggesting that slightly greater proportions entertained thoughts of re-migration – 55 per cent to 42 per cent respectively – thoughts about re-migration were most pronounced (quite obviously) among returnees who assigned a negative rating to their decision to return (82 per cent). Similarly high proportions favoring re-migration were observed among returnees who reported having unfavourable experiences in their respective neighbourhoods (79 per cent), and among returnees who were unemployed – those who sought work (81 per cent) and those who wanted work and were available (65 per cent). Many who did not want to work or see the need to work were more content to stay, probably because they were retired. Involvement in community activities, on the other hand, did not appear to be a convincingly attractive factor encouraging returnees to ignore the prospect or re-migration, since 56 per cent said they harboured re-migration thoughts, compared to 42 per cent who did not.

Table 8.13   Considerations of returnees about re-migration, according to their experiences on returning home

	Percentage distribution	
Consider re-migration to host country	Yes	No
All returnees aged 15 years and over (n=233)	55.4	42.1
Current economic activity		
Student (n = 3)	66.7	33.3
House keeping (n = 43)	51.2	48.8
Self-employed (n = 21)	47.6	47.6
Employer (n = 4)	50.0	50.0
Employee (n = 93)	59.1	39.8
Seeking work (n = 16)	81.3	18.8
Want work and available (n = 17)	64.7	29.4
Setting up business (n = 1)	-	100.0
Do not want work (n = 33)	42.4	54.5
Perception of neighbourhood's response to return		
Favourable (n = 207)	54.6	44.9
Unfavourable (n = 14)	78.6	21.4
Involvement with community groups and organizations	55.8	44.2
Yes (n = 43)	55.9	42.5
No (n = 186)		
Returnee's perception of his/her decision to return	42.8	57.2
Positive (n = 145)	81.7	18.3
Negative (n = 82)		

Source: These estimates were generated using data from the Survey of Return Migrants, Government of Trinidad and Tobago, 1998.

Review of Findings

To be formally considered a returnee in the sample survey used in this study, nationals of Trinidad and Tobago had to have emigrated and spent a period of at least one year overseas before returning with the intention of re-settling permanently in Trinidad and Tobago. The main reason for such emigration was work-related activities or the pursuit of studies; such activities being primary pursuits among returnees aged 20-44 years and 45-64 years. Some interesting demographics have emerged as a result of this examination of the sample survey that targeted returnees, who were residing in Trinidad and Tobago in 1997/1998. While it is not surprising that the vast majority returned from the United States, Canada, England, and to a much lesser extent, the English-speaking Caribbean, there is evidence to suggest that some may have returned from other global regions such as the Middle East and South and Central America. Among returnees, there was a greater proportion that had attained tertiary level education when compared to the corresponding proportion in the national population as a whole. However, the majority of returnees aged 65 years or older had attained no more than primary level education. With respect to returnees' activities while abroad, it is worth noting that an overwhelming majority of at least four in every five returnees engaged in work-related activities, studies or both.

De Souza (1998) noted that more recent cohorts of returnees have often been professionals who possess attributes deemed to be critical in facilitating national development initiatives. He identified four critical motivators as being instrumental in precipitating returnees' decision to return, these being the homeland milieu, family reunification, return obligations and professional fulfillments. From the standpoint of motivation to return, results from the sample survey support the notion of family reunification as a principal factor, especially among returnees aged 45-64 years and 65 years and over. In contrast, professional fulfillment was most likely to be a principal motivator among returnees aged 20-44 years, who generally felt that their academic credentials and training could enhance their prospects of participating in gainful economic activities and making meaningful contributions towards the national development of Trinidad and Tobago. Such aspirations are also consistent with the notion of 'return obligations' as articulated by De Souza.

In terms of their decision to return, the evidence suggests that the returnees were more likely to be self motivated and less likely to be influenced by their relatives and friends. There was also evidence of a systematic accumulation of physical and human capital, although it occurred in variable combinations and quantities to satisfy returnees' desire to fulfill professional and career aspirations, contribute towards national development initiatives, 'bond' with the homeland and enjoy retirement. This was especially true among the eldest returnees aged 65 years or older who, according to the survey, were among the least likely to harbour thoughts of re-migration. Considering that returnees and especially the younger ones have attained a minimum of a secondary level education and given that they primarily migrated in order to obtain work or be educated and trained, their return ought to be lauded as a potential injection into the human capital stock of the country. Some of the progressive life experiences of returnees while overseas plus the noble intentions that in some cases precipitated their return ought to be harnessed in order to compensate for losses due to the 'brain drain' phenomenon. Indeed, this positive consequence of return migration should be better valued when

Trinidad and Tobago considers public outlays in the form of educational investment at earlier stages in returnees' lives.

On returning home, most returnees felt that they had obtained favourable receptions within their residential communities. Relatively low levels of participation in community groups and organizations are an indication that the majority of returnees had not been actively engaged in such activities. However, more often than not, returnees have been observed to express positive rather than negative feelings about their decision to return home. This was observed to be the case despite variable levels of satisfaction with the delivery of public services and was especially evident among the relatively small number of returnees who had been engaged in the activities of community groups and other similar organizations. In contrast, negative feelings outweighed positive feelings among returnees who were seeking work, those who wanted work and were available and the extremely small proportion who had experienced unfavourable responses within their residential communities. An important conclusion arising from this assessment is that returnee's adaptation experiences on returning home both shape and reinforce positive perceptions of their decision to return home. Thus, unfavourable encounters within the job market and within one's residential community constitute consequences that may evoke negative evaluations of his/her return – even prompting thoughts of re-migration. On the other hand, it is worth noting that negative perceptions of a number of critical public services do not seem to be sufficient to evoke overwhelmingly negative evaluations of the decision to return home. The latter might imply that returnees returned home with the expectation that public services would have been sub-standard and have been prepared to make the best of their home-experience irrespective of the quality of public service delivery.

These findings seem to suggest that despite the pleasure that is apparently associated with their return decisions, returnees still seem to believe that they can optimize their prospective life experiences and chances by exercising the option of re-migration. However, among returnees aged 65 years and older, those with tertiary level education and those who spent at least 20 years overseas before returning home, thoughts of re-migration were relatively low. This general reticence appears to be associated with sub-populations of returnees who, by virtue of their old age and limited educational credentials, are more likely to believe that re-migration may have little or no effect upon their prospects in their remaining lifetime. Rather, retirement 'back home' is a final step in their odyssey.

The notion of a 'migrant ideology' as articulated by Philpott (1973: 188) can be operationalized on a continuum as follows:

> 'At one end are migrants whose total commitment and orientation are towards the sending society; at the other end are the migrants whose total commitment and orientation are towards the host society. Empirically, of course, no migrant group falls into either of the extreme positions. But West Indian migrants, for example, would place relatively closer to the former end of the continuum'.

In Trinidad and Tobago, there is no doubt about the relevance of the 'migrant ideology' as hypothesized by Philpott (1973). Some migrants left with intentions to settle indefinitely in the host country and among them, some may have altered their position and returned home upon gaining sufficient foreign experience to weigh 'the pros and the cons'. Others may have migrated with the intention of returning home and opted to stay abroad indefinitely, converting a 'circulation strategy' to an emigration outcome (see Conway 1988). Differences in the form and interpretation of the stimuli associated with living 'in and between two worlds' could potentially alter intentions and in the context of returnees hasten prospective re-migration despite earlier intentions to return home indefinitely. Philpott (1973) made reference to such experiences in the context of Montserratians, who returned and engaged in entrepreneurial pursuits. In Trinidad and Tobago, the preponderance of returnees observed to have entertained thoughts of re-migration is indicative of a level of ambivalence that could be associated with an uncertain future at home. At the same time, however, there have generally been positive feelings about the return decision that appeared to be largely self-motivated and prompted by a similar set of motivations cited by De Souza (1998), mainly familial reunification, professional fulfillment and return obligations.

Conclusion

There have been very few studies examining migratory processes and their consequences in Trinidad and Tobago. Prior to the three studies in this volume (of which this is one), De Souza's (1998) accounts of returnee experiences was the only formal study to have attempted to gain a better understanding of the dynamics underlying return migration in Trinidad and Tobago. Though essentially qualitative, and interpretative by design, De Souza's study was instrumental in identifying critical motivational factors that prompted the return of Trinidadians. This present study, in a complimentary way, has taken a quantitative approach using descriptive tabulations and categorical assessments to produce analyses of a representative sample survey, that reflect the critical importance of motivational factors such as familial unification, professional fulfillment, return obligations and to a lesser extent, identification with the homeland milieu. My study also embraced a demographic approach, focusing on the different experiences of age cohorts; specifically, returnees who were children and teenagers (aged 0-19 years), young adults in the prime of their working lives (aged 20-44 years), adults in the prime of their working lives, those on the verge of retirement (aged 45-64 years) and elderly retirees (aged 65 years and over). Explanations of variation in motivational factors, the consequences of return for returnees and their attitudes towards re-migration could then be specified in terms of roles, expectations and other functional attributes that were associated with specific age cohorts.

This study has examined various aspects of the phenomenon of return migration in the context of Trinidad and Tobago during the 1990s. It has undertaken a quantitative approach that corroborates observations by De Souza (1998) about some of the principal motivational factors influencing the return decisions of nationals of Trinidad and Tobago. It goes a step further to assess the consequences of return for returnees and in particular, their positive or negative feelings about the decision to return, and their consideration of re-migration. In general, greater proportions of returnees rated their return experience as positive rather than negative; except in a few cases that were contingent upon unfavourable experiences in the job market or within their residential communities. On the other hand, over half of the returnees claimed that they had considered re-migration as an option since returning. All cohorts younger than the over 65 group, therefore, view the return move as another transitory step in their migration and life-course strategies. It is neither the final move, nor a permanent return move for many. Retaining 'the migration option' appears to be as much a deeply ingrained, behavioural strategy among Trinidadians and Tobagonians, as it is a region-wide tradition, and practice.

Notes

1. The enumeration districts were primary sampling units that were selected based on the geographical spread of return migrants found through the inquiry.
2. The main geographic domains include the counties of Caroni, St. Andrew/St. David, St. Patrick, Nariva/Mayaro, Victoria, St. George; the cities of Port of Spain and San Fernando; the boroughs of Arima, Chaguanas and Point Fortin and the wards of Trinidad and Tobago.
3. Includes 'A' Level/HSC, Diploma/Certificate, First Degree and Post Graduate Degree.

References

Anderson, Patricia (1985), 'Migration and Development in Jamaica', *ISER Paper #2*, The Institute of Social and Economic Research, The University of the West Indies, Mona, Jamaica.

Berry, Brian J.L. and Alan M. Baker (1968), 'Geographic Sampling', in Brian J.L. Berry and Duane F. Marble (eds), *Spatial Analysis: A Reader in Statistical Geography*, Englewood Cliffs, NJ: Prentice-Hall, Inc., 91-100.

Boodhoo, Martin, J. and Ahamad Baksh (1981), *The Impact of Brain Drain on Development: A Case Study of Guyana*, Kuala Lumpur, Malaysia: Percetakan Intisari.

Byron, Margaret (1994), *Post-War Caribbean Migration to Britain: The Unfinished Cycle*, Aldershot, Hants, UK: Avebury.

Byron, Margaret (2000), 'Return Migration to the Eastern Caribbean: Comparative Experiences and Policy Implications', *Social and Economic Studies*, Vol. 49(4): 155-188.

Conway Dennis (1988), 'Conceptualizing Contemporary Patterns of Caribbean International Mobility', *Caribbean Geography*, 3(2):145-163.

Conway Dennis (2002), 'Gettin' there, despite the odds: Caribbean migration to the U.S. in the 1990s', *Journal of Eastern Caribbean Studies*, 27(4): 100-134.

Cooper, Derek (1985), 'Migration from Jamaica in the 1970s: Political Protest or Economic Pull', *International Migration Review*, 19(4): 728-745.

Davison, R.B. (1962), *West Indian Migrants: Social and Economic Facts of Migration from the West Indies*, London: Oxford University Press.

Davidson, Betty (1967), 'No Place Back Home: A Study of Jamaicans Returning to Kingston, Jamaica', *Race*, 9(1): 499-509.

De Souza, Roger Mark (1998), 'The Spell of the Cascadura: West Indian Return Migration', in Thomas Klak (ed) *Globalization and Neo-Liberalization – The Caribbean Context*, Lanham, Boulkder, New York and Oxford: Rowan & Littlefield Publishers, pp. 227-253.

Ebanks, George E., P.M. George and C.E. Nobbe (1979), 'Emigration from Barbados, 1951-1970', *Social and Economic Studies*, 28(2): 431-449.

ECLAC (1998), *A Study of Return Migration to the Organization of Eastern Caribbean States (OECS) Territories and the British Virgin Islands in the Closing Years of the Twentieth Century: Implications for Social Policy*, United Nations Economic Commission for Latin America and the Caribbean, Sub-Regional Headquarters for the Caribbean, Port of Spain, Trinidad and Tobago.

Girling, Ronald K. (1974), 'The Migration of Human Capital from the Third World: The Implications and Some Data on the Jamaican Case', *Social and Economic Studies*, 23(1): 84-96.

Gmelch, George (1987), 'Work, Innovation and Investment: The Impact of Return Migrants in Barbados', *Human Organization*, Vol. 46, No. 2, 131-140.

Gmelch George (1992), *Double Passage: the Lives of Caribbean Migrants and Back Home*, Ann Arbor: University of Michigan Press.

Harewood, Jack (1994), *The Population of Trinidad and Tobago*, CICRED.

Kritz, Mary, M. (1981), 'International Migration Perspectives in the Caribbean Basin: An Overview', in Mary M. Kritz, Charles B. Keely and Silvano M. Tomasi (ed), *Global Trends in Migration: Theory and Research on International Population Movements*, New York: Center for Migration Studies, Staten Island, Praeger Publishers, 208-233.

Marshall, Dawn (1987), 'A History of West Indian Migrations: Overseas Opportunities and Safety Valve Policies', in Barry B. Levine (ed), *The Caribbean Exodus*, New York: Praeger Publishers, 15-31.

Maunder, William F. (1955), 'The New Jamaican Emigrant', *Social and Economic Studies*, 4(1): 38-83.

Muschkin, Clara (1993), 'Consequences of Return Migration Status for Employment in Puerto Rico', *International Migration Review*, 27(1): 79-102.

Patterson, H. Orlando (1968), 'West Indian Migrants Returning Home', *Race*, Vol. 10(1): 69-77.

Philpott, Stuart. B (1973), *West Indian Migration: The Montserrat Case*, London: Athlone Press.

Plaza, Dwaine (2002), 'The Socio-Cultural Adjustment of Second Generation British – Caribbean Return Migrants to Barbados and Jamaica', *Journal of Eastern Caribbean Studies*, 27(4): 135-160.

Potter, Robert B. (2001), '"Tales of Two Societies": Young Return Migrants to St. Lucia and Barbados', *Caribbean Geography*, 12(1): 24-43.

Potter, Robert B. (2003a), '"Foreign-Born" and "Young" Returning Nationals to Barbados: A Pilot Study', University of Reading, Department of Geography: Geographical Paper, No. 166.

Potter, Robert B. (2003b), '"Foreign-Born" and "Young" Returning Nationals to St. Lucia: A Pilot Study', University of Reading, Department of Geography: Geographical Paper, No. 168.

Potter, Robert B and Joan Phillips (2002), 'Social dynamics of foreign-born and young returning nationals to the Caribbean: outline of a research project', *Centre for Developing Areas Research Paper (CEDAR)*, No 37, 40 pp.

Rampersad, Frank and L. Pujadas (1970), *The Emigration of Professional, Supervisory, Middle Level and Skilled Manpower from Trinidad and Tobago 1962-1968 Brain Drain*, Central Statistical Office, Port of Spain, Trinidad and Tobago.

Richardson, Bonham C. (1983), *Caribbean Migrants – Environment and Human Survival on St. Kitts and Nevis*, Knoxville, The University of Tennessee Press.

Roberts, George W. (1981), 'Currents of External Migration Affecting the West Indies: A Summary', *Revista InterAmericana*, 11(3).

Roberts, George, W. and David O. Mills (1958), 'A Study of External Migration Affecting Jamaica 1953-55', *Social and Economic Studies*, 7(2).

Rubenstein, Hymie (1982), 'Return Migration to the English Speaking Caribbean: Review and Commentary', in William Stinner, Klaus de Alburquerque and Roy S. Bryce-Laporte (eds), *Return Migration and Remittances: Developing a Caribbean Perspective*, Washington, D.C.: Praeger Press, 3-34.

Segal, Aaron (1987), 'The Caribbean Exodus in a Global Context: Comparative Migration Experiences', in Barry Levine (ed), *The Caribbean Exodus*, New York: Praeger Publishers, 67-105.

Smith, T.E. (1981), *Commonwealth Migration: Flows and Policies*, London: Macmillan Press Limited.

Stinner, William, F. and Klaus de Alburquerque (1982), 'Introductory Essay: The Dynamics of Caribbean Return Migration', in William F. Stinner, Klaus de Alburquerque and Roy S. Bryce-Laporte (eds), *Return Migration and Remittances: Developing a Caribbean Perspective*, Washington, D.C.: Praeger Press, pp. xxxiii-xxxvi.

Tidrick, Glenn (1966), 'Some Aspects of Jamaican Migration to the United Kingdom', *Social and Economic Studies*, 15(1): 22-39.

Thomas-Hope, Elizabeth (1985), 'Return Migration and its Implications for Caribbean Development', in Robert A. Pastor (ed), *Migration and Development in the Caribbean*, Boulder, Colorado: Westview Press, 157-177.

Thomas-Hope, Elizabeth (1992), *Explanation in Caribbean Migration*, London: Macmillan Press.

Thomas-Hope, Elizabeth (1999), 'Return Migration to Jamaica and its Development Potential', *International Migration*, 37(1): 183-205.

Chapter 9

A Gendered Tale of Puerto Rican Return: Place, Nation and Identity

Gina M. Pérez

The past decade has witnessed a dramatic shift in migration research. Not only have migration scholars redirected their focus to consider the transnational flows of information, capital, goods and people as well as migrants' transnational practices, feminist scholars have made critical interventions by theorizing the role of gender in these transnational processes (Mahler and Pessar 2001; Hondagneu-Sotelo 1994; Parreñas 2000; Alicea 1997). This gender focus helps to reveal how migration decisions are guided by kinship and hierarchies of power based on gender and generation within households and among migrant social networks which are often the sites of struggle and contestation. In addition, a qualitative variant of this gendered approach also attempts to balance a significant bias in migration research that privileges bodily movement without considering seriously enough what Sarah Mahler and Patricia Pessar have identified as 'the transnational cognitive spaces'. These are cognitive realms of subjective perception, whereby migrants (usually woman) imagine, plan and strategize transnationally, the kind of emotional work that usually prefigures actual migration, and which also may occur without people moving at all (Mahler and Pessar 2001).

This chapter employs a gender focus to consider how the discourse surrounding transnational practices in Puerto Rico – specifically return migration and the social remittances accompanying this movement – is deeply gendered. Furthermore, it tells us a lot about the domains and boundaries of inclusion/exclusion, nation and identity in Puerto Rico based on gender, sexuality, class and race. I begin by discussing various approaches to Puerto Rican migration and then examine the narratives of men and women returning in the 1980s and 1990s to live in San Sebastián, a small largely agricultural town in the northwestern region of Puerto Rico. Many of the women described return migration as *un sufrimiento*, or suffering, as they often found themselves confronting family and community pressure to say at home, a renewed dependency on husbands in order to engage in daily activities outside of the home because of inadequate public transportation, and limited job opportunities, all contributing factors to women's dissatisfaction upon returning to Puerto Rico.

I will then focus on the gendered dimensions of how return migrants are understood in the town and how sexuality – and specifically women's racialized sexuality – informs 'notions of belonging'. As a number of Puerto Rican scholars have documented, return migrants occupy an ambiguous role in the Puerto Rican national imaginary and are frequently the targets of derision, ridicule and contempt (Duany 2002; Zentella 1987; Muntaner 1997). The popular way returnees and their children are referred to, as *los que vienen de afuera* – literally those who come from the outside – clearly reveal their marginal position. It also suggests the uneasy and unresolved relationship between Puerto Ricans on the island and those from the mainland, who not only allegedly refuse to assimilate to life in Puerto Rico, but who dare to transform its cultural, social, and gendered political-economic landscape as well.

Theorizing Puerto Rican Migration

Like many Caribbean nations, Puerto Rico simultaneously experiences high levels of emigration and immigration.[1] Since the 1940s, the island has witnessed the displacement of more than 1.5 million people to the United States, where, according to the 2000 census, nearly half of all Puerto Ricans now reside. This massive displacement from the island is accompanied by substantive immigration to Puerto Rico, with more than 9 per cent of the island's current population classified as foreign born, a statistic that includes children born to Puerto Rican parents abroad as well as Dominicans, Cubans and others.[2] Scholars have employed a variety of metaphors to capture this tremendous volume of mobility, referring to Puerto Rico as 'a nation on the move' or the 'commuter nation', and to Puerto Rican migrants as 'passengers on an airbus'. Such metaphors are useful in highlighting such high rates of mobility, but more importantly, they point to how the geographic displacement of Puerto Ricans, whether forced or voluntary, has come to define Puerto Ricans' social and political identity.

What is the historical root of this heightened mobility and displacement? It starts with the US occupation of Puerto Rico in 1898 and the subsequent consolidation of US agrarian capitalism and shrinking small-scale subsistence cultivation, which helped set in motion population movements to places like Hawaii, Arizona, California and, most notably, New York City. Between 1900 and 1940, more than ninety thousand Puerto Ricans left the island, although many returned after working in New York, Pennsylvania, and New Jersey.[3] As a result of Puerto Rico's state-sanctioned migration strategy and structural changes in the rapidly industrializing Puerto Rican economy, the exodus from the island increased sharply in the early postwar years, peaking in the 1950s. Puerto Ricans went to cities like New York, Chicago, and Philadelphia, where there was great demand for low-wage workers in manufacturing industries. Approximately one-third of the island's population circulated or emigrated to the United States between 1955 and 1970, as Puerto Ricans continued to leave the island in large numbers. By the early 1970s, however, deindustrialization in Northeastern and rustbelt cities resulted in a decline in manufacturing jobs, making emigration a less attractive option for working-class migrants. This continued until the mid–1980s and 1990s, when migration from the island increased once again. And, while cities like New York, Chicago, and Philadelphia continue to serve as home to large Puerto Rican populations, 2000 census data confirm the continued geographic dispersal of Puerto Rican migrants and communities to places like Florida beginning in the 1980s.[4]

Puerto Rican migration, however, has rarely ever been unidirectional. Return migration beginning in the mid-1960s increased dramatically by the early 1970s and in some years even surpassed emigration from the island; a trend that continued through the early 1980s.[5] A number of studies have documented this flow of return migrants and have analyzed its impact on the island, focusing largely on economic and cultural effects of return migration and, to a lesser extent, on its political consequences.[6] Still others emphasize the circular nature of Puerto Rican migration, one in which the flow of goods, people, ideas and capital connect island and mainland communities.[7]

Like many late-twentieth-century Caribbean migrations, Puerto Rican migration has evolved to include a variety of new destinations, multiple movements, and sustained connections among different places, a phenomenon popularly regarded as a 'va y ven' (or *vaivén*) movement. This experience of coming and going is familiar to many Puerto Ricans, and one that has provoked serious debate both inside and outside the academy.

For some scholars, the *vaivén* tradition is a result of economic changes both on the island and on the mainland, and has become a culturally conditioned way for migrants to improve their economic and social position. As sociologist Marixsa Alicea has noted, it is a way for migrants and their families to create and make use of 'dual home bases'.[8] Other writers, however, argue that the continual circulation of Puerto Rican migrants is a key contributor to increased economic immiseration and poverty among Puerto Ricans on the mainland since such movement disrupts families and people's participation in the labor market.[9] More recently, scholars like anthropologist Jorge Duany have re-entered this debate, arguing that circular migration – or 'mobile livelihood practices' – is, in fact, a flexible survival strategy enhancing migrants' socioeconomic status. In response to poor economic conditions on both the island and the mainland, Puerto Rican migrants create, and then make use of, extensive networks; including multiple home bases in several different labor markets. These transnational practices, Duany argues, not only compensate for the fact that economic opportunities are unequally distributed in space, but they also undermine 'the highly localized images of space, culture, and identity that have dominated nationalist discourse and practice in Puerto Rico and elsewhere'.[10] The complicated patterns of migration, return, and subsequent movement present theoretical and methodological challenges to traditional ways of analyzing migration and migrant practices. For this reason, recent exhortations proposing a transnational approach to Caribbean migration are useful in capturing Puerto Ricans' lives both on the island and abroad.[11]

This transnational framework for understanding Puerto Rican migration, however, should not ignore the rootedness of most Puerto Ricans' lives on the island and the mainland. Thus, while 'ir y venir' has become an almost unquestioned cultural trope of the Puerto Rican nation, new work on Puerto Rican communities reveals how 'quedarse y sobrevivir' – remaining and surviving – is perhaps a more appropriate way of understanding Puerto Ricans' experiences.[12] Indeed, most Puerto Rican migrants lead deeply local lives, although they do so transnationally, either by actively maintaining political, economic and social links with another community, or by nurturing affective ties connecting them with other places through ethnic celebrations, cultural events, sports, and even through stories.

That so many Puerto Ricans' lives continue to be bound up with the events, people, communities, and imaginings from 'home-places' they left long ago, or perhaps have never even visited, attests to the profound way in which state-sanctioned migration policies decades earlier have shaped the social, economic, and political landscapes of Puerto Rican communities. The policies guiding Puerto Rico's industrialization program not only succeeded in globalizing the island economy, they also stimulated a variety of migration patterns and social practices that gave initial form to transnational social fields in which people 'take actions, make decisions, feel concerns, and develop identities within social networks that connect them to two or more societies simultaneously'.[13] Over time, these transnational social fields have matured through the active participation of migrants, returnees and even non-migrants, whose daily practices in particular places and in historically determined times intersect with 'transnational networks of meaning and power' in the contested process of 'place-making'. Also deeply structural as well as reflecting the flexibility of human agency, this is a dynamic process whereby local meanings, identities and spaces are socially constructed within hierarchies of power and difference operating at both the local and global scale.[14] 'Place-making' is a critical feature of transnational social fields, grounding them in 'specific social relations established between specific people', and providing a lens for analyzing instances of conflict, resistance, and accommodation among differently situated individuals and social groups that occur within them. For example, community groups, labor unions, and grassroots activists may organize against the presence of global businesses – coffee shops, clothing stores, for instance – that, 'from above', threaten to transform a town's distinctive economic and social life. But these same local grassroots organizations may also actively resist the ways in which migrants, returning with different kinds of cultural knowledge and social remittances, challenge 'from below' ideas of group membership, identity and belonging.[15]

In San Sebastián, this process of contentious place-making has been particularly extreme since the 1960s when both the town and the island experienced high levels of return migration. Between 1970 and 1980, the population of San Sebastián increased by more than 20 per cent; a remarkable demographic shift following two decades of population decline in the 1950s and 1960s. This trend, however, is not surprising and, in fact, reflects population increases throughout the island. While return migration to the island began in earnest in the 1960s, it reached its peak in the 1970s and continued at a steady, although diminished rate, through the 1980s (Rivera-Batiz and Santiago 1996: 55). Over the past three decades, many *pepinianos*[16] returned, although they quickly discovered that returning 'home' is not always easy. After years living *afuera*, many *pepinianos* – particularly women and children – have a difficult time adjusting to their new life in rural Puerto Rico. Return migrants are commonly derided as *Nuyoricans*, a culturally distinct group whose members are usually born and raised *afuera*. While most return migrants eventually overcome such stigma, the idea of *los de afuera* – literally, 'those from the outside' or 'outsiders' – continues to define membership in the Puerto Rican nation, and is used as short-hand to refer to everyone from Dominican immigrants to *Nuyoricans* and 'criminals' allegedly terrorizing the island. This *los de afuera* discourse is also employed in local community politics to resist the ways in which 'progress' (*progreso*) threatens 'authentic' or 'traditional' Puerto Rican culture. Ultimately, these conflicts remind us that place-making within a transnational social field is fundamentally about power – the power to make place out of space; the power to decide who belongs and who does not – and that imagining and forming transnational identities is an historically contingent process circumscribed by power relations operating on the local, regional, and transnational levels.

Return Migration to San Sebastián

The life histories and stories of the residents of *barrio Saltos* in San Sebastián reveal a myriad of economic and non-economic factors informing migration decisions. They also paint a complicated portrait of life in San Sebastián and the ways in which migration, identity, and the politics of place remain emotionally charged issues in everyone's daily lives. The experiences of Carlos and Nilda Arroyo, return migrants who lived in Chicago for many years, provide one example of these processes. Carlos began his migration history by describing his life in San Sebastián before he moved to Chicago:

> 'I was born in San Sebastián, in *barrio Guatemala*, in the sector Central La Plata. We were very poor. The house where we lived, well, we closed the door with only a little rope. We didn't have doors like we now have today. I studied until the sixth grade – until the fifth grade here in Puerto Rico and then in 1957 I went to Chicago. My father took me there supposedly for a vacation which lasted until 1975 ... I was ten years old [when I left]'.

When Carlos and another brother left for Chicago, they went to live with their father and his wife, leaving their mother and other siblings in San Sebastián until 1960. That year, his mother sold their house in Pepino to pay for the airfare to Chicago, and they all moved to a small apartment on Milwaukee Avenue and Racine, then to another apartment on Elizabeth and Chicago Avenue, where they lived until Carlos and Nilda married in 1966.

Like other return migrants I interviewed, both Carlos and Nilda worked in a number of different factories in the Chicago area. In 1965 he joined his father and brothers working for the Merrit Casket Company, but he quickly left to make more money at a nearby rubber factory where he worked for almost four months. On November 11, 1965 Carlos went to work for the Teletype Corporation in Skokie, where he stayed until the factory closed in 1975. When Nilda first arrived in 1965 she was 18 years old and initially worked taking care of her cousins' three children. This arrangement quickly soured, however, when Nilda, like many other immigrants, discovered that the social networks on which she relied provided both 'grounds for cooperation but at the same time breed conflict'.[17] She explains:

> 'The first job I had was, I took care of my cousins' three children while she – she was supposed to find a job for me, right? But time went by and she didn't find anything for me at the factory where she worked. My mom finally told me, "Look, your cousin tricked you. You'd better find another job yourself". So, I went and found a factory job at night. A candy factory, the Holloway Company. I had a shift from 12a.m. to 7 the next day. It was horrible because I, I was always tired. I would also eat some of the chocolates [she laughs] … That was my second job … After that, I wanted to find a day job, because that job was too hard. I went to work at Freddy Hope Lamps, a lamp factory, but I didn't speak English. The girls would ask me, "Do you speak English?" And I said, "Just a little bit" [she laughs again] … When I would go to work, I looked like a model because I would do my hair, put on make-up and I looked like a model. I worked sanding wood, and I worked at a long table with a lot of black women who had liquor and would drink. The owner was an old man who finally took me out of there. He saw me, I was like a flower, he said, so he put me in a cleaner place working with a Japanese man named Frank … and I worked with [him] assembling lamps. It was a really nice job and clean. And I liked it a lot'.

Nilda also worked at a plastics factory, the last job she had before she and Carlos married. Once she married and had children, she worked sporadically in a number of clerical positions.

When I asked Carlos and Nilda why they returned to San Sebastián, they responded with what would soon become a familiar refrain among migrants: 'We always thought we would return here to raise our children. The younger [they are], the easier [it is for them]'. In Puerto Rico, it is a universal truism that the island is a better place to raise children. My first encounter with this virtually unquestioned belief was a conversation I had with a taxi driver the day I arrived to do fieldwork in Puerto Rico. As he drove me from the airport in Carolina to the nearby town of Río Piedras to catch a *carro público* to San Sebastián, he explained to me that Puerto Rico was the best place to raise children. It was okay to live in the United States in order to work and save some money. But the moment a couple has children, they should return to Puerto Rico where life is 'menos complicada y más sana' (less complicated and healthier). When I asked him if he thought San Juan was safer and healthier than other American cities, he said, yes. '(t)here is simply no comparison between the life here and *afuera*. In fact, it would be *irresponsible* for people to raise children *afuera* if they could come back and live in Puerto Rico'. This sentiment was echoed by almost everyone I met and interviewed in San Sebastián. And it was one of the most popular explanations for a family's return.

For example, when Carlos and Nilda decided to return San Sebastián in 1975, Carlos arrived first and began looking for work. And, although it was difficult finding work and a place for his family to live, he believed they were making the best decision for their children. When I asked why they decided to return to San Sebastián, Carlos and Nilda gave the following replies:

> Carlos: 'I believe that my roots called me back here. The family – well, I thought that it would be easier to find work here. We never really considered living in the metropolitan area [San Juan] because we thought that we were going to be faced with the same problems that exist in all big cities, like Chicago'.
>
> I: 'And how did you – Did you have problems when you were living in Chicago? Did that also [influence your decision to leave]?'.
>
> Carlos: 'We didn't have problems per se. But we did have a vision of our children. And we wanted them to study in a better environment. For that reason, we wanted to come to the countryside'.
>
> Nilda: 'Yes, because here they had a place to play and everything'.
>
> Carlos: 'They had more room to run. And that's how it was. They grew up like little wild children. [He smiles] In the open country they could go outside every day and there was no fear and it wasn't even slightly dangerous ['ni estaba calientito'] And little by little, they were able to adjust well'.

Carlos and Nilda's concerns are similar to those of parents in both San Sebastián and Chicago, who use migration as a strategy to protect their children from urban dangers – like gangs, violence, drugs and overcrowded schools – that disproportionately affect residents in poor and working-class neighborhoods. Their decision also highlights the ways in which places like Chicago and San Sebastián are increasingly imagined as mutually exclusive spheres of productive and reproductive labor. As the taxi driver said, you go to the United States to work, but you live and raise your children in Puerto Rico.[18]

Like Carlos and Nilda, Juan and Carmen de Jesús were concerned about potential danger in Chicago and returned to San Sebastián in 1977, after their three children were born. As Carmen described her life in Chicago, she grew increasingly animated and repeated several times how happy she had been there. Her various factory jobs and extensive social network of family, friends, neighbors provided a stable, rich life. But she was suddenly serious when I asked why they moved back. '[We returned] because of the children, you know. It was best for our children'. Once their son was older, she explained, they were afraid that he might have problems with gangs. For that reason, they bought some land in San Sebastián, cashed in Juan's profit shares from the paint company where he worked, and returned to Puerto Rico.

Angie Rubiani, a quiet woman in her early 50s whose husband, Rubén also said their family returned once they had children. She liked Chicago, she explained to me one night as we talked in the living room of her small wood house. The balmy, almost uncomfortably warm evening – pleasantly interrupted by the incessant singing of the *coquís* nearby – provided a fascinating contrast to her stories of the gray, bitter-cold Chicago winters that assaulted her daughters' frail health:

> 'They were always sick … the winter was very bad for them. And it was like, when one wasn't sick, the other was. The winter really affected them, you know? Problems with their throats, their ears. And my oldest even got bronchitis twice … The doctor who treated them – no, not the doctor, it was the pharmacist, I would always see him because of my medical plan, he would see the girls and he told me to come back to Puerto Rico. For the girls' health'.

When I asked her if they would have stayed if her daughters had been healthier, she shook her head, saying that their neighborhood and the local elementary school 'se estaban dañando' – they were deteriorating – and she was scared to send her kids to school. So in 1978 they moved back to San Sebastián, although Rubén returned to Chicago to work for six months before settling permanently with his family in Saltos.

What is striking about these and other narratives of return is the way in which Puerto Rican returnees privilege concerns about raising their children when explaining their decision to leave Chicago. In actual fact, their fuller stories are checkered with myriad reasons for return; including sick relatives in Puerto Rico, housing problems in Chicago, battles with depression exacerbated by long, isolated winters, and, not surprisingly, job loss. Thus, immediately after Carlos and Nilda explained that they wanted to raise their children in Puerto Rico, Carlos also mentioned that he was laid off after ten years working for Teletype Corporation. This was the best time to move, he explained, since the company gave him a severance package that allowed him to ship his belongings and pay for airfare back to Puerto Rico. Juan also confided that even though he worked at a paint shop before leaving Chicago, he was afraid that the paint factory, like other blue-collar jobs in Chicago at that time, might relocate. These narratives of return, however, are consistent with a migration ideology that anticipates people will return to Puerto Rico to live a better life than before. According to this logic, migration is an economically motivated decision to '*mejorarse y progresar*' – better oneself and progress – while return is largely informed by 'place utility' and sentimental attachments to home and nation. The assumption is that one returns to Puerto Rico to retire or to enjoy life after many years working hard *afuera*.[19]

Many *pepinianos* explained they never adjusted to life *afuera*. In fact, they often compared themselves to the coquí, the tiny, melodious frog native to Puerto Rico that popularly represents 'Puerto Rican-ness'. Like the coquí, I was repeatedly told, *true* Puerto Ricans cannot thrive outside of Puerto Rico. They might be able to live, *pero no cantan afuera* (they don't sing outside). For that reason, many proudly reminded me, to really live your life, you must return to your native land. But for many, returning to San Sebastián was not easy. Although most of the people I interviewed have lived there at least fifteen years, they recalled in vivid detail the difficulties they faced adjusting to life in San Sebastián again. Women and children were particularly clear about how they suffered during this return transition. As I demonstrate below, these narratives reveal the ways in which migration is often a conflictual process in which women and children contest and resist new gender and generational ideologies and culturally prescribed behaviors.

The Suffering of Return

Elena Rodriguez is an energetic pastor of an Evangelical church, who has lived in San Sebastián for more than twenty years. Born in Aibonito in the southeastern region of Puerto Rico, Elena was raised in Chicago with her nine brothers and sisters. At the age of 17, she married her husband, Lolo. Shortly after they were married, they went to live in San Sebastián. When I asked her if she wanted to return to Puerto Rico, she smiled and said, 'This is where the story begins! You cannot even imagine how much I suffered in that change!'. When she first arrived in Puerto Rico, she lived with her mother-in-law for a month, while her husband remained in Chicago with plans to eventually join her there:

> 'I came to Puerto Rico before [him] to set things up. And in that month, well, for me, that month was terrible for me because there weren't the same facilities you could find [in Chicago]. When I began to stay in my mother-in-law's house and she would say to me, "There's no water", ... and I saw that the stores were far away, that there weren't any stores close or pharmacies close ... and when one left the doctor's office worse than when they arrived ... I saw all this and I wrote a letter to my husband saying, "Send me a ticket home fast because I'm leaving. I am not going to live here" ... Well, he sent me [and our two children] tickets and I left'.

Shortly after Elena returned to Chicago, however, her sister-in-law died, and her husband decided it was best for them to return to Puerto Rico to take care of his mother. Because they didn't have their own home, they lived with Lolo's mother which only made a difficult transition even worse:

> 'You know the saying, "*Que se case, casa quiere*" (Everyone who marries wants a home). And living in another's house, no matter how well they treat you, you want your own home. I

lived in a room on the patio (*marquesina*) and I already had two children, and I felt uncomfortable in my mother-in-law's home. And seeing that there was never any water – I had to go with buckets to get water from the neighbors. And carrying that water in order to do something to help my mother-in-law. And my husband was only making $60 a week. [With that] we had to pay the bank, feed our children – It wasn't easy. And I'm telling you, for me it was so traumatic. I cried every night because I wanted to leave. Every night. And I told my *compañero*, "If you want to stay, you stay. But I'm leaving. I can't take this anymore". Because I saw how if my children got sick here – in Chicago [the doctors] took care of them quickly. But here I had to wait to be called, sometimes almost three hours while my children suffered from fever, pain, and they still made me wait ... I finally wrote to my father ... and I told him, "Papi, send me tickets because I'm leaving. I'm going to leave my husband because life here is full of suffering (*muy sufrida*)" ... This is how I lived. It was so traumatic. And honestly, honestly, I cried to leave [Puerto Rico] for ten years'.

Elena's husband begged her to stay, and he eventually convinced her to use the money her father sent for tickets back to Chicago to begin building a house. She stayed in San Sebastián, but regretted her decision for a very long time.

Like Elena, Nilda also vividly recalls her difficult transition to life in San Sebastián. Even though she disliked the weather in Chicago, she preferred living there because life was easier in the city and there was more to do. They lived near museums and parks, and they had access to public transportation. She adapted to her new life slowly, she explained, and she was much happier once they had their own home. She still complains, however, about the town's lack of public transportation, a common lament among almost all the women in the town. Unless one owns a car and knows how to drive, one has to depend on the town's *carros públicos* which run irregularly until 2pm and wait until they have a full car of five or six people before they go on their routes. Women complained bitterly about this new dependence on husbands, friends, and family and were frustrated that these transportation problems limited their movement, confining them largely to their homes.

Women's isolation was further heightened by their failure to work outside the home, a remarkable difference from their lives in Chicago. When I asked women about their jobs on the mainland, they were extremely animated and provided great detail about all the places where they worked, what they enjoyed about their jobs, the people with whom they worked, and frequently the race/ethnic politics of the workplace. In San Sebastián, their labor history changed dramatically. Nilda worked irregularly when she first arrived, but eventually she decided to collect unemployment once she began to have back problems. Carmen also stayed at home collecting unemployment while her husband worked in a series of different factory jobs in San Sebastián, Isabela, and Aguadilla. Angie has been employed sporadically since her return and currently cares for an old man, cleaning his house and cooking his meals. For some women, working outside the home became a necessary condition for agreeing to stay in Puerto Rico. Elena, for example, became very involved in her local church and eventually studied to be a pastor and now leads her own community. And Yahaira, a young woman who, like Elena, was born in Puerto Rico but raised in Chicago, decided to stay in San Sebastián once her husband agreed to help her open a small cafeteria in the town.

On the surface, most women did not see their new domestic roles as contradictory or conflictual. They often explained that they enjoyed staying home to care for their children. But they did express frustration with how much more difficult domestic tasks were in San Sebastián. In addition to the problem with water – which continues to be a problem even today – women complained that their more isolated lives meant they had to cook more often, take more time to buy food because of poor public transportation, and there were fewer entertainment venues. The women also noted that because their husbands earned less money and the cost of living was higher in Puerto Rico than in the United States, they had to find creative ways to stretch their money. These new responsibilities increased women's *sufrimiento* and prompted some to advise their sisters to resist their husbands' efforts to return to Puerto Rico. Elena explains:

> 'It wasn't easy. After living in Chicago where you have your good job, and you would eat out on Fridays and maybe Saturdays too…and to come to Puerto Rico I had to get used to cooking breakfast, lunch, and dinner … It wasn't easy … After one has lived in Chicago, it's not easy to adjust to life here … I would never tell anyone to come to Puerto Rico [to live]. No one, no one, no one. When my sisters would come to visit, [I would take them aside and tell them], "Don't come to Puerto Rico to live. Leave me here, it's okay since I'm more settled here now … My husband is here and I have a house…But I would advise you not to and live in Puerto Rico"'.

Return migration influenced different domestic arrangements, thereby making housework less egalitarian. While none of the women explicitly discussed their husbands' willingness to share in domestic chores, they implied that housework was primarily their responsibility and expressed great discontent with the amount of housework they had to do, and how it usually involved more labor-intensive chores than in Chicago. In Chicago, most of the women worked outside the home and had more disposable income, allowing them some reprieve from cooking duties. Their lack of wage labor in San Sebastián, as well as their husbands' reduced earnings, circumscribed the household's disposable income and heightened women's feelings of being overwhelmed by domestic tasks.

Even though women like Elena, Yahaira, and Nilda admit they eventually adjusted to life in San Sebastián and, in retrospect, believe it was the best move for them, they also describe their children's difficulties with adjusting to life in San Sebastián. Language problems, different norms for dress, and the stigma attached to being *de afuera* alienated migrant children from their peers, especially in school.

Yahaira worried about her son's performance in school, because of his language difficulties. When I asked her if Tito had problems with other students, she assured me that he got along with everyone. Overhearing our conversation, Tito politely corrected his mother saying in Spanish, 'I like it here, I like being in the open air and everything. But at school they bother me. They call me the gringo. Well, they used to call me that, because I didn't speak Spanish, I spoke English too much. I used to get really mad'. Yahaira was surprised by her son's response and sympathetically added that there were certain words *she* still couldn't pronounce correctly since she is English-dominant. Then, slipping into English, she complained, 'There are some things that I'll never get used to here. People are so *nosey* here. Everybody *quiere saber la vida de uno*' (Everyone wants to know about your life).[20]

Gendered Dimensions of *Los de Afuera*

In addition to the difficulties involved with adapting to a new life in Puerto Rico, women like Elena, Yahaira and Nilda are also enmeshed in the punitive national discourse which blames outsiders or *los que vienen de afuera* for almost everything imaginable. Narrowly defined, '*los de afuera*' refers to Puerto Ricans who currently reside on the island after many years living in the United States. More broadly, the label popularly applies to Dominicans, *pillos* (criminals), *asesinos* (murderers), or anyone else who does not properly belong in the imagined circle-of-the-we of the Puerto Rican nation.[21] Like the term *Nuyorican*, *los de afuera* is a racialized discourse usually employed pejoratively, connoting a culturally distinct group whose values, behaviors, language, and dress directly challenge dominant understandings of 'authentic' Puerto Rican culture. This stigmatizing label also homogenizes an extremely diverse group allegedly responsible for contaminating, polluting, and corrupting national identity.[22] On both a local and a national level, *los de afuera* is used to define membership in the immediate and national community. It is also strategically deployed in local politics to resist the transnational flow of ideas, people, and capital because they are believed to destabilize traditional understandings of community, identity, and place by the reactionary, the conservative and the insular in Puerto Rico's barrios.

In San Sebastián, young Puerto Rican men and women who do not fit the expected norms of behavior, dress, and linguistic performance are immediately labeled *de afuera,* regardless of whether or not they have actually lived outside of Puerto Rico. Multiple piercings, hip-hop styles for men, and 'provocative dress' for women are commonly regarded markers of one's outside status and usually provoke pointed remarks by older barrio residents. On several occasions, *la tienda* – local store – became the arena for heated intergenerational conflicts between young and old men who consistently demonstrated their disdain for each other. Because Cristina's store was directly across from the basketball courts, the young men playing basketball or hanging out around the courts would often intermingle with the old men drinking and playing dominoes at the *tienda*. These men – most of whom had lived *afuera* for varying lengths of time – consistently complained about how the young men dressed, the music they listened to, and their multiple piercings; attributing these behaviors to the negative influences they received while living in the United States. One evening as I worked in the store, I teasingly told one young man with multiple piercings that he shouldn't smoke because he'll die young. One of the regulars looked up from his beer and sneered, '*Que se muera, no más! Mira cómo es! Déjalo que se muera!*' [Let him die! Just look at him! Just let him die!]. The young man looked surprised at the outburst, said something under his breath and left the store. Some men playing dominoes witnessed the exchange and laughed as one commented, '*La juventud de hoy está perdida. Vienen de afuera, vienen para acá y no hacen na*' [The youth of today is lost. They come from the outside, they come here and they don't do anything.].

The idea of '*la juventud perdida*' was a common lament. Older community residents would pray for them – '*Roguemos por los jóvenes, tan metidos en el vicio, en la droga, en la prostitución*' [Let us pray for our youth, completely lost in vice, in drugs and prostitution] – describe how they are vastly different from how they were when they were younger – '*la juventud de hoy no es lo mismo de antes. Son perdidos, y no saben vivir la vida*' [Today's youth is not the same as before. They are lost, and they don't know how to live their lives] – and would accuse them of being lazy, unmotivated and corrupted by life *afuera*. One day over a game of dominoes, the topic emerged again as two older men tried to convince me and another woman, Beba, that today's youth was worthless. Severino commented, '*Chacho! Estos jóvenes – yo diría que solo un 25% de la juventud hoy en día son buenos. Jovenes que valen la pena y que no están metidos en problemas – en las gangas, en la droga*'. [Man! These young people—I'd say that only 25 per cent of today's youth are good. Young people who really are worthwhile and are not involved in problems – in gangs, in drugs.] '*Sí*', Carlos added, '*la juventud está perdida*'. [Yes, the youth are lost.] When Beba protested that this surely was not true, that there had to be more than 25 per cent of the youth who were worthwhile, she was corrected by Severino. '*No, no. Mira, los jóvenes de hoy no trabajan! No saben trabajar. Cogen el welfare, los cupones. No saben trabajar como nosotros ... por eso los jóvenes son bancarrota. Es el modernismo – mundo moderno. Y es lo mismo que allá [en los Estados Unidos]*'. [No, no. Look, today's young people don't work! They don't know how to work. They collect welfare, food stamps. They don't know how to work like us ... that's the reason why the youth are a lost cause. It's modernism – the modern world. And it's the same over there in the United States.]

Young men, according to these narratives, do not know the value of work, the honor (*honra*) of making money, and living an honest life. Moreover, unlike the critics who also lived *afuera* for varying lengths of time, these young men have allegedly internalized the cultural values attributed to urban 'ghetto' living and bring them to the island: Gangs, drugs, and increased violence in Puerto Rico are blamed, in large part, on *los que vienen de afuera*.

Notions of '*la juventud perdida*' are also extremely gendered. While young men are derided for their apparent refusal to be productive members of the community, women *de afuera* are blamed for their domestic failings. They are immodest, '*locas*', [wild] '*fiesteras*', [partiers] '*enamoradas*', [easily-seduced] and responsible for failed marriages and the moral deterioration of the town. Young women's dress is one arena of struggle and an important terrain upon which ideas about the nation, gender, and identity are contested.[23] When I first arrived in San Sebastián, for example, one of my great-aunts took enormous pride in introducing me to her friends and neighbors. When she invited me to meet the men at the bakery across the street from her home, she pointed out that although I was born and raised *afuera*, I was not '*una de esas locas de afuera. Miren cómo se viste* [one of those wild girls from the United States. Look at how she dresses]', she insisted, as she passed her hand along the front of me, highlighting my appropriately short skirt and conservative blouse. '*Es como si fuera criada aquí, como en los tiempos de antes*', she concluded. [It's like she was raised here, like in the old days.]

By contrast, young women born and raised on the island, who do not fit expected norms of dress and behavior, are criticized for acting as if they were *de afuera*. '*[Mi nieta] se viste como si fuera de Nueva York*' [[My granddaughter] dresses as if she were from New York], a middle-aged woman complained to four of her friends while visiting from Bayamón one afternoon in Cristina's store. The comment provoked an intense conversation among the five middle-aged women, each of whom had lived at least ten years in Brooklyn, Newark, New Jersey and Chicago and returned in the 1970s to live in San Sebastián and the San Juan metropolitan area only to find that the 'American' lifestyle and values they hoped to leave behind now reach and corrupt young island women. '*Nueva York*' and '*de afuera*' are part of a racialized sexual vocabulary of contamination and impurity used to describe women who defy culturally prescribed roles of dress, sexual behavior, and modesty. Like many young Chicanas and *mejicanas*, young Puerto Rican women are constantly watched, their bodies, dress, and sexuality policed by family members and neighbors.[24] An essential component of this policing is the 'semi-private, semi-public talk' of gossip which, as anthropologist Roger Lancaster cleverly demonstrates, 'can take up almost any subject and imbue it with significance'. In the case of return and transnational migrants, however, gossip often serves a punitive function of further marginalizing those groups already stigmatized and racialized as 'other' and 'outside' the Puerto Rican nation.[25]

Thus, when Melinda, a flashy young mother of three currently living in New York, made several trips to San Sebastián within six months, it was rumored that she was there because she had AIDS and needed to rest in the quiet of San Sebastián. Although no one ever really *knew* whether or not she had AIDS, the fact that people believed she was dying from the still-stigmatized disease – she was very thin, pale (she arrived in the middle of New York winter), and reportedly used drugs – fueled more gossip about her sexual liaisons with *los jovenes* and married men in the neighborhood. On many occasions, angry wives would fight with their husbands at Cristina's store, accusing them of having sexual liaisons with '*esa loca de afuera que tiene SIDA* [that wild girl *de afuera* who has AIDS]'. Melinda emphatically denied the accusations and blamed the town's provincialism for perpetuating the rumors in the absence of any evidence. Regardless of its veracity, the gossip (and marital strain) Melinda's presence provoked highlights again how women's bodies are the site of struggle and contestation, and how transnational migrants are not only blamed for culturally polluting the Puerto Rican nation, but are popularly regarded as diseased and capable of physically contaminating the body politic with AIDS and other diseases as well. Alberto Sandoval Sánchez notes that the metaphor of an 'air bus' or an 'air bridge' facilitating the ease with which Puerto Ricans can move between Puerto Rican and U.S. cities, has also become 'a space of continuity and contiguity that makes possible the passage of those condemned by Puerto Rican society: the sick, the infected, the contaminated, the marginal'.[26]

HIV/AIDS is unquestionably one of the most pressing problems facing Puerto Rican communities: The Center for Disease Control reports Puerto Rico registers one of the highest rates of HIV/AIDS in the United States, and that Puerto Rican women are one of the fastest growing sectors infected with HIV/AIDS in the US Yet how this public health crisis that transcends international boundaries gets reframed as a problem imported by migrants and those already on the margins, speaks to how ideas of purity – linguistic, sexual, cultural and racial – are marshaled to 'uphold the frontier allegedly separating the island from the United States, and islander elites from US/lower-class Puerto Ricans ... [in order to] distinguish between 'real' Puerto Ricans from imposters ...'.[27]

These racialized metaphors of contamination, sexual pollution and corruption of the nation have also been applied to immigrants in the United States, particularly to Mexican immigrants who, as Leo Chávez demonstrates in his study on popular images of US immigration, have been 'represented almost entirely in alarmist imagery' of crisis, bombs, disease, and invasion since the 1960s. In both instances, women's bodies, sexual practices, and reproductive potential are the objects of disdain, fear, and even punitive social policy measures like California's Proposition 187 in 1994 which sought to deny publicly funded health care and education to undocumented immigrants and their children. Such draconian policy measures have not been adopted in Puerto Rico, but the discourse enveloping return migrants is of the same xenophobic cloth, although with an important caveat. As Frances Negrón-Muntaner explains, cultural, linguistic, and sexual 'impurity' implies hybridity and ultimately replacement, since 'given the nation-building narratives' concern with reproduction, a hybrid cannot produce; it is sterile'.[28]

In addition to their sexual behavior, Puerto Rican young women are also criticized for being out too much, relying too often on relatives to care for their children, and not attending to their husbands properly. Invariably, people blame these problems on the unsavory influence of American popular culture and distinct gender ideologies threatening 'traditional' Puerto Rican family values. As feminist scholars have aptly noted, women are expected to signify cultural tradition and are the contested terrain upon which nationalist projects and ethnic identity and self-respect are constructed, resisted, and reconfigured. For some Filipino immigrant parents in California, for example, their young daughters' bodies and sexual behavior are critical to 'assert cultural superiority over the dominant group' and to resist the ways in which their racialized sexuality is disparaged by white society. Research among Indian immigrant families in New York City uncovers similar dynamics of how ideas of 'tradition' and the 'authentic immigrant family' are contingent on young women's chastity, and how these ideas are produced transnationally. Sociologist Yen Le Espiritu notes that while such strategies are, indeed, attempts for racialized groups to assert moral superiority over oppressor groups, they are problematic since they simultaneously reinforce 'masculinist and patriarchal power' and unfairly circumscribe women's lives since they 'face numerous restrictions on their autonomy, mobility, and personal decision making'.[29] When Puerto Rican women fail to live up to constructed norms of behavior or dress, they are punished and labeled *de afuera*, a process that betrays the ways in which a glorified Puerto Rican past rests on racialized constructions of women and motherhood.

Community concern with women's private and public behavior also reveals important intergenerational conflict about shifting cultural values regarding gender and sexuality embedded in regional political economies. As ethnographer Caridad Souza has shown in her research among working-class *puertorriqueñas* in New York City, young women's reproductive work is critical to households whose survival depends on family and community solidarity through the exchange of goods, services, and the pooling of resources across multiple kin and non-kin households. In this context, young women are expected to be *en la casa* – inside the home or family – for important material and cultural reasons: Without their reproductive work, adult women – mothers, grandmothers, and aunts – are easily overwhelmed by (and unable to meet) the reproductive and productive demands within their own households and their kin networks. Being *una muchacha de la casa* also means abiding by the cultural norms of respectability, chastity, and family honor valued by the community. To be *de la calle* – outside the home, or literally, on the streets – is to be transgressive, sexually promiscuous, and dangerous.[30] Because this good versus bad girl dichotomy also maps onto dominant understandings of those who are *de acá* and *de afuera*, young women from the mainland are usually regarded suspiciously and as *de la calle* until they prove themselves to be otherwise.

In short, the term *de afuera* reproduces a particular understanding of community that is static and smoothes over its heterogeneity. It also stigmatizes those people whose economic marginalization obliges them to move as a survival strategy. As many of the town's residents quietly lament, jobs are scarce in San Sebastián. The closing of Central La Plata, the town's sugar refinery, in May 1996, as well as recent factory relocations and layoffs *left* thousands of men and women unemployed. And despite organized attempts to resist plant closings, San Sebastián's unemployment rate hovers around 25 per cent; a statistic that does not include workers underemployed in low-wage service sector jobs.

Global economic restructuring has had a tremendous impact on San Sebastián. The repeal of Law 936 in 1998 awarding tax exempt status to North American companies, as well as increased reliance on temporary workers in factories, have heightened laborers' vulnerability. When I asked *pepinianos* if one could find work in the town, they answered tentatively, saying that people could find *trabajitos* – or little jobs – here and there. But as Rubén, one of the most vocal critics of *la juventud perdida* pointed out, they are usually dead-end jobs and even young people with college degrees – including his own daughters, trained in the sciences but unable to find employment in the area – will almost certainly have to *irse pa' fuera*.

Conclusion

Young people often make convenient scapegoats for larger political-economic problems.[31] Older adults regard them as significantly different from earlier generations – usually lazier, unwilling to make sacrifices or delay gratification – and guided by a qualitatively different value system than their parents or grandparents. Much of today's intergenerational conflict in San Sebastián is certainly a result of generational distance and misunderstanding. Yet the past two decades have also witnessed key political-economic changes throughout Puerto Rico in general – and in San Sebastián in particular – which have prevented many young women and men from participating in the formal economy and leading the kind of life their forbearers allegedly led, usually, as migrants themselves. Some scholars have described Puerto Rico as a 'post-work society', whereby the American government promotes a political economic system that discourages wage labor and provides, instead, federal transfer payments as a way to help people survive while maintaining the political status quo.[32] Others have carefully documented how the various stages of Puerto Rico's industrialization program have differentially incorporated some workers and marginalized others since the 1950s.[33] Anthropologist Helen Safa, for example, has also noted in her research among female factory workers in western Puerto Rico that most believe it is easier for women than men to find employment. For most young *pepinianos*, unemployment, migration, varied domestic arrangements, and marital strain have very little to do with poor values, being *de afuera*, or *de la calle*. These problems are largely the product of many Puerto Ricans' economic marginalization both on the island and the mainland. In San Sebastián, however, the *los de afuera* trope reveals not only disdain for outside influence, but deep ambivalence concerning the consequences of a transnational existence.

Notes

A longer version of this chapter appears in Gina M. Pérez, *The Near Northwest Side Story: Migration, Displacement and Puerto Rican families* 2004, University of California Press.

1. Segal 1987; Richardson 1989; Olwig; Duany 2002.
2. Duany 2002, 13.
3. History Task Force (1979); Sánchez-Korrol (1994).
4. Bonilla 1994; Ellis et al. 1996; Rivera-Batiz and Santiago (1996).
5. Meléndez (1993, 15–17).
6. For studies of the socio–economic impact of return migration see José Hérnandez–Álvarez's pioneering work (1967) as well as Ashton 1980; Bonilla and Campos 1986; Bonilla and Colón Jordán 1979; Cordero-Guzmán 1989; Enchautegui 1991; Muschkin

1993; and Torruellas and Vázquez 1982. Zentella (1990) and Negrón-Muntaner (1997) provide important analysis of language and culture and return migration and Vargas-Ramos (2000) provides a thorough discussion of return migration and political participation in Aguadilla.

7. The idea of circular migration was first advanced by Juan Hernández Cruz (1985) and employed by other scholars of Puerto Rican migration including Rodríguez 1993; Ortiz 1993; Torre et al. 1994; Duany 2002; Ellis et al. 1996; Kerkhof 2001.

8. Alicea (1990) writes that many Puerto Rican migrants create dual home bases both on the island and the mainland, a process that allows them to 'maintain social and psychological anchors in both the United States and Puerto Rico' and belong simultaneously to several different dwellings at one time (14). See also Rodríguez 1993; Meléndez 1993; and Ortiz 1993 for a discussion of circular migration.

9. Tienda and Díaz 1987; Chávez 1991.

10. Duany (2002, 235). See also Alicea 1990, 1997; Ellis et. al 1996; and Rodríguez 1993 make similar arguments about the use of migration as an important survival strategy. See also Olwig and Sorenson (2001) for a more elaborate discussion about mobile livelihood practices among global migrants.

11. Examples of this work include Glick Schiller et al; 1992; Levitt 2001; Smith and Guarnizo 1998.

12. I would like to thank Isa Vélez and Patricia Zavella for helping me to clarify this point. Work by Alicea 1997; Souza 2000; Ramos-Zayas (2003); Rúa 2001; Whalen 2001; Stinson-Fernández 1994; Bourgois 1995; Glasser 1997 are examples of research focused on specific Puerto Rican communities on the mainland.

13. Glick Schiller, Basch and Blanc-Szanton, 1992, 2.

14. Michael Peter Smith 2001,106, 144.

15. Smith and Guarnizo, 1998

16. *Pepinianos* is a popular way to refer to people hailing from the town of San Sebastián de las Vegas del Pepino, or simply El Pepino.

17. Menjívar 2000, 174. Menjívar's research among immigrant women from El Salvador provides important analysis of the conflicts, cleavages and limits to immigrant social networks, the ways in which gender shapes them, and of the prevailing 'structures of opportunities' in which these social networks are embedded. Works by Hondagneu-Sotelo 1994; Kibria 1993; and Mahler 1995, 1999 also provided critical attention to the conflicts surrounding immigrant social networks.

18. Alicea (1997: 619) makes similar observations explaining how Chicago Puerto Ricans regard the island as a site for investment – buying land and a home, for example – recreation, and a safe place to raise children, while American cities are productive sites for work and reliable social services. Using Goldring's work (1992) among Mexican transnational families as an important case for comparison, she explains how this differentiation of social space is largely a result of global and political restructuring.

19. Ellis et al., (1996) describe four categories for return/circular migration: those who return because of the labor market, 'tied' migration (a move precipitated by a partner's migration or family need), to improve place utility (better place to live due to improved housing or preferred climate and culture), and various other reasons such as religion or political reasons.

20. Many *pepinianos* spoke passionately about the relationship between language, cultural authenticity and identity. See Pérez, chapter 4 for more on these discussions. Many scholars have discussed the role of language and Puerto Rican identity see Negrón-Muntaner 1997, 259; Urciuoli 1996, 47-48; Zentella 1990, 85; Flores 1993; Kerkhof 2001.

21. The conflicts arising from return migration provide an interesting counterpoint to Anderson's notion of an imagined community (1983) and point to the different ways in which this imagined community is constructed differently over time. Here I have borrowed from David A. Hollinger's notion of 'the circle of the "We"' (1993) to discuss how different groups are included and excluded in popular understandings of the nation.

22. Many scholars have carefully documented the ways in which returnees and Spanglish-speaking Puerto Ricans on the island and the mainland are charged with corrupting Puerto Rico's linguistic integrity. Citing island writers, intellectuals, and government officials, Zentella (1990) reveals widespread concern for the deterioration of the Spanish language. Puerto Rico's distinctive colonial history with the United States has reinforced the 'consistent identification of Puerto Rican identity with the Spanish language'. As a result, 'many of the island's intellectual and others believe that English has had a continuously deteriorating effect on the Spanish of Puerto Rico and that, as a result, Puerto Rico's national identity itself is being threatened'(85). Negrón-Muntaner (1998) masterfully maps the gendered politics of language onto enduring debates regarding Puerto Rican identity, nationalism and migration. For some island academics and politicians, bilingualism is often a metaphor for 'ambiguity, cultural disorders, and political passivity'. Similarly, defenders of the Spanish First legislation – an attempt to make Spanish the official language of government in Puerto Rico – regard bilingual Puerto Ricans on the island and the mainland as 'a race of *tartamudos* [stutterers], unable to communicate either in English or Spanish'. She writes, 'For many intellectuals on the island, US Puerto Ricans serve as a 'futuristic' projection of what all Puerto Ricans will/have become: culturally 'impure' or hybrid, racially *mestizo* and bilingual (that is, having two 'national' loyalties). The notion of 'hybridity' is important since given the nation-building narratives' concern with reproduction, a hybrid cannot reproduce; it is sterile. The possibility that the elite's destiny will be explicitly tied to the US diasporas (the *hampa*) or be displaced by the 'lower classes' partly fuels these groups' writing off of two-thirds of the Puerto Rican population' (279). Kerkhof (2001) makes similar observations about the struggle over language and Puerto Rican return migrants as well.

    In addition to linguistic corruption, return migrants are also popularly regarded as diseased, physically contaminating the body politic with AIDS and other diseases. A controversial *New York Times* article described the migration between Puerto Rico and New York as an 'air bridge' transporting sick and polluted migrant bodies. 'New Yoricans [*sic*]', according to the article are blamed for importing AIDS from the mainland, further cementing their marginal status (in Sandoval Sánchez 1997, 203). Sandoval Sánchez writes 'the metaphorical construct of an 'air bridge' constitutes a space of continuity and contiguity that makes possible the passage of those condemned by Puerto Rican society: the sick, the infected, the contaminated, the marginal (IV drug users, homosexuals, gay tourists, prostitutes)' (203).

23. Goldring (1998) argues, for example that migration often leads to 'disagreements over meanings' within a transnational community and that women's dress is an important site of struggle that is contested both directly and through gossip.

24. Zavella 1997, 396.

25. Lancaster 1992, 71-72. In his hilarious account of how community gossip successfully uncovered the true origin of his sudden illness, Lancaster demonstrates the form and function of gossip in a Managua neighborhood. It is both useful and playful, he writes, and binds people together in a web of economic and non-economic relationships. Gossip in San Sebastián functions quite similarly to Lancaster's accounts in Managua, although it also marginalizes those groups who are the popular objects of gossip.

26. Sandoval Sánchez 1997, 203. Sandoval Sánchez also notes that the high incidence of AIDS among Puerto Rican men and women has created a new kind of migration: 'The pattern is the following: Those who get infected in Puerto Rico come to New York and other US cities for treatment. As American citizens they can get on a plane and qualify for Medicaid in the United States. And those who are in a terminal state go back to Puerto Rico to die with their families and to be buried in their native land' (1997, 202).

27. Negrón-Muntaner 1997, 280. Center for Disease Control, Division of HIV/AIDS Prevention, HIV/AIDS Surveillance Report, December 2001, V. 13, nu. 2.

28. Negrón-Muntaner 1997, 279. Chávez 2001, 260. For more on how *mejicanas* are implicated in discussion of immigration, reproduction and nativism, see Hondagneu-Sotelo 1995.

29. Espiritu 2001, 415, 416; Das Gupta 1997, 574. Other important examples of feminist analysis of nationalism and nation-building projects include di Leonardo 1984; Chatterjee 1989, 1993; Kaplan, Alarcón, and Moallem 1999; Grewel and Kaplan 1994.

30. Souza, 2002, 35. Behar (1993) also discusses this dichotomy of female respectability and women's attempts to resist and subvert such designations to claim power and place on the street. This duality of street and home and female respectability exists throughout Latin America and among US Latinos as well.

31. Zentella 1990, 91.

32. López (1994) argues that as a result of the most recent technological innovations and global restructuring, an important shift has occurred in which 'those who can work, work too much, entering into forms of multiskilling and double shifts, while the majority become deskilled, disqualified, overly-trained and redundant' (112). In this context, Puerto Rico has, once again, become another living laboratory to examine the 'strategies of coping with a post-work society and experiments in ambivalent postmodern sensibility' (ibid). This post-work society, she argues, is the product of a colonial political-economic relationship with the United States that originally promoted export-led industrialization, but, more recently, has 'ceased to be oriented toward economic growth and turned instead toward a redistributive role in sustaining basic consumption and the standard of living of the population' (116). López notes in 1991 US transfer payments totaled almost $5 billion and the gross national product was a little more than $5.5 billion, figures buttressing unconfirmed reports that only 1 in 10 islanders does not receive transfer payments (ibid). While the idea of a post-work society is certainly a debatable one, her arguments do underscore how different the island's economy is now compared to the 1950s and 1960s when many of the old-timers migrated and returned from working *afuera.*

33. See Safa 1995; Acevedo 1990; and Ríos 1993 for a discussion of gender and Puerto Rico's industrialization program.

References

Acevedo, Luz del Alba (1990) 'Industrialization and employment: Changes in the patterns of women's work in Puerto Rico', *Demography* 18(2): 231-255.

Alicea, Marixsa (1997) '"A chambered nautilus": The contradictory nature of Puerto Rican women's role in the social construction of a transnational community', *Gender & Society* 11: 597-626.

Bonilla, Frank (1994), 'Manos que sobran: Work, migration and the Puerto Rican in the 1990s', in *The commuter nation: Perspectives on Puerto Rican migration*, Carlos Antonio Torre, Hugo Rodriguez Vecchini and William Burgos (eds), Rio Piedras: Editorial de la Universidad de Puerto Rico.

Bonilla, Frank and Ricardo Campos (1986), *Industry and idleness*, New York: History and Migration Task Force, Centro de Estudios Puertorriqueños, Hunter College.

Bonilla, Frank and Héctor Colón Jordán (1979) '"Mamá, Borinquen me llama!" Puerto Rican return migration in the 70s', *Migration Today* 7(2): 1-6.

Bourgois, Philippe (1995), *In search of respect: Selling crack in El Barrio*, Cambridge: Cambridge University Press.

Chatterjee, Partha (1989), 'Colonialism, nationalism and colonized women: The contest in India', *American Ethnologist* 16, 622-633.

Chavez, Linda (1991), *Out of the barrio: Toward a new politics of Hispanic assimilation*, New York: Harper Collins.

Cordero-Guzmán, Héctor (1989), 'The socio-demographic characteristics of return migrants to Puerto Rico and their participation in the labor market, 1965–1980', Master's thesis, University of Chicago.

Das Gupta, Monisha (1997), '"What is Indian about you?" A gendered, transnational approach to ethnicity', *Gender and Society* 11(5): 572-596.

di Leonardo, Micaela (1984), *The varieties of ethnic experience: Kinship, class and gender among California Italian-Americans*, Ithaca: Cornell University Press.

Duany, Jorge (2002), *The Puerto Rican nation on the move: Identities on the Island and the mainland*, Chapel Hill, NC: University of North Carolina Press.

Ellis, Mark, Dennis Conway and Adrian Bailey (1996) 'The circular migration of Puerto Rican women: Towards a gendered explanation', *International Migration Quarterly Review* 34(1): 31-64.

Enchautegui, María (1991), *Subsequent moves and the dynamics of the migration decision: The case of return migration to Puerto Rico*, Ann Arbor: Populations Studies Center, University of Michigan.

Espiritu, Yen Le (2001), '"We don't sleep around like white girls do": Family, culture, and gender in Filipina American lives', *Signs: Journal of Women in Culture and Society* 26(2): 415-440.

Flores, Juan (1993), *Divided borders: Essays on Puerto Rican identity*, Houston, TX: Arte Público Press.

Glasser, Ruth (1997), *Aquí me quedo: Puerto Ricans in Connecticut*, Connecticut: Connecticut Humanities Council.

Glick Schiller, Nina, Linda Basch and Cristina Szanton Blanc (1992), *Towards a transnational perspective on migration: Race class, ethnicity and nationalism reconsidered*, New York: New York Academy of Sciences.

Goldring, Luin (1992), 'Blurring borders: Community and social transformation in Mexico-US transnational migration', Paper presented at New Perspectives on Mexico-U.S. Migration Conference, University of Chicago, October 23-24.

——— (1998) 'The power of status in transnational social fields', in *Transnationalism from below*, Michael Peter Smith and Luis Eduardo Guarnizo (eds), New Brunswick: Transaction Publishers.

Grewal, Inderpal and Caren Kaplan (eds) (1994), *Scattered hegemonies: Postmodernity and transnational feminist practices*, Minneapolis: University of Minnesota Press.

Hernández-Álvarez, Jose (1967), *Return migration to Puerto Rico*, Berkeley: University of California Press.

Hernández Cruz, Juan (1985), 'Migración de retorno o circulación de obreros boricuas?', *Revista de Ciencias Sociales* xxiv: 81-109.

History Task Force (1979), *Labor migration under capitalism: The Puerto Rican experience*, New York: Monthly Review Press.

Hollinger, David A. (1993), 'How wide the circle of the "We"? American intellectuals and the problem of the ethnos since World War II', *American Historical Review* 98 (April): 317-37.

Hondagneu-Sotelo, Pierrette (1994), *Gendered transitions: Mexican experiences of immigration*, Berkeley: University of California Press.

Kaplan, Caren, Norma Alarcón, and Minoo Moallem (eds) (1999), *Between woman and nation: Nationalisms, transnational feminisms, and the state*, Durham, NC: Duke University Press.

Kerkhof, Erna (2001), 'The myth of the dumb Puerto Rican: Circular migration and language struggle in Puerto Rico', *New West Indian Guide*, 75 (3 & 4), 257-288.

Kibria, Nazli (1993), *Family tightrope: The changing lives of Vietnamese immigrant communities*, Princeton, NJ: Princeton University Press.

Lancaster, Roger (1992), *Life is hard: Machismo, danger, and the intimacy of power in Nicaragua*, Berkeley: University of California Press.

Levitt, Peggy (2001), *The transnational villagers*, Berkeley: California University Press.

López, Maria Milagros (1994), 'Post-work selves and entitlement "attitudes" in peripheral postindustrial Puerto Rico', *Social Text* 38 (Spring): 111-133.

Mahler, Sarah J. (1995), *American dreaming: Immigrant life on the margins*, Princeton, New Jersey: Princeton University Press.

Mahler, Sarah J. and Patricia R. Pessar (2001), 'Gendered geographies of power: Analyzing gender across transnational spaces', *Identities* 7(4): 441-459.

Meléndez, Edwin (1993), *Los que se van, los que regresan: Puerto Rican migration to and from the United States, 1982–1988*, Centro de Estudios Puertorriqueños, Political Economy Working Paper Series #1.

Menjívar, Cecilia (2000), *Fragmented ties: Salvadoran immigrant networks in America*, Berkeley: University of California Press.

Muschkin, Clara G. (1993), 'Consequences of return migrant status for employment in Puerto Rico', *International Migration Review* 27(1): 70–102.

Negrón-Muntaner, Frances (1997), 'English only jamás but Spanish only cuidado: Language and nationalism in contemporary Puerto Rico', in *Puerto Rican jam: Essays on culture and politics*, Frances Negrón-Muntaner and Ramón Grosfoguel (eds), Minnesota: University of Minnesota Press.

Olwig, Karen Fog and Nina Nyberg Sorenson (eds) (2001), *Work and migration: Life and livelihoods in a globalizing world*, New York: Routledge.

Ortiz, Vilma (1993), 'Circular migration and employment among Puerto Rican women', Special issue on Puerto Rican migration and poverty *Latino Studies Journal* 4 (2): 56-70.

Parreñas, Rhacel Salazar (2000), 'Migrant Filipina domestic workers and the international division of reproductive labor', *Gender & Society* 14: 560-80.

Pérez, Gina M. (2004), *The Near Northwest Side Story: Migration, displacement and Puerto Rican families*, Berkeley: University of California Press.

Ramos-Zayas, Ana Yolanda (2003), *National performances: The politics of class, race and place in Puerto Rican Chicago*, Chicago: University of Chicago Press.

Richardson, Bonham (1989), 'Caribbean migrants, 1838-1985', in *The modern Caribbean*, Franklin W. Knight and Colin A. Palmer (eds), Chapel Hill: University of North Carolina Press.

Rios, Palmira (1993), 'Export-oriented industrialization and the demand for female labor: Puerto Rican women in the manufacturing sector, 1952-1980', in *Colonial dilemma: Critical perspectives on contemporary Puerto Rico*, Edwin Meléndez and Edgardo Meléndez (eds), Boston: South End Press: 89-102.

Rivera-Batiz, Francisco L. and Carlos E. Santiago (1996), *Island paradox: Puerto Rico in the 1990s*, New York: Russell Sage Foundation.

Rodriguez, Clara (1993), 'Puerto Rican circular migration. Special issue on Puerto Rican migration and poverty', *Latino Studies Journal* 4(2): 93-113.

Rúa, Merida (2001), 'Colao Subjectivities: PortoMex and MexiRican Perspectives on Language and Identity', Theme issue, 'Puerto Ricans in Chicago', *CENTRO: Journal of the Center for Puerto Rican Studies* XIII(2): 117–133.

Safa, Helen (1995), *The myth of the male breadwinner: Women and industrialization in the Caribbean*, Boulder: Westview Press.

Sandoval Sánchez, Alberto (1997), 'Puerto Rican identity up in the air: Air migration, its cultural representations, and "cruzando el charco"', in *Puerto Rican jam: Essays on culture and politics*, Frances Negrón-Muntaner and Ramón Grosfoguel (eds), Minnesota: University of Minnesota Press: 189-208.

Segal, Aaron (1987), 'The Caribbean exodus in a global context: Comparative migration experiences', in *The Caribbean exodus*, Barry B. Levine (ed), New York: Praeger.

Smith, Michael Peter (2001), *Transnational urbanism: Locating globalization*, Malden, Massachusetts: Blackwell Publishers.

Smith, Michael Peter and Luis Eduardo Guarnizo (1998), *Transnationalism from below*, New Brunswick, NJ: Transaction Publishers.

Stinson Fernández, John H. (1994), 'Conceptualizing culture and ethnicity: Toward an Anthropology of Puerto Rican Philadelphia', Ph.D. diss., Department of Anthropology, Temple University.

Souza, Caridad (2002), 'Sexual Identities of Young Puerto Rican Mothers', *Diálogo*, Special issue on Latinas. Winter/Spring (6): 33-39.

Tienda, Marta and William Díaz (1987), 'Puerto Ricans' special problems', *New York Times*, August 28.

Torre, Carlos, Hugo Rodríguez Vecchini and William Burgos (eds) (1994), *The commuter nation: Perspectives on Puerto Rican migration*, Río Piedras, PR: Editorial de la Universidad de Puerto Rico.

Torruellas, Luz M. and Jose L. Vazquez (1982), *Los puertorriqueños que regresaron: un análisis de su participación laboral. [El movimiento migratorio de retorno en el périodo 1965-1970 y su impacto en el mercado laboral]*. San Juan, Puerto Rico: Centro de Investigaciones Sociales.

Urciuoli, Bonnie (1991), 'The political topography of Spanish and English: The view from a New York Puerto Rican neighborhood', *American Ethnologist* 18: 295-310.

Vargas-Ramos, Carlos (2000), 'The effects of return migration on political participation in Puerto Rico', Ph.D. diss., Department of Political Science, Columbia University.

Zavella, Patricia (1997), '"Playing with fire": The gendered construction of Chicana/Mexicana sexuality', in *The gender/sexuality reader: Culture, history, political economy*, Roger N. Lancaster and Micaela di Leonardo (eds), New York: Routledge.

Zentella, Ana Celia (1990), 'Returned migration, language and identity: Puerto Rican bilinguals in dos worlds/two mundos', *International Journal of Social Language* (84): 81-100.

Chapter 10

Landscaping Englishness: The Postcolonial Predicament of Returnees in Mandeville, Jamaica

Heather A. Horst

Returning residents[1] have captured the imagination of the Jamaican public over the past decade, particularly in the town of Mandeville now considered 'The Returning Resident Capitol of Jamaica'. Although exact numbers are difficult to determine due to the transient nature of 'returning resident' status and the fact that only one returnee must register per household for customs purposes, 20,085 individuals registered for returning resident status between 1993 and 2003 (Economic and Social Survey 2004, Ministry of Foreign Affairs and Trade 1997). While only 250 individuals registered with Mandeville's two Returning Resident Associations in the year of 2000, there are enough returning residents in Mandeville to form at least six active organisations.[2] Moreover, the 250 registered members does not count spouses or the returning residents who did not wish to formally register but often participate in the many activities organized by the association(s) (Personal Communication 2001). Certainly the large concrete homes dotting the landscape visually attest to the social and economic presence of returnees in Mandeville.[3]

While Mandeville has always been a choice retirement location for those Jamaicans and expatriates who could afford to purchase property in the town, it has not always been viewed as a sanctuary for returning residents. In fact, it was only with the relatively large remigration of individuals who lived in the United Kingdom (8,634 between 1993 and 2003) that the prominence of Mandeville increased, an association that Jamaicans presume reflects the town's uniquely British heritage (Economic and Social Survey 2004, Ministry of Foreign Affairs and Foreign Trade 1997). Indeed, many scholars note that there is something delightfully peculiar about the tendency of individuals who migrated to the United Kingdom (henceforth returnees) to return to the most English town in Jamaica (See Nettleford 1998). As Harry Goulbourne (1999, p. 164) describes the situation,

> '... the hill town of Mandeville has acquired the reputation of being a desirous destination for returnees who create a prosperous ghetto characterised by some English pastimes: tea in the afternoon, the cultivation and display of well manicured lawns and gardens ordered for more aesthetic pleasure than

> practical use, which stand in sharp contrast to the utilitarian
> kitchen and fruit gardens of rural Jamaica. Some would see an
> irony here because the town of Mandeville in the parish of
> Manchester, like Simla in the Himalayan foothills, used to be
> the retreat for British Administrators in the colonial past
> during the hottest months'.

As Goulbourne suggests, returnees choose Mandeville because they have been transformed by their experience in the United Kingdom (UK), becoming Englishmen and Englishwomen in words and action. In fact, many Jamaicans actually refer to returnees as 'the English'.

In this chapter, I consider how the idea of Mandeville as an 'English place' resonates with *returnees'* sense of being an 'English people'. Recognising place as both a geographical location as well as a particular location within a social hierarchy, I begin by tracing the production of Mandeville as an English place, describing Mandeville's origins as a British hill station and its' subsequent development. I then discuss how returnees, who are referred to as 'the English', interpret and resituate their own sense of being English as well as Jamaican. Research was based upon ethnographic fieldwork in Mandeville in 2000 and 2001 which included (but was not confined to) participation in Returning Resident Association meetings and functions, living in a returning resident neighbourhood as well as in-depth interviews with twenty returnees concerning their return and the material culture of home.

The English Heritage of Mandeville

Located 628 meters above sea level on the Manchester plateau, Mandeville and its surrounds remained isolated until the English and Spanish contest[4] over the island. H.P. Jacobs (1994) marks the Spanish arrival into the present boundaries of Manchester at Porus, located twelve kilometres from the centre of Mandeville, in 1656. Despite the early Spanish presence, Manchester remained relatively uninhabited until 1814 when the freehold landowners of the parishes of Vere, St. Elizabeth and Clarendon appealed to the House of Assembly for the creation of a new, centralised administrative centre. The measure (approved in Act 55 George III C 23) resulted in the establishment of Manchester, named after the governor of the island, the Duke of Manchester, Williams Montagu[5] (Brathwaite 1971). The parish remains the youngest parish in Jamaica.

At its outset, Manchester comprised 250 landed proprietors, 750 free people of colour and 15,000 slaves (Sibley n.d.). Two years later Mandeville, named for Montagu's eldest son Lord Mandeville, was appointed the capital. In order to purchase land in Mandeville, the vestry (six vestrymen and two magistrates) determined that a person should hold British Nationality and must own at least ten slaves or alternatively earn a salary of £160 per annum in 1819 (Grant 1946, p. 11). The next year the vestry raised the requirements to £200, or 20 slaves, a condition which continued to increase annually.

Once the capital was established, the vestry planned four buildings in the town: the courthouse, parsonage, Gaol/Workhouse and church. The courthouse, noted today as an historic monument, was completed in 1817. Based on a Georgian design and ornamented with Doric columns and a double staircase, slaves built the courthouse out of limestone bricks. The courthouse still stands today as a symbol of the law and order established under British rule. Across the village green (now a park named after the prominent mayor Cecil Charleton) stands St. Mark's Anglican Church. Completed in 1820, one of the church's more unique features is the lynch gate, 'a roofed gate in the churchyard under which the bier traditionally rests during the initial part of a funeral' (Bowen 1986). It also possesses a gleaming cross, constructed out of local alumina. The accompanying rectory, the first official house built in Mandeville, was rented out as a tavern by the first rector, the controversial Reverend George Wilson Bridges. In the mid-nineteenth century a number of English troops living at the garrison in the town centre were buried in the parish churchyard after a yellow fever outbreak.

Unlike the rest of the island which was dominated by large sugar (and later banana) estates, the parish of Manchester became known for the presence of small coffee plantations,[6] established after the prohibitions were lifted on coffee importations to Britain (Hall 1959, Braithwaite 1971, Jacobs 1994, Higman 1995). Due to the rocky limestone soil and the cooler climate enjoyed throughout Manchester, coffee thrived in the Carpenter Mountains, the May Day Mountains and in the area north of Mile Gully. Although Higman (1995) notes that scholars have argued that picking coffee remains a difficult task, most scholars agree that procuring coffee involved less intensive physical work as well as shorter days and seasons than the labour required on sugar estates. In addition, Higman (1995, p. 26-27) suggests that 'the organization of labour was less strictly regimented' and more flexible occupationally on the coffee estates, which in Manchester were often diversified with livestock used to power the coffee mills due to the 'very few streams and rivers present from which water power could be accessed' (Monteith 2002, p. 125). Nonetheless, coffee remained the primary source of income (Cf. Delle 1998).

The structure of coffee plantations resulted in two other distinct features of colonial Manchester. In contrast to the predominance of absenteeism on sugar estates, two-thirds of the proprietors lived on or near their coffee plantations. Shepherd (2002) attributes this pattern to the small size of the plantations as well as the expense of employing others to oversee the property, which made living abroad less feasible for the coffee proprietors who did not share the wealth, prestige or political influence of the absentee sugar estate owners who could afford to return to Europe. Shepherd further adds that the white non-sugar plantation and pen owners occupied a lower social position in Jamaica due to their residence on the island as well as their smaller land ownership. However, coffee plantation owners possessed greater status than pen owners due to their ability to export the coffee (Shepherd 2002, p. 159). This residence pattern resulted in closer supervision of the slaves by the estate owners and many of the owners married and brought their wives to the area. In addition, coffee plantations in Manchester (as well as Port Royale) primarily used young African slaves (i.e. slaves born in Africa and transported to

Jamaica), rather than the more established Creole population of slaves born on the island, which dominated the larger sugar estates by a two-thirds majority. Higman (1995, p. 77) suggests that the near majority of African slaves (49.2 per cent) present in the newly settled parish influenced the process of creolisation on the plantations, although the degree to which this influenced integration or division varied. For example, Brathwaite (1971) contends that the newly imported African slaves were more prone to rebellion than the Creole populations, who were often preferred due to their knowledge of the plantation system and previous contact with whites. In contrast, Jacobs (1994) suggests that the proprietors encouraged their African slaves to conduct themselves in an English manner by introducing marriage, European family patterns and participation in religious life. He further observes that in 1950 the parish of Manchester showed the third lowest illegitimacy rate on the island, behind the Kingston-St. Andrew Metropolitan area and the parish of St. Ann (Jacobs 1994). After emancipation in 1838, the large upheavals occurring across Jamaica between slaves and planters were relatively absent in Mandeville and Manchester,[7] and many of the freed slaves became independent farmers who grew coffee and other small crops.

During the Crown Colony government, Manchester developed citrus products and Mandeville continued to be a leisurely retreat for British officers, planters and wealthy Kingstonians. In 1875, the former officer's quarters and Mess Hall were transformed into the Waverley Hotel which, under the guise of the Brooks Hotel, became the centre of social life in Mandeville. Marshall's Pen, a nature preserve noted for its large number of endemic birds and animals, became a guest house alongside the Mandeville Hotel, the King Edward Hotel and the Newleighly Hotel. In addition, the Manchester Horticultural Society was founded in 1865 with 27 members. One of the oldest horticultural associations in the world, in 1927 the Society became affiliated with the Royal Horticultural Society of Great Britain. The Society remains very prominent in Mandeville today, hosting an annual flower show in May which is regularly attended by sellers, growers and buyers across the island. The Manchester Club also opened in 1895. The club's 9-hole, 18-par course, built to imitate St. Andrew's Golf Course in Scotland, is one of the oldest in the Caribbean. For many years, the course hosted tournaments that were well attended by wealthy Jamaicans and others visiting the island and they later added squash and tennis courts.

Presently over 161,000 people live in the parish of Manchester, with just under one-third of Manchester's residents residing in greater Mandeville. Mandeville continues to maintain its' reputation for civility, education and order. Alongside the more established schools, such as Manchester High School, Bel-Air Academy serves the expatriate community as well as the children of the town's wealthy elite. Bible colleges recruit young men and women to the area, as does Northern Caribbean University (NCU) which is the first University on the island located outside Kingston. Mandeville also enjoys its own dance teacher, music teacher and a bustling children's fine arts program. Local Jamaican artists' work appear in Bloomfield Great House, a renown steakhouse and former coffee plantation, as well as an annual art fair sponsored by the Catholic Centre. The charity event and opening night dinner feature Mandeville's most prominent citizens.

Corresponding with this English heritage is a narrative of wealth and status. Mandeville possesses a large population of highly educated citizens, many individuals with foreign degrees and honours. It also is the wealthiest parish on the island due to the development and influence of the bauxite industry.[8] As a result, the town offers a wide range of services, such as supermarkets, small grocers, pharmacies, hardware stores, florists, banks, insurance and investment services as well as local and foreign fast food chains and numerous shopping malls. Many of these shops, supermarkets and groceries have taken advantage of the presence of the 'new English' in Mandeville by importing lamb, British-style baked beans and special tea biscuits. In fact, one store stocks a range of Tesco (a British supermarket chain) products, including oil and bath soap.

Returnees in Mandeville

Although bauxite has added a North American flavour to Mandeville, the English heritage of Mandeville remains evident, particularly in the town's architecture and town planning. On the morning walks in *Ingleside* ('English side'), a local neighbourhood, returnees enjoy identifying the beauty of the large, modern homes alongside the vestiges of the English past, such as bricks and chimneys. Returning residents often remark upon the green hills of the surrounding countryside and delight in the fog of the winter mornings which remind them of their days in England. In fact, returnees' interpretation of the landscape is not unlike the British troops who transformed the area into a hill station and summer retreat in the 1800s because it reminded them of the green rolling hills of the English interior. But did returnees make the very profound decision to uproot themselves and disrupt their lives in England only to return to the very place they left? Did returnees move back to Jamaica to (re-)create a 'little England'?

Throughout my research, I talked at varying lengths with one-hundred returnees about their return and why they decided to move to Mandeville. For those returnees I interviewed in detail, three-quarters lived in the Greater London area for an average of 37 years, returning to Jamaica around 1996 which reflects the fact that returnees from the United Kingdom represented the largest single group of returnees between 1994 and 2001 (See Table 10.1). Eighty per cent of the individuals I interviewed and talked with returned to Jamaica as married couples in their mid-60s, although the number of widows and widowers continues to increase with time.

The motivation behind the decision to live in Mandeville could be broadly lumped into three, often related, categories. The first category was the possibility of a retirement lifestyle. Most returnees felt they worked very hard all their lives and deserved to retire. Jamaica represented a place that, unlike England, was warm year round making it possible to enjoy themselves. Because many returnees

Table 10.1   Returning residents in Jamaica, 1993-2003

Year	UK returnees	Total
1993	919	2493
1994	1145	2417
1995	1007	2353
1996	995	2349
1997	995	2094
1998	821	1875
1999	793	1765
2000	594	1282
2001	531	1177
2002	417	1113
2003	417	1167
Totals	**8634**	**20085**

Note: Figures reflect individuals who registered for returning resident status.

Source: Adapted from Ministry of Foreign Affairs and Foreign Trade figures reported in Social and Economic Survey 2004 and Ministry of Foreign Affairs and Trade Website 1997.

endure arthritis, a condition complicated by the cold English climate, Jamaica also improved their health and ability to stay active. For example, the Thompsons represent a typical example of people who made decisions to relocate in Mandeville based upon lifestyle. When they first started making preparations to return to Jamaica, the couple contemplated living in Hanover where Mrs Thompson's family originated. In Hanover, they envisioned a life at the sea side, enjoying the ocean breeze surrounded by extended family members. The couple also considered living in the hills of Kingston where they had easy access to shopping, cultural events and aspects of urban life they had become accustomed to in London. However, they disliked the need for extensive security systems and the traffic and pollution in Kingston itself. The Thompsons eventually decided upon Mandeville because it possessed cultural events, was only a short (two to three hour) drive from Kingston and had all of the modern conveniences such as healthcare, water and electricity, without the crime of Kingston. In addition, and after living so many years in England, they became accustomed to the climate and found that the north coast and Kingston made their hands and feet swell, an uncomfortable physical side effect of the heat which could potentially restrict their ability to enjoy life in Jamaica.

Proximity to family and the parish of birth helped to determine where other returnees relocated. Thomas-Hope (1992) notes that in the 1950s and 1960s residents living in the rural areas of Manchester emigrated in large numbers[9] and, not surprisingly, half of the individuals I interviewed at length were from the parish of Manchester, a pattern which Chevannes and Ricketts (1997) also observe in their study of returnees in Kingston and Jamaica's north coast. In fact, individuals who lived in Manchester as children referred to Mandeville as 'my town' and could tell childhood stories about visiting the large market or shopping over the Christmas season. Related to this familiarity, individuals also relocated to areas where their family house or property remained, although many were unable to build on family land due to inaccessibility and lack of facilities. Other returnees moved in or near their home parish and recreated a version of family land in Mandeville. For example, three siblings returning from England built homes on individual plots adjacent to each other while their other two siblings moved five minutes away on adjacent streets. While the direct proximity between the siblings remained somewhat unique, most returnees thought it ideal to live in the same town where they could easily visit their brothers, sisters, cousins, aunties and uncles.

Finally, returnees opted for Mandeville because they learned that there was a relatively large returning resident community. For example, Mr and Mrs Brown went back to their home parishes in eastern Jamaica only to discover that very few of their family or their friends were still living there. Mr Brown, who was more reluctant to return to Jamaica than his wife, worried that he would feel lonely and trapped in the parish of their birth. He also managed to convince his wife that Mandeville would be a better place to move because they could make friends with other returnees who had shared their experience of living in England. In addition, two of their close friends moved to Mandeville as they were contemplating their return. The ease of attaining land and a building contract while living in London solidified the Brown's choice.

The decision to return to Mandeville appears to be based upon familiarity with place and people as well as the ability to obtain a particular lifestyle, a lifestyle of leisure which reflects an English desire tending to home and garden in retirement (See King, et. al. 2000). Yet to my surprise what remained noticeably absent from the discussions of returnees' relocation to Mandeville was the English factor popularly touted as an explanation for their choice by scholars and other Jamaicans. In fact, when I directly asked returnees if they moved back to Mandeville because it was so much like England (or alternatively described Mandeville's noted English qualities), returnees appeared quite puzzled by the association between a decision to return to Jamaica and the idea that they were moving back to some version of England. Other returnees asserted that while certainly there were English aspects of Mandeville that remained, Mandeville simply was not England.[10] Noting the dramatic changes in the town over the last thirty to forty years, returnees contended that what remained were mere vestiges of English life captured in the chimneys and buildings of old Jamaica. While no one I interviewed wanted to live anywhere else in Jamaica and enjoyed Mandeville's modern conveniences, they mentioned that Mandeville was not as 'nice' or 'cool' as it used to be due to the concrete buildings, reduction in trees, increased traffic and poor roads. At least half of the returnees I talked to claimed that the bauxite industry had not helped the situation and caused environmental damage to the area which affected the water and air quality in the town. One-third of the returnees went so far as to say that Mandeville has become a 'little America' with all of the shopping malls and fast food chains, the quaint country town destroyed by modern consumerism.

In addition to critiquing the notion of English Mandeville, returnees consistently denied that the idea that Englishness played any role in their decision to return to the town. Many returnees were even offended by the suggestion that they were at all English. Furthermore, and although they admitted to being somewhat amused in the beginning, returnees complained about how 'locals' insisted upon calling them 'English', thinking it ridiculous that an English accent or drinking tea changed who they felt they were. But if Mandeville is not English and calling returnees 'the English' causes offence, what does it mean to be English? In the following section, I trace how the concept of Englishness has been transformed over the past 40 years, in order to discern how returnees view Englishness and, by extension, the importance of Mandeville's English heritage today.

Shades of Englishness

When most returnees left for England, a white bias permeated Jamaican society, particularly within the aspirant classes. Henriques (1953) describes the valuation of individuals considered 'fair', a term associated with light skin colour, European features and straight hair (See Hoetink 1985). Variations upon the dichotomy of *good* and *bad* (or 'European' and 'African') physical characteristics, such as European features and kinky African hair or an individual with 'good' hair but with a dark, African complexion were prevalent. As Chevannes (1995) characterises the situation before the introduction of the Black Power Movement in Jamaica in 1968:

> 'Ideologically speaking, to this day, hair in Jamaica is either *good* or *bad*. *Good* hair is described as *pretty*, not soft or fine, but *pretty*, or sometimes *nice and straight*. *Bad* hair is *nati-nati* (knotty). These ideologically laden terms are often used purely descriptively, and do not necessarily reflect the outlook of the user ... The matter did not end at the level of values. Society produced for the grooming not of "bad" hair but of "good" hair. The purveyors of hair pomades and oils assumed that everybody either had or wanted to have "good" hair, and the combs manufactured or imported into Jamaica were designed for grooming only "good" hair'.
> (pp. 105-106, *author's emphasis*)

Women's internalisation of this valuation of European images of beauty was made particularly evident through the painful process of straightening the hair with the application of hot oil, creams and use of a steel brush as well as lightening the skin with creams and face powders, in order to 'lift up' themselves within society through the erasure of colour and texture on their skin and bodies. Henriques (1953) notes that fair women were particularly concerned with any permutations in their skin colour, such that women of fair complexion might avoid the sun for fear of darkening. This white bias extended to the practices involved in social reproduction, such as marriage and education, wherein many men sought fairer women for marriage. In addition, families with restricted budgets granted children of lighter colour greater access to education and other avenues than their darker-skinned children. As Austin-Broos (1997, pp. 150) argues, in Jamaica 'culture, class and race do not merely coincide. They merge as phenotype is rendered through culture; inheritance made potent through environment and experienced inscriptions on the body'. To think of oneself as English was therefore one step closer to being white.

There is clear evidence of scepticism concerning the acceptance of the colour-continuum and the valuation of whiteness, such as through the adult suffrage movement which fought to give black Jamaicans the right to vote and self-govern and movements led by Marcus Garvey, Claudius Henry and the Rastafarians in the first half of the twentieth century (Chevannes 1994, 1995). Yet, many returnees I spoke with admitted that when they travelled to the 'Mother Country', at some level they carried these positive views of whiteness (See Deakin 1970). Seeing white people who were poor, living on the streets and begging jarred many returnees' worldview when they first arrived in England. They never imagined that they would be sitting next to a white person while working in a factory. These images were not what returnees learned about and not how they were taught to view the English.

Most returnees' education about being English and the English way of life came through the work environment and on the street. Nurses (which comprised one-quarter of the people I interviewed) offered particularly poignant stories about how, for instance, a patient with a stab wound who was bleeding profusely protested and refused treatment when he saw a black nurse coming to attend to him. The man almost died before the doctor told him that he had no choice but to deal with her because a white nurse was unavailable. Even in nursing school when the UK was desperate for their services, female returnees reported being made to share the smallest boarding rooms, work the worst hours and carry out the dirtiest duties, such as bedpans. They endured verbal abuse not only from their fellow students but also their supervisors who sometimes made jokes about how the black nurses needed to use *Sno Mo*, a bleaching detergent, to become clean enough to touch the patients. Even the high achievers and individuals of more privileged classes acknowledged that they were often passed over for promotions. One returnee noted that she received the highest marks in her nursing class, an honour which normally accorded the student a prestigious award. The year that she won, the award was mysteriously cancelled.

Men characterised their early years in England as filled with tension and antagonism. A man living in one of the neighbourhoods I frequented sat down one evening and told all the neighbours about his life in England as 'the fighter'. In the early years, he was dating an English woman from a working class English family. No one liked it. Her family did not like the fact he was black and his family thought that he was in for trouble trying to date a white woman, particularly if he had any intention of marrying her or starting a family. Even when they went out on the street he was accosted by strangers who launched racial slurs at the couple and tried to coax his girlfriend into leaving him. One night he recalled dropping his girlfriend off at her door and being beat by white youths as he made his way back to the bus stop, simply for the audacity of dating a white woman. He managed to give them 'a few good licks' and ran home, but he also said that every time he went on the street as a young man in England, he took on the role of 'the fighter'. As Winston James (1993) argues, what is most significant about the daily encounters with whites (particularly working class whites) throughout 1950s and 1960s was the ways in which West Indian conceptions of colour and race changed, resulting in disillusionment with the esteem accorded to whites and conceptions of whiteness learned through their socialisation in the Caribbean.

In the 1970s, returnees' attention shifted to the plight of their children growing up and making their way through the British education system in an increasingly conservative climate (typified by Enoch Powell's infamous 'Rivers of Blood' speech in 1968). The growth of the conservative movement, which sought to restrict the development of the black population, was compounded by massive unemployment across the country. Gilroy (1987) argues that this environment fostered the growth of a new form of racism which reflects a transformation from a biological view of race and nation to a cultural approach to nationhood. As Gilroy (1987, p. 60) describes:

> 'Alien (i.e. black) cultures have been introduced into this country with disastrous effect: "the indigenous population perceives its own predicament as that of physical pressure and attack". The increased competition for limited resources and the variety of disruptive behaviours introduced by the immigrant population create problems for the national community. The most profound difficulties are uncovered by trying to dilute our nationhood and national culture so that they can accommodate alien interlopers and their formally but not substantively British children'.

Mrs Taylor, who returned to Jamaica in 1994, often describes the 1970s and raising her four children as the most difficult time of her life, to the extent that the family contemplated moving back to Jamaica or re-migrating to the United States. Mrs Taylor spent five nights per week working as a nurse while her husband came home every evening, cleaned up from dinner and put their children to bed. Each morning Mrs Taylor came home from work, prepared breakfast and sent the children to school. However, one of her sons had a learning disability. Mrs Taylor spent a days travelling back and forth between home and school trying to fight for her son's right to learn. In the end she felt that the schools underestimated his potential and sent him to a school for the 'dumb' because he had a learning disability and was black. Twenty-five years later, she still wonders what would have happened if she had sent him back to Jamaica, but realistically Mr and Mrs Taylor never had the option because 'the money wasn't there'. All she was left to do was pray, but because she was so busy raising her children she never even had a chance to attend church.

Returnees also felt that the 1970s and early 1980s were difficult for their children who, although born in England, were denied full access to their rights as British citizens. Unlike their parents, the second generation lacked the attachment to their West Indian heritage that played a significant role in their well-meaning parent's negotiation and tolerance of racism (Gilroy 1987). Mrs Grant, for example, could recall the disappointment and shame the first time she received a call from the police saying that her son (normally a 'good boy') had a run-in after attending a club and Mrs Tulloch remembers constantly worrying about her son getting 'mixed up' with the wrong crowd, the ones involved in ganja smoking. Cashmore (1995) argues that during this time Rastafarianism became a particularly cogent symbol of resistance for disenfranchised British black youths.[11] Coupled with the global popularity of Bob Marley, he contends that while the US-based Black Power movement which 'in its overtly political form might have captured a few minds ... Rastafari, by contrast, phrased its critique in a religious form' (Cashmore 1995, pp. 184-185). Rastafarianism explicitly addressed the ideological use of Christianity by white European culture, rejecting the concept that one must endure through hardship because salvation will come in the afterlife. In terms of racial consciousness, particularly the valuation of blackness, Rastafarianism turned away from the painful use of the comb and, similar to the development of the Afro in the Black Power movement, allowed the hair to grow naturally into dreadlocks. Moreover, the features, colour and texture of hair associated with blackness were recognised as beautiful rather than something to be erased or hidden (Chevannes 1995).

Getting Your Colour Back

The new experiences and developments within the UK's West Indian community prompted many returnees to interrogate their assumptions about the meaning of being black. Mrs Clarke, a 50 year-old returnee reflecting on the change in racial consciousness, told me how shocked she was to hear one of the older returnees she went to church with lamenting over her 'brown' grandson's choice of a 'black' wife, openly questioning how the woman could make such a statement after the years of mistreatment due merely to skin colour in England. For Mrs Clarke, there was a general acknowledgement that blackness and being black could be positive, reversing (or at the very least altering) the 'white bias' of the 1950s and 1960s. As Goulbourne (1998) describes the situation:

> 'The paradox is that black and brown people often share the
> same culture with whites who determine the nature of
> incorporation, but the racial affinity is nearly always stronger

than cultural ties … race relations pivoted upon the divide of
the colour-line remains relevant in Britain' (pp. 152-153).

Clearly, the experience of racism in England transformed the Jamaican returnees'
conceptions of what it means to be English, or more specifically for the post-WWII
generation migrants, what it means *not* to be English. Although most returnees lived
over half of their lives in England, they continued to be asked where they were from.
If the returnee responded like the English and mentioned their locality, such as East
London or Croydon, they were met with the response 'no, where are you *really*
from'. For this generation, however, where 'yu barn an' gro' shapes one's identity
and sense of self. For this reason, a question concerning their origins did not disturb
returnees, particularly as their plans to return concretized.[12] Many returnees
acknowledged that these issues would be much more difficult for their children,
some returnees advising their sons and daughters that 'the English can be funny' and
that England remains 'a strange country in as much as you were born there', one
more indication that returnees did not feel that they were English.

In addition to altering returnees' ideas of Englishness, the experience of racism
in England also changed the meaning of being Jamaican. For example, when Mrs
Cole, Sister B and I were talking about a recent visit to England, Sister B looked up
and flung her arms out to embrace the sun, proclaiming how wonderful it was 'to
be back in the Jamaican sun and get my colour back'. Hearing Sister B's comment,
Mrs Cole started chuckling about how fair she had become sitting in her job in a
factory in England, so fair that every time her English workmates returned from
holiday they felt compelled to compare their new tans with her sun-deprived brown
skin. Noting how ironic it was that her white work mates were so keen to brown
their skin, but were reticent about accepting brown or black people, Mrs Cole
acknowledged that she too felt, and indeed welcomed, the idea that she regained
her colour once she returned to Jamaica (they often described their skin colour in
England as gray or 'ash-y'). Mrs Cole and Sister B's discussion reveals none of
the sun avoidance and bleaching measures described by Henriques and further
suggests that pigmentation change, particularly turning darker appears to indicate
for many returnees that they are not only back on the island but also *belong* in
Jamaica. There is therefore a curious symmetry with the arrival in England turning
Jamaicans into blacks based on colour and not background and the re-arrival in
Jamaica turning them into 'the English' but which returnees deny because being
English now represents a colour category. Calling returnees 'the English' negates
this physical, mental and emotional transformation.

Conclusion

Mandeville has long claimed the title of 'the most English' place in Jamaica.
Noting the prevalence of 'fair', 'coloured' or 'upper class' residents living in
Mandeville, Fernando Henriques (1953) described Mandeville as 'Jamaica's
Cheltenham' and historian H.P. Jacobs (1994, p. 1) observed that 'Manchester was
the only area in which the English ever came anywhere near to achieving their

original idea of establishing a tropical New England'. Like the British Virgin Islands which accentuates the English heritage of the islands to lure investors into their offshore financial centre (Maurer 1997), Jamaicans emphasise Mandeville's clean, orderly streets and cool climate. Moreover, they often translate the orderliness of the town into the character of its people, with Mandeville's residents being noted for their law-abiding nature and cool, but civilised detachment, thus sustaining the myth of Mandeville's Englishness in the face of change.

Returnees, on the other hand, associate Mandeville's importance with the climate, sense of community as well as economic prosperity and status. Living in the salubrious hills of Mandeville signals achievement and a feeling of accomplishment, and returnees are keen to realise a lifestyle associated with Mandeville, particularly by maintaining a proper house as well as participating in the community through church, voluntary and charitable associations. In other words, and in contrast to earlier suggestions that returnees move to Mandeville because 'it is England' (Nettleford 1998, pp. 396-397), returnees no longer equate Mandeville with Englishness. In fact, what becomes evident is that, unlike many of the other towns where returnees might resettle, the association of Mandeville with England exacerbates rather than resolves their desire to return and be recognised as Jamaicans. In moving back to Jamaica, returnees want to experience the positive aspects of Jamaican culture which they remember and imagine while abroad, aspects of Jamaican culture which were ignored or denied in British depictions of Jamaican culture (Gilroy 1987, Skelton 1998, Small 1998). Carrying the material and cultural symbols of success resulting from their migration, returnees arrive in Jamaica only to find that Jamaicans who stayed behind possess the authority to define who they are which effectively leaves the returnees in the position of being cultural outsiders in the very culture they identified with and attempted to retain while living abroad.

Notes

1. 'Returning resident' is a government category for a person who leaves Jamaica for five years and claims the title for customs purposes upon moving back to the island. In this chapter, I use 'returning residents', 'returned migrants' and 'returnees' interchangeably to refer to individuals who migrated to the United Kingdom in the 1950s and 1960s and remained there in the 1990s when they opt to retire in Jamaica, unless otherwise specified.

2. The two associations are the Central Manchester Returning Resident Association, which was established in 1999, and the Mandeville Returning Resident Association, established in 1989.

3. I estimate that there are at least 2,000 returnee households in Mandeville based on interviews with returnees in various communities around Mandeville. Watson-Thomas (1996) noted that in some of the new subdivisions in Mandeville 94 per cent of the residents were returnees, a pattern I also observed but was not equipped to verify.

4. Before the Spanish landed in Jamaica in 1494, the island was inhabited by Arawak Indians who arrived from South America between AD 600 and AD 900 (Sherlock and Bennett 1998).

5. Montagu was a governor of Jamaica between 1808 and 1827.
6. For a comparison of sugar plantations, see Austin-Broos 1997 and Harrison 2001.
7. There is some controversy over the extent to which the large-scale rebellions were experienced throughout Mandeville and Manchester. Hall (1959) argues that the Moravian churches of Manchester actually initiated what later became known as the Baptist Revival. While Hall presents the spread of the revival throughout the island as reasonable, he is perplexed by its Manchester origins. As Hall queries, 'It is difficult to explain why the movement began where it did. Manchester contained no sugar estates. It was a relatively prosperous parish of small settlers. It is unlikely that the religious feeling arose as a reaction to any peculiar economic distress. In part, it may be described as a desire to break away from the rather hum-drum routine of daily labour. Social amenities and recreational facilities were limited. The chapel was usually the social centre. An intense emotional appeal by the local preacher might well sway a congregation, and the response, as well as the appeal, might prove infectious' (p 237).
8. Bauxite strongly redefined the dynamics of life in Mandeville after WWII. The bauxite industry commenced in 1942 when Mr. R. F. Inncs, the Jamaican Government Senior Agricultural Chemist, informed the government of the rich alumina content of the soil. Alcan Aluminium Ltd (Montreal) subsequently surveyed the soil on behalf of the Jamaican government. After the content of the soil was formally confirmed, Alcan carried out a second survey shipping 2,500 tonnes of bauxite ore to North America in October of 1943. When the Jamaican government decided to commercially develop the land, they started building a site at Kirkvine, just outside of Mandeville, in 1950 under the name of Jamaica Bauxite Limited, which was a fully owned subsidiary of Alumina Canada Limited. By March 1952 the company began commercially mining the land, changing the name of the company to Alumina Jamaica Limited or Alcan, as it is known today. In December of 1952, the first alumina was produced and shipped in January 1953 (Austin 1975). Alcan's Kirkvine plant supplied a significant volume of bauxite and alumina for the United States, whose demand for alumina products stemmed from extensive military involvement in Korea. By 1957 Jamaica was the largest supplier of bauxite to the United States and a number of local men were employed in the initial phases of plant construction; peak employment in the industry occurred between the years of 1969 and 1971 (Kaufman 1985, p 27). In addition to employment within the bauxite industry, a number of Jamaicans moved from rural towns to the Mandeville area to obtain employment in the new service-oriented industries. They also improved the conditions of roads in greater Mandeville and participated in the Land Exchange Program, relocating families to company owned and, in some instances, built housing in exchange for their bauxite-rich land. This movement was also encouraged by the Jamaican government which mandated that the bauxite companies remove the topsoil, mine the bauxite underneath and replace the topsoil when finished. The aim of this law was, at minimum, to maintain the virility and fertility of the soil. Alcan's Kirkvine operation centre planted various crops, attempted to improve the water supply, maintained cattle for beef and dairy products as well as produced and sold the ortanique, a unique mix of the orange and tangerine discovered in the 1920s.
9. Peach (1968) notes that 11.2 per cent of Jamaica's population migrated to the UK between 1956 and 1960.
10. This is not to suggest that returnees were not equally critical of Jamaica. Many returnees lamented about the crime and violence in Jamaica and the 'indiscipline' of Jamaicans generally. Some returnees even suggested that Jamaica was 'better off when the queen mum ruled Jamaica'. Their disappointment and the idea that they were moving back to live with other returnees could suggest that returnees wanted to return

to a little England. However, they did not wish to live with the wider English population. They wanted to live with people who shared their experiences and perspectives.

11. Sutton and Makiesky-Barrow (1987, p. 96) attribute transformations in racial consciousness in Britain during this era to the 'rhetoric and symbols…[of] the Black Power movement in the United States' and state that 'Black American forms have been transplanted and reworked in a new environment' and 'provid[e] a ready-made vehicle for expressions of Black consciousness'.

12. However, James (1993) observes that for the elderly who could not return, the fact that they would never be seen as belonging, or as being English, was quite painful (See also Byron 1999).

References

Austin, D.J. (1975), 'Jamaican Bauxite: A Case Study in Multi-national Investment', *Australian and New Zealand Journal of Sociology*, 11, 53-59.

Austin-Broos, D.J. (1997), *Jamaica Genesis: Religion and the Politics of Moral Orders*, Chicago: University of Chicago Press.

Bowen, G. (1986), 'An Intriguing 170 years', *The Jamaica Gleaner*, Sunday, November 16, 1986, 11A.

Brathwaite, E. (1971), *The Development of Creole Society in Jamaica, 1770-1820*, Oxford: Clarendon Press.

Byron, M. (1999), 'The Caribbean-born Population in 1990s Britain: Who Will Return?', *Journal of Ethnic and Migration Studies*, 25, 285-302.

Cashmore, E. (1995), 'The De-Labelling Process: From "Lost Tribe" to "Ethnic Group"', in *Rastafari and Other African-Caribbean Worldviews*, (ed) B. Chevannes, London: Macmillan Press in association with ISS, The Hague, 182-195.

Chevannes, B. (1994), *Rastafari: Roots and Ideology*, Syracuse, NY: Syracuse University Press.

Chevannes, B. (ed) (1995), *Rastafari and Other African-Caribbean Worldviews*, London: Macmillan Press in association with the Institute of Social Studies, The Hague.

Chevannes, B. and Ricketts, H. (1997), 'Return Migration and Small Business Development in Jamaica', in *Caribbean Circuits* (Ed, Pessar, P.zR.) Center for Migration Studies, NY, 161-196.

Deakin, N. (1970), *Colour and Citizenship in British Society*, London: Panther Books.

Delle, J.A. (1998), *An Archaeology of Social Space: Analyzing Coffee Plantations in Jamaica's Blue Mountains*, New York and London: Plenum Press.

Expo Central Supplement (1995), 'A Look at Manchester', *Mandeville Weekly*, Tuesday, February 28, 1995, 6-7.

Gilroy, P. (1987), *'There Ain't No Black in the Union Jack': The Cultural Politics of Race and Nation*, Chicago: University of Chicago Press.

Goulbourne, H. (1998), *Race Relations in Britain Since 1945*, Social History in Perspective, New York: St. Martin's Press.

Goulbourne, H. (1999), 'Exodus? Some Social and Policy Implications of Return Migration from the UK to the Commonwealth Caribbean in the 1990s', *Policy Studies* 20, 157-172.

Grant, J.T.G. (1946), 'Address on the Early History of Manchester (1816-1838)', Mandeville, Jamaica.

Hall, D. (1959), *Free Jamaica, 1838-1865: An Economic History*, New Haven: Yale University Press.

Harrison, M. (2001), *King Sugar: Jamaica, the Caribbean and the World Sugar Economy*, London: Latin American Bureau.

Henriques, F. (1953), *Family and Colour in Jamaica*,London: Eyre & Spottiswoode.

Higman, B.W. (1984), *Slave Populations of the British Caribbean, 1807-1834*, Baltimore: Johns Hopkins Press.

Higman, B.W. (1995), *Slave Population and Economy in Jamaica, 1807-1834*, Kingston: University of the West Indies Press.

Hoetink, H. (1985), '"Race" and Colour in the Caribbean', in *Caribbean Contours* (eds) S.W. Mintz and S. Price. Baltimore and London: Johns Hopkins University Press, 55-84.

Jacobs, H.P. (1994), 'Manchester – Roots and Branches', *Mandeville Weekly*, June 2, 1994, 8.

James, W. (1993), 'Migration, Racism and Identity Formation: The Caribbean Experience in Britain', in *Inside Babylon: The Caribbean Diaspora in Britain* (eds) W. James and C. Harris, London and New York: Verso Press, 231-288.

Kaufman, M. (1985), *Jamaica Under Manley: Dilemmas of Socialism and Democracy*, London: Zed Books.

King, R., T. Warnes and A. Williams (2000), *Sunset Lives: British Retirement Migration to the Mediterranean*, Oxford and New York: Berg Press.

Maurer, B. (1997), *Recharting the Caribbean: Land, Law and Citizenship in the British Virgin Islands*, Ann Arbor: University of Michigan Press.

Ministry of Foreign Affairs and Trade (1997), 'Number of Returning Residents and Sources', http://www.mfaft.gov.jm/jod/number_of_returning_residents.htm. (May 30, 2003).

Monteith, K.E.A. (2002), 'Planting and Processing Techniques on Jamaican Coffee Plantations during Slavery', in *Working Slavery, Pricing Freedom*, (ed) V. Shepherd. Kingston and Oxford: Ian Randle and James Currey Publishers, 112-132.

Nettleford, Rex (1998), Interview with Professor Rex Nettleford, in 'The Irrisistible Rise', Phillips, M. and T. Phillips, *Windrush: The Irrisistible Rise of Multi-Racial Britain*, London: Harper Collins, 391-402.

Peach, C. (1968), *West Indian Migration to Britain: A Social Geography*, London: Oxford University Press.

Personal Communication (2000), Interview with President of the Central Manchester Returning Resident Association, May 20, 2000.

Planning Institute of Jamaica (2004), *Economic and Social Survey Jamaica 2003*, Kingston: The Planning Institute of Jamaica.

Planning Institute of Jamaica (2001), *Economic and Social Survey Jamaica 2000*, Kingston: The Planning Institute of Jamaica.

Shepherd, V. (2002), 'Land, Labour and Social Status: Non-Sugar Producers in Jamaica', in *Working Slavery, Pricing Freedom* (ed) V. Shepherd, Kingston and Oxford: Ian Randle and James Currey Publishers, 153-180.

Sherlock, P. and H. Bennett (1998), *The Story of the Jamaican People*, Kingston: Ian Randle Publishers.

Sibley, I.K. (no date), *The History of the Parish of Manchester*, Manchester Public Library.

Skelton, T. (1998), 'Doing Violence/Doing Harm: British Media Representations of Jamaican Yardies', *Small Axe* 2(1):27-48.

Small, G. (1998), 'Do they mean us? A Reflection on the Making of the Yardie Myth in Britain', *Small Axe* 2(1): 13-25.

Sutton, C.R. and S.R. Makiesky-Barrow (1987), 'Migration and West Indian Racial and Ethnic Consciousness', in *Caribbean Life in New York City: Sociocultural Dimensions* (eds) C.R. Sutton and E.M. Chaney, New York: Center for Migration Studies of New York, 86-107.

Thomas-Hope, E. (1992), *Explanation in Caribbean Migration*, London: Macmillan.

Watson-Thomas, Sheila (1996), 'Coming Home to Mandeville', *Jamaica Herald Mandeville Supplement*, December 1, 1996, 3-4.

Chapter 11

Transatlantic French Caribbean Connections: Return Migration in the Context of Increasing Circulation between France and the Islands

Stephanie Condon

French Caribbean Migrations, as Part of the Wider Caribbean Migration Experience

As citizens of the French Republic, Martinicans and Guadeloupeans who move to continental France are considered internal migrants.[1] It is principally for this reason that academic research on immigration to France has rarely included Caribbean migration. At the same time, studies of internal migration have focussed solely on movements within mainland France. Yet, as specific studies have shown, the dynamics of emigration from the French Caribbean displays similarities with both internal movements and post-colonial immigration (Domenach and Picouet, 1992). Furthermore, several authors have stressed the relevance of this migration within the history of post-1945 labour migration to France (Anselin, 1979; Constant, 1987; Condon and Ogden, 1991). This national context within which the movements have taken place – and continue today – goes some way to explain differences in relation to Caribbean migration to Britain or the Netherlands (Brock; 1987; Cross and Entzinger, 1988), although numerous similarities result from the broader socio-cultural context of emigration from this region. Although the aim of this chapter is not to explicitly compare aspects of migration between the French Caribbean islands and mainland France – this latter referred to as the metropole- those which differ strikingly from the British Commonwealth case will be highlighted. The heuristic value of comparing French Caribbean migrations with others in the region is undeniable (Brock, 1987; Byron and Condon, 1996; Peach, 1991), as other comparisons have shown (Foner, 1979; Richmond, 1987).

The migration cycle – emigration and return – has been shown to be increasingly inappropriate for understanding much of Caribbean migration, as movements between the metropoles and the Caribbean are frequent and of a varied nature (Thomas-Hope, 1986). This is certainly true in the French case. The purpose of this chapter is to discuss the extent to which thousands of people born in the islands of

Guadeloupe and Martinique have emigrated to France, yet live out their lives in a transatlantic space (Condon, 1996). Their multiple attachments and circulation of individuals between the islands and metropolitan France bear similarities to the dynamics of the much discussed Caribbean transnational experience (Basch et al., 1994; Goulbourne and Chamberlain, 2001; Grasmuck and Pessar, 1991; Olwig, 2003; Sutton, 1987). Given the absence of national boundaries in the French West Indian case, one could hypothesise a greater intensity and fluidity of mobility, return and circulation, as well as of information flows. Nonetheless, various obstacles to movement do exist and a host of factors related to gender, socio-economic position, life course stage, family attachments and obligations, health and personal biographies come into play; refuting simple generalizations and complicating explanations of the contemporary situation.

Migration to the Metropole

A Post-World War II, Labour Migration

It was during the 1950s and 1960s that most of today's Caribbean retirees migrated to metropolitan France. Despite not crossing a national boundary and thus not being considered foreign immigrant workers, Caribbean migrants in many respects played a key part in post-war labour migration to France. Several aspects of their integration into metropolitan society over the subsequent two decades (low skilled jobs, minimal training, housing segregation, discrimination) indicate a similar status to migrants in contemporary labour movements to other Western European destinations. Moreover, in contrast to the British case, the launching of central state policy to encourage emigration from the islands further emphasised the national uniqueness of this particular transatlantic, migration system. This 'encouragement' policy, put into effect in 1961, constituted a major strand of state policy towards the former island colonies, which had been integrated into the nation state as full administrative regions in 1946. Post-war governments' consistent policies have been central both to defining the constitutional status of the islands and encouraging migration flows, as a specific solution to labour shortage in metropolitan France's public sector (where French nationality was required) and to the perceived potential for demographic and political explosion in the islands (Condon and Ogden, 1991a). A crucial element in economic and political planning came in 1962, with the creation of the state agency Bumidom (*Bureau pour les migrations intéressant les départements d'outre-mer*), which was responsible for the rapid acceleration of out-migration. French state policy equally encouraged emigration of women and men of working age, with gender specificities with regard to employment. The policy included a family reunification process. A further strand of policy was to encourage the emigration of young women of child-bearing age, in order to permanently remove future mothers from these 'overpopulated' islands (Condon, 2004). Explicit in the discourse of this 'encouragement to migrate' was the anticipated goal of permanent settlement of migrants in the metropole.

The movement that took off from the early 1960s bore the socio-demographic traits of a labour migration – young, working age, low-skilled individuals (Condon and Ogden, 1991). In the early years, as in the decade prior to the state's orchestration of the migration, there was a strong presence of married couples and young families; sometimes, with the man leaving the islands first to be joined later by the spouse or family. Military service too had an important influence on flows, on the proportion of men in the movement and on the characteristics of their settlement in France. As time went on, there were an increasing number of single persons in the emigration stream. Many women, including young single mothers, initially left children behind with their mother or parents. Divorced or separated women – often wishing to distance themselves from their ex-partner (Condon, 2004) – and widows – often grandmothers joining a daughter or son and their family – were more numerous from the 1970s onward.

There were a host of life course situations within which migration plans and trajectories could vary substantially. Whilst many among the earlier generations of emigrant islanders talked of an original plan to only stay four or five years, others had vaguer plans and some planned to leave the islands for the long-term. For many who had hoped to return to the islands after a few years, plans changed because of distance, cost of transport, low wages, commitments to sending remittances, and their stay was prolonged. As discussed below, the migration project – and its temporal and spatial dimensions – has changed in a number of ways since the 1960s.

The Establishment of a Caribbean Community in France

The Caribbean-born population living in metropolitan France grew rapidly from the 1950s. There was a doubling in numbers at each census until the turn of the 1980s (Table 11.1). The onset of economic crisis in the metropole, combined with the halting of the organised migration, however, led to a slowing of the movement; particularly from the 1980s.

Table 11.1   Growth in the Caribbean-born population resident in metropolitan France, 1954-1999

Census year	Caribbean-born
1954	15 620
1962	37 591
1968	61 160
1975	115 465
1982	182 728
1990	211 550
1999	211 140

Source: French census, 1954-1999.

An increasing population of descendants of Caribbean migrants, who were born in metropolitan France was evident by the 1982 census; at this time, 89,068 such descendants[2] were recorded. By 1999, over 100 000 such people were living in mainland France. At the same time as this rapid growth in numbers, there has been a considerable geographical concentration of this population in the Paris region, where roughly three-quarters of the Caribbean population live. Housing market forces – combined with the nature of the organised migration and massive recruitment into the public services – led to a concentration in particular sectors and areas, notably suburban public housing estates (Condon and Ogden, 1993). Similarly, from the outset, there has been employment concentration in a few occupational sectors, and grouping of migrants in certain workplace locations (hospitals, post offices etc.).

The economic crisis led to a series of employment restrictions and institutional changes. State policy regarding emigration from the Caribbean changed direction and the island populations were widely informed of unemployment and reductions in public service employment. Although effectively reducing flows, young men and women continued to emigrate. However, migrants arriving in metropolitan France during the early 1980s found access to public employment much more limited than previously (Constant, 1987, Condon and Ogden, 1991). Cutbacks in the public sector affected men more than women: amongst arrivals between 1982 and 1990, less than a third of women and fewer men gained posts in this sector (Marie, 1996). At the same time, unemployment levels throughout metropolitan France were increasing dramatically.

Census data in the 1990s provides evidence that the young and women were more negatively affected than men. For example, at the 1999 census, in the 25-29 age group, 24 per cent of women on the labour market and 19 per cent of men were unemployed, average rates for the whole migrant population being 13 per cent for men and 14 per cent for women. Whilst a certain proportion of these young unemployed are recent arrivals, many are in fact former child migrants who have spent several years in the metropolitan education system. In addition to the reduction in recruitment in the public sector, young people seeking work are confronted with stereotypes relating to the occupational position of Caribbean migrants in France and are expected to seek work in certain sectors; for example, for women, childcare and hospital work. This specific 'hybrid form' of discrimination combines with others, notably those based on gender or age, and with forms of racism experienced by descendants of some foreign migrant groups, to severely disadvantage the access to employment of Caribbean migrants.

From 1980, Diminishing Emigration and Increasing Circulation

The change in status of islands from colonies to *départements d'outre mer* (administrative regions of the French state) led to a gradual modernisation of their infrastructure, housing and health care, and a rather slower introduction of social welfare benefits. However, provision of further education and vocational training programs in the islands remained insufficient. So, whilst outflows have decreased substantially since the 1970s, migration to France has remained an attractive alternative to looking for employment in the islands. Unemployment trends as well as family or individual ambitions have encouraged increasing numbers of young people to further their education, both in the Caribbean and in metropolitan France. It is very common for Caribbean-born students, and increasingly women, to emigrate to attend universities in Paris or other French cities, often with the intention of returning after completing their studies. For these, and young metropolitan-born people of Caribbean origin, other options are being explored, with alternative destinations sought – particularly for those having chosen to study foreign languages (usually English or Spanish) – in Europe or the Caribbean.

Apparently trapped within a dual labour market on either side of the Atlantic, young people continue to try their luck in the metropole. Such mobility can be compared with the strategic flexibility described by Ch. Carnegie (1987). Sometimes this leads to back and forth movement as several attempts are made. Family networks are relied upon heavily for accommodation and assistance (Urunuela, 2002). Within this increasing circulation, more and more young people of Caribbean origin, but born in metropolitan France, are returning to the islands. Similarly, young men and women who migrated with their parents to the metropole are returning alone, on leaving school or after further education. They are part of the movement summarized in census figures of Caribbean-born returnees. At the same time, these returnees, often with further education degrees and diplomas provide stiff competition in the island labour markets for people who have not migrated (Chanteur, 2002). For example, in 1999, nearly half (43 per cent) of returnees aged 25-45 have such qualifications, as compared to only 23 per cent of non-migrants; the former are also less often unemployed. This imbalance not only creates tensions within island labour markets and their societies, but it highlights the continuing insufficient investment in education and training in the islands. Moreover, those individuals outside family networks partly located in mainland France may be at a disadvantage in relation to opportunities to further their education.

Return Migration Trends

Analysis of census enumerations in the French Caribbean reveals a substantial return movement, as the figures in Table 11.2 indicate. Note, however, that the census estimates must be assessed with some degree of circumspection. First, they rely on respondent's reporting of place of permanent residence at the previous census; second, children's places of residence may be recorded as that of their parents (whether or not this is accurate); thirdly, other movements may have taken place in the intervening period.

Table 11.2   Number of residents in the islands of Guadeloupe and Martinique who have moved from metropolitan France during each inter-censal period

Intercensal period	Returnees
1969-1974	3542
1975-1982	29786
1982-1990	25018
1990-1999	36000

Sources: Domenach and Picouet, 1992; Guengant, 1993; Marie and Rallu, 2004.

As with most aspects of the post-1970s migration streams, return flows to each island are remarkably similar. Prior to 1975, the figures reveal a substantial amount of family return migration; with 12 per cent of the returnee population aged under 15 years and 25 per cent aged 35-44 years (Domenach and Picouet, 1992, pp. 216-17). Approximately one-quarter of the returnees were aged over 55. Slightly more men than women were returning in each age group; reflected the gender balance of emigration in the 1950s and 1960s. However, from the 1970s, there were several changes within the migration system. Comparing migration during the periods prior to 1975 and 1975-82, Domenach and Picouet (1992) found an increase in the number of returnees; with the proportion of younger returnees having grown (20-39 year olds increased between the two periods from one third to 60 per cent of returnees) and with a decrease in the proportion of men reflecting the feminisation of migration to the metropole in the 1960s. In the 1980s, in addition to the continuing return flow for retirement, a considerable proportion of people in the 25-40 age group were returning (Condon and Ogden, 1996; Rallu, 1997). Between 14-17 per cent of the 28-32 age cohort were estimated as having returned between 1982 and 1990, and 12-13 per cent of 33-37 year-olds, in contrast to 8-9 per cent of 53-57 year-olds (Rallu, 1997, p.709). By the 1990s, whilst return flows of the young age groups remained similar, those of the retirement age group increased considerably; around 14 per cent of 59-63 year-olds returned between 1990 and 1999 and 16-18 per cent of 64-68 year-olds (Marie and Rallu, 2004, table 3). There is no gender difference apparent in the retirement age groups. Interestingly, however, the younger age groups up to 39 contain proportionally more women returning than men. This trend was already apparent in the 1980s and, despite continuing high rates of unemployment in the islands, is a likely reflection of more job openings for women, in what have become 'female employment sectors' (secretarial, retail, tourism) in the Caribbean. It has been observed that, whilst job instability and frequency of part-time work prevail in some sectors, women have higher rates of employment, especially among those over the age of 40 (Marie and Rallu, op.cit.).

The returnees 'captured' at each census may be long-term returnees planning to stay in the islands definitively, or they may be at the beginning or the end of a short stay before emigrating again to the metropole. A definite feature of this most recent period is the increasing complexity of migration biographies. And, an important aspect of these biographies is the prevalence of short visits to the Caribbean.

Circulation: From Visits to Longer Stays – and Return

Return visits – when affordable – had always occurred from the early days of this economically-motivated, labour migration; by individuals, or whole families. The motives for such visits ranged from holidays, family events, fetching offspring left with parents in the initial stages of migration, sending offspring to grandparents for holidays, or longer for schooling in the Caribbean, to return visits to search for jobs, land or housing (Byron and Condon, 1996). Modernisation of transport, with the introduction of commercial air travel, greatly facilitated journeys across the Atlantic from the 1970s (Atchoarena, 1992; Nicolas, 2001). A number of factors made this back and forth circulation easier in the French case than, for example, British experiences: financial advantages for workers in some public services, possibility of grouping together holidays, and reduction in travel costs. Such short stays are not included in return statistics yet they are important to examine since they represent a crucial part of the transnational process for maintaining links and perpetuating information flows. Analysis of a survey question put to 1,112 Caribbean-born migrants living in metropolitan France in 1992 gives a strong indication of the volume of short return visits (Condon, 1996). Around one third of interviewees stated that they returned once every three or four years, while a further 30 per cent said they went back every two years. A particularly privileged group (16 per cent) went back to the islands every year (with the proportion being 38 per cent amongst the higher socio-economic groups), whilst 13 per cent said they had never returned.[3]

The climatic and environmental advantages of the islands make them attractive places for holidays for metropolitan-born offspring of migrants – and not purely a visit to grandparents and other relatives. It is thus that Jacqueline[4] talks of the arrangements she has made so that her children can maintain a link with family in Guadeloupe. Her partner was born in French Guyana, but his parents were Guadeloupean. He has been quite happy to spend a holiday in Guadeloupe with Jacqueline and the two boys every three years and, since the boys were aged around 7-8, to send them to stay with their grandparents. According to Jacqueline (aged 38), the boys enjoy their stays there, going to the beach and other outside activities. She is really happy about this, as she feels it is important for the family to be able to see the boys grow up and, she hopes, it leaves the option for one or both of them to settle in Guadeloupe one day: 'They need to go regularly to keep up with local life, to make friends, to be able to speak Creole. That way, if they want to go back, there would be no problem.'

Mirroring these movements are the visits to metropolitan France made by individuals permanently residing in the Caribbean. In addition to student and labour migration, shorter stays and visits are made: for health reasons (visit to specialists, hospitalisation for an operation), for holidays (often in the case of returned migrants) and visits to relatives. These movements too are important links in the transatlantic social field and contribute to shaping the context of emigration and return.

The Social and Economic Context of Return Migration to the French Caribbean

Given the variety of age groups concerned, returning migrants are positioned differently in relation to the labour market as well as the housing market. Age, combined with a number of other factors, influences their ambitions in the return move. We can examine the various factors intervening in the decision to return to the Caribbean by grouping them into three broad categories: economic, socio-demographic and socio-cultural (Thomas-Hope, 1993; Byron, 1999). Gender relations intersect with most of these issues relating to return.

The main considerations with respect to *economic factors* are those of having a source of income in the Caribbean (earnings or pension payments) and possessing a home or having the resources to acquire one in the Caribbean. There are two broad categories of returnees; those who return for retirement and those who return at working age. For both groups, those returning to work and those to retire, the different costs of living in the metropole and the islands are an obvious comparative task. In several respects the cost of living is higher in the islands, owing to the cost of imported goods.[5] The benefits of being able to reintegrate a network of close relatives and long-standing friends within which services can be exchanged (for example assistance in building a home), being able to grow one's own fruit and vegetables, or no longer having to pay for heating are weighed against the high cost of manufactured goods and the need to purchase one or more private vehicles. Other costs involved in return are, of course, those incurred during the move (transport of possessions, insurance, legal fees, among others ...).

Generally, migrants interviewed in the Paris Region are of the opinion that life after a return to the Caribbean will be less costly – as long as a satisfactory pension or a reasonably stable job can be procured – since there are various ways of supplementing one's formal income and benefiting from services within one's networks. Those migrants most in favour of returning have continuously participated in their social network, sending money or gifts to family in the Caribbean, offering lodgings to siblings, cousins or their offspring, making return visits to attend christenings, weddings or funerals. They thus feel at liberty to ask for lodgings or other assistance during the first months after return. Nor do they expect their needs to be overlooked when family land is shared out following the decease of a parent or grandparent (Urunuela, 2002). Housing intervenes in the return decision depending upon the possibilities and ideals of migrants. Migrants who have inherited a small plot of land may have been able to start building a home over the years, working on the project during holidays in the island.[6] Some stay with relatives on their return while completing the building work. Others plan to purchase a house after a short period in rented accommodation. However rising land prices and rents mean that many are not able to save the capital necessary to achieve their objective.

In the case of return to the French Caribbean around *retirement age*, the fact that movement takes place within the same national space is important.[7] Nonetheless, the diversity of pension levels means that some migrants are in a more comfortable position to choose where to spend their retirement than others. The level of pension is of course related to employment experience in metropolitan France, its stability and level of remuneration. Public sector workers, whatever the grade, thus are assured of a regular pension. Outside this sector, only the minority of people falling within the upper socio-economic groups benefit from substantial occupational pensions. As a group, amongst the older generations, many women have been less financially secure than have men, owing to interrupted working careers or to not having benefited from social security cover (particularly in domestic employment).[8] In retirement, they will be particularly dependent upon their partner's pension. Women who are separated from their partner or single may have considerably lower household pensions and find themselves dependent upon relatives should they wish to return. Unlike the British case, where men bore the brunt of mass redundancies in industry (Modood, 1997, p.93-94), Caribbean men in metropolitan France have been largely protected since a majority worked in the public sector or nationalized industry.

Pre-retirement return concerns people at various stages of their working life: young people who have completed their studies, couples with young children, single parents who have spent fifteen or twenty years working in the metropole, and public sector employees who have obtained a transfer to the Caribbean in the latter stages of their career. Problems arise for some couples employed in the public sector when one partner obtains a transfer and not the other. For public sector employees not having obtained a transfer there is the possibility of requesting a release from duties for a period up to three years, which some use to seek employment locally in the islands. Outside this sector, returnees may attempt to find work in the tourist industry or in other services. Interviewees spoke of relatives or friends who have become self-employed, for example working as taxi drivers or in the building trade (usually men), or setting up a small restaurant (often women). In addition to the necessity of skills and social networks, a certain amount of savings are necessary for setting up a small business. Younger people returning from metropolitan France having completed studies in medicine or law are fairly numerous among those setting up private practice. However, they face competition from metropolitan French doctors and lawyers who choose to settle in the Caribbean (Urunuela, 2002). Those migrants trained in fields such as computing, business studies or languages are well received in firms in the Caribbean and are seen by some to be at an advantage over people who have not left the islands (Chanteur, 2002). They are better protected from unemployment than are their compatriots who have not migrated. Both young and more mature migrants thinking of a return move are aware that finding stable employment and an adequate salary will not be easy in a context of persistently high unemployment.[9]

Those who return with skills or experience, have a reasonable knowledge of local labour market changes and are linked into a strong social network will be more successful in securing satisfactory employment.

Age, along with marital or family status, is a key *socio-demographic factor* in the return decision. It of course influences a range of characteristics: as well as the distinction of being of working age or retired, age also conditions the likelihood of having dependent children or not, of having ageing parents in the Caribbean, health status, and so on. Age is also linked to the generational experience of migration, since labour market opportunities have varied over time, education levels have improved, means of communication and transport between the islands and the metropole have increased and their cost(s) have dropped. So, as the older generation with lower levels of education moved into a labour market with fairly abundant low-skilled, often stable jobs, the younger generations who have benefited from a longer time in the school system, have been moving into a labour market where jobs are more unstable, many requiring high level qualifications, and where the number of lower-skilled jobs in the public services is decreasing.

With regard to 'marital' and family status, if a migrant has a partner, return plans have to be negotiated between partners who may have conflicting interests (professional, family or social), or have a more personal feeling of attachment either to the Caribbean or to metropolitan France. For couples where each partner originates from a different island, which island the couple returns to is an important question. If one partner is metropolitan-born of non-Caribbean origin, a decision for the couple to settle permanently in the Caribbean will depend very much on the relationship this partner has built up with the Caribbean as well as the degree to which the Caribbean partner has maintained links with his or her place of origin. Generally, as found for other migrant groups (Cribier, 1992; Leite, 1997), interview and survey evidence tends to suggest a weaker desire amongst women – and often those who are married to Caribbean men – to return permanently to the Caribbean (Condon and Ogden, 1996). Although those who do not live as a couple and wish to return may not have to negotiate with a partner, the presence of close relatives in the islands may be a decisive factor in their residential choice. Young women raising children alone may choose to return to be close to parents or siblings and become more involved again in their family network of exchange of services; widowed or divorced persons may move for similar reasons.

If there are children who are still dependants, the question of schooling – in addition to their own feelings toward return – have to be taken into consideration. Older children may be left in the care of an uncle or aunt in metropolitan France in order to complete their education and join their parents during the summer holidays. When relatives are not available, parents may find lodgings for offspring until they are financially independent. It was thus for François when at the age of 45 he unexpectedly received the long-awaited offer of a transfer of his job as postman to Guadeloupe. Prior to this, he had been resigned to finishing his working life in metropolitan France and to remain there while his daughters completed their training. He arranged the rental of a small apartment for the girls, in the eastern suburbs of Paris near an aunt. His wife, a school canteen assistant, obtained a transfer six months afterwards. They had bought previously a piece of land in their village of origin (they had been childhood friends) and begun building for their retirement. When François moved back to take up his post in Guadeloupe, he occupied a room at his parents' home. This house had been extended and improved over the years with the help of François, his sister who had migrated and the two sisters who had remained in Guadeloupe.

Gender combines with age and other characteristics to intervene as a positive or negative factor in the return decision. The growing literature on gender relations in the Caribbean attests to the different opportunities and constraints experienced by men and women from childhood (e.g. Senior, 1991, pp. 166-84). In some respects, migration can bring about increased power for women within the couple or the family. After having gained independence through regular, waged work, control of their earnings, more equally shared domestic duties (Foner, 1975; Byron, 1998) some women are afraid of losing their independence once they return to the Caribbean. In other cases, for example among Puerto Rican women, traditional gender roles however have been maintained in the destination and this has contributed to a strong desire to return to social networks in their island of origin (Ellis et al., 1996).

In the French case, a crucial element of acquired independence referred to by women is the access to public transport enjoyed in the Paris region, in particular, enabling them to take up work where they wish, to participate in activities outside work, travel to large shops, public amenities and so on. Amongst the earlier generations, it is much more common for women rather than men not to hold a driving licence[10] and those in this position fear that they will be dependent on others in their daily activities if they return. As Véronique (aged 40) said, 'I'm used to moving around a lot, I don't like to feel stuck in a place and since I never took my driving test, I'd have to rely on other people for taking me to places'. This feeling combines with a certain reluctance among women to leave behind other convenient public amenities shown to be important factors in the return decision within metropolitan France (c.f. Cribier, 1992 p.95-96). During visits to the Caribbean for holidays, people thinking about return take note of such restrictions on their access to services and amenities and these observations are often passed on to other Caribbean migrants living in France.

However, an aspect of gender relations in the Caribbean seen by many as a cultural norm – and a highly resistant one – is the tendency for some men to have relationships with more than one woman at the same time and for such behaviour to be considered as an acceptable facet of masculinity. One preoccupation that emerged from interviews was a deep concern for the stability of couples' ties on return. As Rosette (aged 47) stated 'many women I know say that they are afraid that their husband will be tempted away once they are back in the Caribbean. I know a few couples who have split up after returning'.

Nonetheless, although these factors may interrupt, delay, or even pre-empt the return move amongst older migrants, younger women living in the metropole – particularly those who migrated at an early age, as well as the metropolitan-born – have had a different experience of gender relations. Although inequality between women and men persists in many aspects of work and home life in France, norms of what is acceptable behaviour within couples are changing. These women have also had access to higher levels of education than their mother's generation and have enjoyed relative freedom of movement within the urban environment. In addition, many have had the opportunity of acquiring a driving licence. Those who wish to return and perceive openings in the job market corresponding to their skills see themselves as playing an active role in all realms of the island society when they return.

The Central Role of the Social Field

Certainly, the most important factor influencing the decision to leave the metropole to settle in the Caribbean is the presence in the islands of family and friends; i.e., being able to access a *social field*. Family networks are usually the core of the social field and very definitely contribute to the transnational identity and experience of many individuals. The strength of links maintained by migrants with the Caribbean are largely based on continued participation in family affairs and on strong emotional attachment(s) to individual members (Philpott, 1977). The networks are the theatre of various dynamics: real or perceived obligations, loyalties, conflicting positions or roles in their network, emotional proximities, and preferences (Byron, 1994). The presence of ageing parents or other relatives in the Caribbean may encourage return migration, especially when these relatives do not wish to, or cannot, join the migrant in metropolitan France. Women tend to be the ones generally considered to be at the centre of such decisions, given their ascribed roles of caring for children as well as older relatives. However, French Caribbean men have been shown to play an active role in the 'pivotal' generation between young adults and retired parents, supporting the former and caring for the latter (Attias-Donfut and Lapierre, 1998) – and not only in the absence of female siblings. This was the case for François, who had been very involved in helping his parents, including improving their house over the years and with repairs to the home after 'Hurricane Hugo' damage. Being close to his parents in their old age was one of the main factors in his desire to return. On the other hand, concerns regarding one's own welfare in old age negatively influence the decision to return, especially when no close relatives are present in the Caribbean and the familial support networks are thus more French metropole-oriented. Many migrants hope – or even expect – that those for whom they have cared and often made sacrifices will in turn assist them after their retirement (Byron, 1999).

An important component of the migrant's life-world, encompassing the place lived in during childhood and youth and the one in which she or he has elected residence – as well as perhaps other places – is the flow of information across borders, between places and among those involved (Basch et al., 1994). These information flows often acquaint migrants with minute changes in the organisation of the daily lives of friends and family, with opinions about such changes and proposed solutions to problems. The flows circulate, influencing migrant decisions on both sides of the Atlantic. It is thus that an accumulation of knowledge is built up, with the metropoles becoming less and less unknown societies for individuals thinking of emigrating, and eventually moving off the island. For the person contemplating return migration, such flows informing her or him of changes in norms and practices in the islands 'back home', can influence the decision to return. This 'migration knowledge' takes the form of experiences of return recounted by individuals concerned or those close to them, or is a transfer of knowledge about, as well as stories of, the behaviour of returnees. Such information commonly circulates amongst migrants in metropolitan France (Condon and Ogden, 1996, p.45).

Different types of information work in favour of, or against a wish to, return. Thus concerns about being able to 'reintegrate' into the local community are very often expressed. Those who feel that they have acquired what they consider a more 'modern' way of thinking, different values and attitudes to life, feel that they will no longer be able to communicate. As Rosine explained: 'We've got used to doing things differently here in France. We hurry more, we work more quickly. They don't understand why.' And Lucy: 'I'm no longer used to that mentality, jealousies and so on, everyone knowing everything about you.' Many also fear that envy will be expressed toward them. For migrants thinking of retiring to the islands, but who have not been able to return frequently and feel out of touch with changes – similar to the circumstances of most migrants of the older generations in Britain (Thomas-Hope, 1986) – there is the fear of not being able to adapt. This is a process made all the more difficult because they are that much older and such attitudes towards adaptation-problems are shared by men and women.

Such pro- and con- issues are weighed, debated and negotiated within families, when plans to return to the Caribbean are being drawn up. What some see as an insurmountable obstacle to return, others will weigh that 'inconvenience', against the advantages they see in such a move. Sylvianne had heard people who had returned talk of problems with settling back into Guadeloupe, and of the reactions of colleagues who felt that they were trying to impose metropolitan working habits. She had been given the impression that it took longer to adapt 'back' into Guadeloupean society, than it had been for her to adapt to life in metropolitan France: 'Of course it's difficult to begin with, the time to realise that you're no longer on holiday and for people to get used to your ways (…) I was happy, always busy in Paris but the weather, I was often ill during the long winter months (…) and my family and friends were here. So when the children were coming to the end of their schooling, I came over on a period of extended leave but three months after arriving, I organised a detachment to another post here'. Sylvianne went back to Guadeloupe ten years after emigrating, taking her five children with her. Each of them later went back to Paris to pursue further education.

Discourses of 'Belonging' and Return

Back home, *'chez nous'*, is an ever-present image in the minds of most Caribbean people (Gmelch, 1992; Western, 1992). It is the island in which they grew up, where places became familiar, where their role as member of extended family and friendship networks was consolidated. It's the place that symbolises many aspects of their identity, a place that previous generations struggled to build as their own, freeing themselves from the domination by plantation society (Chivallon, 1998). Regular communication with people 'back home' and return visits have usually been used in numerous migration studies as an indicator for predicting the likelihood of (permanent) return. Although these aspects are important in preparing a future resettlement, they can coexist with a plan to stay in the place to which they have emigrated and settled; what they may see in some ways as their 'new, metropolitan home'. Various kinds of movement are thus intertwined; shorter and longer stays at each side of the Atlantic, shorter stays which are prolonged, stays that were planned as long-term which are foreshortened, visits for special events or holidays, among other permutations. At the same time, these movements maintain, feed into and develop the social field of which individuals and families are a part. Surrounding these various plans, stays, movements, are a number of discourses of belonging, some reflecting a firm desire to return, others reflecting dilemmas, or 'simply' dreams.

Of course the place 'the islands' hold in the heart of Caribbean migrants living in the metropolitan France evolves throughout their stay there and discourses evolve from one generation to another. Younger and older migrants talk of returning 'home' in various ways – for Yolande, Elise, Gilberte, this represents a return to their roots, for Laure and François, a return to their own people, whilst for Jacques, it's the place to which he could always return if need be, where 'people know you'. Concomitant with the expression of this desire is the need to be able to defend their 'right to return' – as encapsulated in the word *'rentrer'* (to go back to whence one came). Continuous contact and regular visits to the islands are necessary to keep up with family and friends, Creole language,[11] local knowledge of current events and affairs.

The Desire to Return to Work

For some, as in the case of Laure (33 years of age, living in furnished rooms in Paris), the desire to return is very strong. After twelve years away from Guadeloupe and having recently obtained the qualification of nurse's aide, she feels it is her duty to return to Guadeloupe: 'I've had my training. I've worked in the hospital. Now, I want to go home to care for my own people.' In other cases, this desire to return can be prompted and realised when an opportunity arises. For example, Jacques (aged 47) took up the offer of work made while he was in Martinique on holiday by the local mayor in his home town. He had recently changed jobs in the south of France, following his divorce and health problems associated with work as a lorry driver. His passion was swimming and he had qualified as a life guard and swimming instructor; and this was the job offered to him near Fort-de-France. The offer was too tempting: 'So in 1988, I packed my bags and came home after twenty years in France.'

'Return' of Young People who Migrated at an Early Age, with their Parents, or Who were Born in France

Elise M was determined to pursue her further education in the Caribbean with a view to re-settling permanently there again. She had migrated to Paris at the age of seven. Her mother had left her with her grandmother while she followed a secretarial training course, her mother's expectation being that she would then work in Guadeloupe. After deciding to go into hospital work and to look for a job in Paris, Elise's mother went to fetch her daughter. Eleven years later, when Elise told her mother of her desire to go back to Guadeloupe, her mother, who had a permanent post at a Parisian hospital arranged to take unpaid leave for a maximum period of three years. Elise then registered at the faculty in Martinique and, after a few years of short contracts, obtained a steady job in agricultural development.

Yolande emigrated to France at the age of eight with her four sisters. Her parents had left six years previously; the youngest sister was born in the metropole. Yolande had been brought up by her grandmother from the age of two; leaving her was a traumatic experience. She says that all her good childhood memories were those in Martinique. Her parents had made one attempt of settling back in the island but had returned to western France. For Yolande, the desire to go back to the island increased as the years went by. Part-way through her university course, she decided to return and try her luck.

Desire to Return for Retirement

For many of the early labour migrants, returning on retirement is an unquestioned decision. For Gilberte (50 years of age): 'And when I've finished here, I'll go back home to retire under my coconut trees.' This is clearly an idealised view but reflects what her Caribbean home symbolises: rest, calm, a feeling for being at home. Another migrant, Francine (46 years) expresses a real, physical need to go back to the Caribbean to see her brother and other relatives: 'I really need to go back often, I can't stay here in France more than three years without going, I need to go back to bathe in the sun. I am a Caribbean plant, I get my strength there. Otherwise I get depressed. I need to see my brother, eat the local food. I couldn't stay here for my retirement. I have to go home, to really make the most of life. Here it's good for work, because there's none in Martinique, but otherwise, that is where I want to be.'

Issues relating to health are thought about, of course; but there always remains the possibility of travelling to the metropole for medical treatment. Separation from children and grand-children remains, however, an issue. Sylvianne's aunt was faced with this dilemma as retirement approached and her husband went ahead with his plans to return and live in the house they had built there. 'She said, though, that her place was not in Guadeloupe, that her children and her grand-children were in France. So she goes back and forth now, to see her husband. She stays in Guadeloupe from December to March. He wanted peace and quiet on retirement, now he lives alone most of the time!'

'Home' is in the Caribbean, and in France

The attachment to the Caribbean can remain strong, whilst at the same time the desire to remain in metropolitan France can outweigh any future return intention. Gabrielle had no intention of returning to Martinique when preparations were being made, it was only her husband who was so set on the idea: 'Here is where my friends are, I've got my routine. Then there are my grandchildren. But my husband thinks about nothing else.' He accepted a transfer of this customs' officer's post to French Guyana three years before retirement, after which they would move back to Martinique. Gabrielle conceded, since her youngest son, who had experienced difficulties during schooling and was keen on training in forestry, was prepared to move with them. Gabrielle firmly intended to go back to France each year to see friends – particularly one best friend – and family.

Whilst such a return project is realised by many, it remains a dream for many others. In some cases, it is renounced following changed circumstances, or a family crisis. In others, where emigration was enacted to make a clean break with the past, and leave the island 'once and for all', return was never considered from the outset. It was thus for Murielle, who left after she was abandoned by her partner for another woman. He was the father of her baby and had also been her employer. She left Guadeloupe in 1977 with her child and declared having no intention of ever returning.

Conclusions

The social and economic context of migration and the return decision in the late 1990s and early 2000s is radically different from that prevailing in the 1950s and 1960s. Mass unemployment in the Caribbean is still seemingly impossible to diminish, but unemployment also is high in certain sectors and for certain age groups in metropolitan France. Meanwhile, transport and telephone costs are constantly decreasing and internet communication plays an increasing role in forging and maintaining links between mainland France and the French Caribbean. Migration decisions thus are taken in a rapidly changing technological context in which social fields can be consolidated via new forms of communication. Retirement migration strategies are undertaken with more knowledge about opportunities and drawbacks. Many skilled young people try out both metropolitan and island labour markets before settling for a longer period. Even when they are unemployed, other life events – or the decisions of offspring or other close relatives – may lead some people to choose one place over the other. At the same time, there is a population of women and men, not necessarily unskilled but often marginal, or loosely integrated, to mainstream society, circulating between Paris and tourist areas in the French Caribbean, entering and exiting the informal sector, adventuring and risk taking – 'experiencing the wider world'. Migrants at all stages of the life course evaluate the usefulness of their individual skills or of their network relations before moving from one place to the other in the hope of improving their living standards and quality of life.

All commentators on French Caribbean migration agree on the increasing importance today of the phenomenon of circulation – the back and forth movement – rather than the migration cycle of long-term emigration and return. But rather than always being performed through choice, for many it constitutes a search for what individuals deem to be the best solution. Unemployment, low skill levels, discrimination, racism, violence, as well as individual problems linked to health and personal status generate such behaviour for many. Circulation by choice corresponds to mature, adequately well-off people who can afford a form of 'double residence' or at least regular visits; or young, highly qualified people experimenting both places, life styles and labour markets, often perceiving an ideal in both places, having multi-layered ambitions. In an increasingly globalised world, transatlantic circuits between island and metropolitan homes have become well entrenched and enduring aspects of French Caribbean life.

Notes

1. Only recently have problems of integration and discrimination experienced by Caribbean migrants in metropolitan France been discussed before a wider audience in the field of immigration studies, *Espace, Populations et Sociétés*, special issue 'Immigrés et enfants d'immigrés', 1996 n°2-3.
2. The census identified individuals having at least one parent born in Martinique or Guadeloupe, unmarried and living in a parent's household, whatever their age, as 'offspring in Caribbean families'. They are added onto the population of Caribbean-born to form the population of '*originaires des Antilles*'. Excluded from the Caribbean 'minority' thus are those descendants of migrants living outside their parents' households.
3. This survey was conducted by the national statistics institute (INSEE), using the 1990 census as the sampling base. The interviewees were of all age groups and various lengths of stay. Thus this last group included both very recent migrants and people who had lost contact with the islands. Analysis presented here was conducted by this author.
4. One of the 40 in-depth interviews gathered in the Paris Region during ethnographic research conducted between 1990 and 1996. Several interviewees have been re-contacted a number of times since. 25 further interviews were conducted in Guadeloupe and Martinique in 1996. Translation of interview extracts was done by the author.
5. For example, in the mid-1980s prices (excluding rent) were over 16 per cent higher than prices in the metropole. Domenach and Picouet, 1992, p. 124.
6. Interviewees for the 1992 INSEE survey who said they wished to return to live in the islands were asked about the type of housing envisaged on their return. Only one in ten already owned a home (most were aged over 50), 14 per cent owned a plot of land and 11 per cent were saving to buy a plot. One-quarter planned to rent lodgings. Over one-third planned to stay with their family (particularly the younger age groups) (results published in Condon S, 1999, 'Politiques du logement et migrations de retour. Retour des migrants antillais de la France métropolitaine', in d'Armagnac, J. et al., *Démographie et aménagement du territoire*, 10th CUDEP National Demography conference, Paris, PUF, pp. 297-305).
7. In contrast with most situations within the Caribbean, since French Caribbean migrants returning to the islands remain within the same national boundaries, they do not lose

any pension rights. Moreover, retirement pensions in the French overseas départements (DOM) are at a higher level than in the metropole. Unfortunately, this topic is scarcely documented.

8. In comparison to many other migrant women in France, Caribbean women are currently in a more favorable position as retirement approaches since they have more often been in stable, full-time work (particularly in the public sector). Thus trends are similar to those observed in Britain (see Holdsworth and Dale, 1997, 'Ethnic differences in women's employment', *Work, employment and society*, 11(3), pp. 435-457). However, pension prospects of the younger generations are less optimistic.

9. Returnees to the French Caribbean who do not find stable work after resettlement can claim unemployment benefit.

10. Furthermore, military conscripts had the opportunity of taking their driving licence.

11. The link between the use of creole in the home in families in metropolitan France, the maintenance of family ties and the desire to return has been explored by the author elsewhere.

References

Anselin, A. (1979), *L'émigration antillaise en France*, Paris, Anthropos.

Atchoarena, D. (1992), 'Les communications: évolutions et conséquences', in Domenach, H. and Picouet, M., *La dimension migratoire des Antilles*, Paris, Economica, 139-167.

Attias-Donfut, C. and Lapierre, N. (1998), *La famille providence. Trois générations en Guadeloupe*, Paris, La Documentation Française (avec la CNAV).

Basch, L., Glick-Schiller, N. and Szanton-Blanc, C. (1994), *Nations unbound: Transnational projects, post colonial predicaments and deterritorialised nation states*, London, Gordon and Breach.

Brock, C. (ed) (1986), *The Caribbean in Europe: aspects of West Indian experience in Britain, France and the Netherlands*, London, F. Cass.

Byron, M. (1994), *Post-war Caribbean migration to Britain: the unfinished cycle*, Avebury, Aldershot.

Byron, M. (1998), 'Migration, work and gender: the case of post-war labour migration from the Caribbean to Britain', in M. Chamberlain (ed), *Caribbean migration, globalised identities*, London, Routledge, pp. 217-231.

Byron, M. (1999), 'The Caribbean-born population in 1990s Britain: who will return?', *Journal of Ethnic and Migration Studies*, 25 (2) 285-301.

Byron, M. and Condon, S. (1996), 'A comparative study of Caribbean return migration from Britain and France: towards a context-dependent explanation', *Transactions of the Institute of British Geographers*, 21, 91-104.

Carnegie, C. (1987), 'A social psychology of Caribbean migrations: strategic flexibility in the West Indies', in Levine B.B. (dir.) *The Caribbean exodus*, Praeger, NY, pp. 32-45.

Chanteur, Bénédicte (2002), 'Les jeunes adultes de retour au pays: Partir multiplie les chances de réussite', *Antiane Eco*, n°52, mai 2002, 19-21.

Chivallon, Christine (1998), *Espace et identité à la Martinique. Paysannerie des mornes et reconquête collective (1840-1960)*, CNRS Editions.

Condon, S. (1996), 'Les migrants antillais en métropole: un espace de vie transatlantique', *Espace Populations et Sociétés* (n° spécial: *Immigrés et enfants d'immigrés*), n° 2-3, 513-520.

Condon, S. (2004), 'Gender issues in the study of circulation between the Caribbean and the French metropole', *Caribbean Studies*, vol. 32 (1), 129-159.

Condon, S.A. and Ogden, P.E. (1991b), 'Afro-Caribbean migrants in France: employment, state policy and the migration process', *Transactions, Institute of British Geographers*, 16 (4) 440-457.

Condon, S.A. and Ogden, P.E. (1993), 'The state, housing policy and Afro-Caribbean migration to France', *Ethnic and Racial Studies*, vol. 16, 256-297.

Condon, S.A. and Ogden, P.E. (1996), 'Questions of emigration, circulation and return: mobility between the Caribbean and France', *International Journal of Population Geography*, vol. 2(1), 35-50.

Constant, F. (1987), 'La politique française de l'immigration antillaise de 1946 à 1987', *Revue Européenne des Migrations Internationales*, vol. 3(3), 9-29.

Cribier, F. (1992), 'La migration de retraite des Parisiens vers la province et ses transformations récentes', in Lelièvre, E. and Lévy, C. (eds), *La ville en mouvement: habitat et habitants*, Paris, L'Harmattan (collection Villes et entreprises).

Cross, M. and Entzinger, H. (1988), *Lost illusions: Caribbean minorities in Britain and the Netherlands*, Routledge, London.

Domenach, H. and Picouet, M. (1992), *La dimension migratoire des Antilles*, Paris, Economica.

Ellis, M., D. Conway and A.J. Bailey (1996), 'The Circular Migration of Puerto Rican Women: Towards a Gendered Explanation', *International Migration*, 34(1), 31-64.

Foner, N. (1979), 'West Indians in New York City and London: a comparative analysis', *International Migration Review*, 13, 284-297.

Gmelch, G. (1992), *Double passage: the lives of Caribbean migrants abroad and back home*, Ann Arbor, University of Michigan Press.

Goulbourne, H. and Chamberlain, M. (eds) (2001), *Caribbean families in Britain and the Trans-atlantic world*, London and Oxford, Macmillan (Warwick University Caribbean Studies).

Grasmuck, S. and Pessar, P. (1991), *Between two islands: Dominican international migration*, Berkeley, University of California Press.

Guengant, J.-P. (1993), 'Migrations : moins de départs, plus d'arrivées', *Antiane-Eco*, n° 22, 32.

Leite, C. (1997), 'Femmes immigrées autour des projets de 'maison'', in B. de Varine (ed), *Les familles portugaises et la société française*, Mâcon, EditionsW/ Paris, Interaction France-Portugal, 207-222.

Marie, C.V. (1996), 'Femmes antillaises outre-mer', *Espace, Populations, Sociétés*, n°2-3 (special issue *Immigrés et enfants d'immigrés*), 521-528.

Marie, C.V. and Rallu, J. L. (2004), 'Migrations croisées entre DOMs et métropole: l'emploi comme moteur de la migration', *Espace, Populations et Sociétés* 2004 n°2, pp. 237-252.

Nicolas, T. (2001), 'La circulation comme facteur d'intégration nationale et d' 'hypo-insularité': le cas des Antilles françaises', *Les Cahiers d'outre-mer*, oct.-déc., n°216, 397-416.

Olwig, K.F. (2003), '"Transnational" socio-cultural systems and ethnographic research: views from an extended field site', *International Migration Review*, vol. 37 (4) 787-811.

Peach, G.C.K. (1991), *The Caribbean in Europe: contrasting patterns of migration and settlement in Britain, France and the Netherlands*, Research Paper in Ethnic Relations n°15, University of Warwick, Centre for Research in Ethnic Relations.

Rallu, J-L. (1997), 'La population des départements d'outre-mer. Evolution récente, migrations et activité', *Population*, n°3 1997, 699-727.

Richmond, A. (1987), 'Caribbean immigrants in Britain and Canada: Socio-demographic aspects', *Revue Européenne des Migrations Internationales*, 3 (3), 129-148.

Sutton, C.R. (1987), 'The Caribbeanisation of New York City and the emergence of a transnational socio-cultural system', in C.R. Sutton and E.M. Chaney, *Caribbean Life in New York City: Socio-Cultural Dimensions*, New York, Center for Migration Studies, 15-30.

Thomas-Hope, E. (1986), 'Transients and settlers: varieties of Caribbean migrants and the socio-economic implications of their return', *International Migration*, 24, 559-71.

Thomas-Hope, E. (1993), *Explanations in Caribbean migration*, London, Macmillan.

Urunuela, Yvan (2002), *Dynamiques migratoires et développement d'une petite économie insulaire: le cas de l'émigration des guadeloupéens en France métropolitaine*, Thèse de doctorat, Université de Versailles Saint-Quentin-en-Yvelines.

Western, J. (1992), *A passage to England: Barbadian Londoners speak of home*, Minneapolis, University of Minneapolis Press.

Chapter 12

Expressions of Migrant Mobilities among Caribbean Migrants in Toronto, Canada

David Timothy Duval

Much has been written on the Caribbean migrant experience from the perspective of adaptive mechanisms in the host country and the complex relations maintained with former homelands (e.g., Chamberlain 1997; Foner 1997; Pessar 1997). The impacts of migration on the receiving countries have received considerable attention; either generally empirically-quantified and examined in terms of economic success, or, using a more intrepretivist or realist approach, the meaning of migration (and return) from the perspective of the migrant has been the center of interest (e.g., Gmelch 1992).

For some time, a growing research community has been interested in the transnational networks of relationships that migrants hold, and retain, in both the social fields of origin and destination. Deserving more exploration, however, is the means and mechanisms by which such connections and networks are forged, consolidated and maintained. For example, it is well-established that technological innovations in international communications have often facilitated communication among migrant families (i.e., cheaper telephone rates, email) (e.g., Parham 2004; Vertovec 2004; Waldorf 1995). Furthermore, the ability of 'transnational migrants' to ascribe a sense of belonging to more than one physical place has been considered a form of 'transnational livelihood' (Olwig and Sørensen 2002), while Guarnizo (2003) has offered the concept of 'transnational living' as a livelihood strategy which recognizes the unintended, as well as intended, economic consequences of migration.

There are persuasive arguments for adopting a transnational approach to characterizing migrant mobilities, although some doubt has been cast on the anthropological approach championed by Glick-Schiller et al (1998, among others), because it fails to illuminate migrant behaviour, and does not at all address the material bases of day-to-day existence (see, for example, Bailey 2001; Conway 2000). One issue to be more fully explored, for instance, is the daily, monthly and seasonal interactions of migrants with multiple 'homes'. Missing from transnational explanations of societal changes, multi-local network development and consolidation, hybrid identity formation and the like, are the sequential, repetitious, short- and longer- term set of mobility strategies employed by migrants as they maintain ties to former homelands. Urry (2003) argues that these strategies represent a manner in which corporeal, shared experiences are manifested, and compliment, if not transcend, advances in technology that, one might argue, should make communication easier. To this end, one question we can ask is: why do migrants continue to engage in face-to-face encounters with family and friends 'left behind' in former homelands?

In essence, the connectivity dynamics of migrant networks invariably involves migrants' homelands and other places of significant social meanings and importance. Furthermore, it speaks to Harvey's (1989) concern that peoples' lives are undergoing time-space compression in today's global, interactive world. Therefore, an investigation of transnational linkages and conduits of social practice, involving family and friends in multiple localities, would appear to be essential (see also Appadurai 1991). Moreover, with the range of roots following routes (Coles 2004), it is now, more than ever, possible to identify the multiple social realms that migrants occupy. By extension, it should be possible to understand global social formations of identity that are linked in numerous ways, not the least of which are acts of 'co-presence' (Urry 2003).

The broad purpose of this chapter is to focus upon one way in which migrant identities are fostered across distances through physical contact between migrants and former homelands (Duval 2003, 2004). More specifically, I undertake an assessment of the role of the return visit, as a potentially-important part of transnational practice in the return migration event. I wish to argue here that serious consideration should be given to the *function* of the return visit as a contributing factor in return migration. Qualifying this position, the return visit may, itself, not necessarily be the sole causal factor in determining future return migration. Rather, it is conceptualized as one contributing factor to such decisions, in large part because it is an activity through which social relationships are maintained by person-to-person contacts and personal assessment, and in this way has some degree of influence on the reasons being considered for returning. To show how the return visit facilitates social interconnectedness and, at the same time, how return visits play a more than cursory role in return migration, a transnationalism framework is particularly useful (Basch et al. 1994). Goulbourne notes that transnationalism:

> '...may involve groups of people who are connected in some significant ways to maintain the close links across the boundaries of nation-states; it may involve the enjoyment of more than one citizenship or belonging to more than one society; it may also involve the easy movement of people across states and societies'. (Goulbourne 2002:6)

In short, a transnational framework incorporates, 'ordinary people' (Goulbourne 2002:6), who have the propensity to be mobile and commonly share various social elements that span borders. The examination of return visit-return migration connections in this chapter builds upon results obtained from exploratory ethnographic fieldwork conducted by the author among Commonwealth Eastern Caribbean migrants living in Toronto, Canada. I argue that migrants may use the return visit to retain ties to their former homeland and to support in their social re-integration upon permanent return (also see Rodman and Conway, this volume).

Anastomotic Social Networks

Transnationalism works as a conceptual representation of 'cross-border social conduits through which migrants situate themselves and their respective social identities' (Duval 2004:51). Fittingly, Ward (1997) draws a rough comparison between the networks established by migrants and the anastomotic characteristics of water networks in the study of geomorphology. Just as fluvial networks can be altered for any numbers of reasons (e.g., natural and human obstacles, topography), social networks can be altered by financial restrictions, the strength of the network, and the motivation to maintain a flow of social capital. Transnationalism has been used to explain the social ties that individuals maintain in multiple localities, almost as a response to the 'objective categories of analysis' (Bailey 2001:415) formulated in previous studies of international migration (*cf.* Waldinger and Fitzgerald 2004). As a conceptual framework, it helps provide, to some extent, a functionalist interpretation of the meaning behind the return visit in the context of return migration.

Transnationalism has also been used to explain interconnected social experiences that help shape and influence migrant social spheres (Basch et al. 1994; see also Goulbourne 2002; Yeoh et al. 2003). As a conceptual framework, it has been used to tease out such 'deterritorialised' activities as remittances, health networks, and cross-border movement relating to education and employment (Appadurai 1991). Many of these transnational iterations point to the multiple relationships maintained by migrants irrespective of political-national demarcations (Kennedy and Roudometof 2002). Such interconnected experiences are represented by identity structures and social networks that, as noted, often span geopolitical borders. Indeed, the notion of social-derived boundary space (see Longhurst 1997) from geography fits nicely into a transnational approach with a realist orientation toward migration. Glick Schiller and Fouron (2001), for example, suggest that young Haitian adults live in a social field that encompasses both the USA and Haiti, so much so that their lives are influenced by transnational/border-irrelevant, yet typically 'Haitian', migrant experiences:

> '... young people of Haitian descent living in the United States, although much more familiar and comfortable with the pace and outlook of daily life than their parents, often seek ways to identify with Haiti. They are reclaiming Haiti by strengthening their ties with their ancestral land and reaffirming their Haitian identity'. (Glick Schiller and Fouron 2001:157)

Migration scholars certainly recognise the social networks that migrants maintain across boundaries, or perhaps more perceptively, irrespective of boundaries (Vertovec 2001). Of particular interest is how such transnational behaviour is operationalised. In other words, investigations into transnational migrant behaviour often assume that such connections exist, but do not elaborate on how the transnational connections are maintained. Again, while the degree to which electronic connections have been fostered speaks to the extent to which networked interrelationships are made possible through IT, the actual physical movement of migrants back and forth between multiple localities, and its influences and consequences, still deserves further exploration.

While transnationalism research has fostered many empirical studies of migrant identities, it has not promoted the same volume of examinations of the social meanings behind return migration and, by extension, return visiting as a means by which return migration may be facilitated. This transnational theme, that conceptualises return visits as a definitive feature of social mobility generation, is the focus here.

A Transnational Interpretation of the Return Visit/Return Migration Interface

Return migration, as an actor-based process, has been described in a number of ways, including 'reflux migration, homeward migration, remigration, return flow, second-time migration, [and] repatriation' (Gmelch 1980:136). The term is generally used to refer to the movement of migrants going back to their previous homeland (Gmelch 1980). Within the context of migrant communities, there has been a conscious effort to focus on the reason(s) for return, including investigations of adaptive mechanisms employed by migrants and the propensity to return to the external homeland. Extensively considered in the literature are economic hardships and realities in receiving countries, pressure from family members still resident in the external homeland, and problems with social adjustment (e.g., Boyd 1989; Gmelch 1983; King 1978).

To explain and justify the return visit, it is possible to use the same arsenal of social motivations that, themselves, have come to clarify return migration. It is to be expected that the extent to which transnational ties exist will very much determine whether a return visit takes place. As a similar actor-centred process to that of return migration, the return visit can be positioned within the framework of 'social mobilities' (e.g., Larsen 2001; Urry 2003) and, to a large extent, migration (Bell and Ward 2000) and tourism-related mobilities (Coles et al. 2005). In her research, Baldassar (2001, 1998, 1997, 1995) used the concept of *campanilismo* (roughly referring to spatially-determined identity among individuals) to understand how 'home' is constructed by both Italian migrants in Perth and residents of the small village of San Fior in Italy. In the context of migrants' construction of home, Baldassar (2001) found that, for some, their external homeland (in particular the village of San Fior) was the central source for cultural identification, with the return visit offering special meaning for all travellers:

> 'The emigrants' visit "home" is a secular pilgrimage of
> redemption in response to the obligation of child to kin,
> townsperson to town. In the case of the second generation
> migrant, the visit is a transformatory rite of passage brought
> about by the development of ties to one's ancestral past'.
> (Baldassar 2001:323)

Practically, the return visit is a means by which migrants travel temporarily to previous homelands, or a locality from which they had previously emigrated. For example, given the multi-local nature of Caribbean migration itineraries, a return visit of a Canadian West Indian born in Trinidad of Grenadian and Trinidad parents could be to Britain, where she had lived previously, or the United States, or a return to the island home(s) of her youth, or her parents. Conceptually, the return visit has the potential for being a social demonstration of the social and cultural ties that migrants may have to particular physical (and social) places. As an expression of connectedness, the return visit can also be considered a social practice allowing migrants to maintain multiple identities in both the host country and their former homelands. Rightfully, Sørensen and Olwig (2002; see also Sørensen 2003) argue that such connections are not necessarily restricted to one receiving society and one singular homeland, but rather involve multiple livelihoods across multiple levels (local, national, international). It is this multiplicity of localities that offers strong support for examining return visit/return migration relationships in terms of the transnational nature of migrant identities.

Return visits are certainly influenced socially, but are also subjected to, for example, economic considerations and issues of accessibility. More broadly, however, any conceptualization of the return visit and its social and cultural worth requires consideration of the actor's world view, but in particular as it applies to, at the very least, two social spaces: (1) their place of birth and (2) their current city, country or region of residence. Combined, the two speak concisely to Brah's (1996:208) 'diaspora space', which for Brah represents 'intersections of diaspora, border, and dis/location as a point of confluence of economic, political, cultural, and psychic processes'.

As outlined above, transnationalism has allowed for the recognition of multiple social fields linking numerous and dynamic social spaces. One concern, yet to be resolved however, is how to conceptually locate the return visit within such a dynamic and open-ended framework. Elsewhere, I (Duval 2004) have suggested that the return visit is an example of a 'transnational exercise' through which multiple social fields are linked. By 'transnational exercise', I mean an activity through which transnational relationships are fostered and nurtured via the social networks that are incorporated in a migrant's multi-national sense of belonging. As a context of migrant mobility, it represents an expression or even demonstration of the 'multiple ties and interactions linking people or institutions across the borders of nation-states' (Vertovec 1999:448). As a transnational activity, I view the return visit itself as one manifestation of the numerous social conduits between former homelands and new homelands. As such, it is in good company with other conduits through which contact is maintained (e.g., by courier services, fax, telephone and email).

If, however, the return visit acts as a particular expression of migrant *transnationality*, a further question can be identified. How, does the return visit, as a transnational exercise bridging identities and facilitating the maintenance of social networks and transnational spaces, facilitate return migration? If both the return visit and return migration are actor-centred processes by which migrants demonstrate their social connectivity, can they, in some way, be linked conceptually? More specifically, can the return visit serve to enhance return migration? Data from an ethnographic case study of Commonwealth Eastern Caribbean migrants living in Toronto are used to explore these postulates empirically. First, however, a brief review of the Canadian-Caribbean migration system sets the contextual scene.

Migration from the Caribbean: Canadian Perspectives

A substantial number of migrants from the Caribbean arrived in Canada in the 1960s, after the United Kingdom in 1962 severely restricted the process of auto-entry; a privilege previously afforded to immigrants entering the United Kingdom from a colony state or dependency (Carter et al. 1993; Fraser 1990; Richardson 1992). Changes to the Canadian Immigration Act in 1967, which saw the introduction of a points-based system of judging the suitability of applicants and the decreasing emphasis placed on the United Kingdom as a source of migrants, greatly impacted the patterns of immigration by individuals from the Caribbean (Anderson 1993; Anderson and Grant 1975; Richmond and Mendoza 1990). This change ultimately led to a sharp increase in the total number of immigrants from the region in the 1970s.

Migrants from the Commonwealth Eastern Caribbean have traditionally sought residency in Canada for reasons of education and economic success (e.g., Pool 1989), although some migrate to be close to family members. As Chamberlain (1997) points out:

> 'The motive for migration may have had as much to do with maintenance of the family and its livelihoods, with the enhancement of status and experience, within a culture which prized migration per se and historically perceived it as a statement of independence, as to do with individual economic self-advancement'. (Chamberlain 1997:27)

By 1992, over 400,000 people in Canada identified a specific country in the Caribbean as their birthplace (Henry 1994). Throughout the 1990s, the annual number of migrants from the Caribbean to Canada has decreased substantially from a high of just over 16,000 in 1993 to 6,700 in 1998 (Statistics Canada 2001a). In the 2001 census of the Metropolitan Toronto region, over 167,000 people identified the broader Caribbean (including Bermuda) as their country of birth

(Statistics Canada 2004). In the 1996 census, 64,305 individuals living in the Greater Toronto Area indicated the Commonwealth Eastern Caribbean region as their place of origin (Statistics Canada 2001b).

Caribbean Return Visits and Return Migration Interfaces

The importance of transnational connections in Caribbean migration experience is reasonably well established in the literature (e.g., Basch et al. 1994; Byron 1994, 1999; Glick Schiller and Fouron 2001; Olwig 1993; Soto 1987). As Byron (1999) points out:

> 'So much of the Caribbean has now been transposed onto the streets of London, Birmingham, Toronto and New York that many migrants simply do not see a permanent return to the Caribbean as the ideal retirement condition. Being 'in touch' is more important'.
> (Byron 1999:299)

While a more focused and wider discussion of the role of the return visit in the context of Caribbean identities can be found in Duval (2003) and Duval (2004), what is considered here, once again, is the extent to which return visits play a role in return migration prior to the actual return migration event. The data utilized are from an exploratory study that consisted of eleven months of ethnographic fieldwork in 2000 and 2001 among members of specific (convenience-based) social networks within the Commonwealth Eastern Caribbean community in Toronto. Two methods of data collection were utilised. The primary method consisted of unrecorded, informal discussions and observations with migrants throughout the wider Toronto community. Importantly, this meant building relationships, establishing trust, and integrating myself into the community. The second method of data collection involved a series of focused semi-structured interviews with twelve migrants which were designed to build on the information gleaned from the unstructured interviews. The specific, verbatim comments from migrants featured below stem from the latter method, with the broader discussions from the former.

Similar to the wider population from which this ethnographic sample was derived, the migrants I spoke with were quite diverse in terms of demographic characteristics. Some held full-time, 'white collar' positions, while others were studying full-time. In terms of spatial settlement – their address and community location -several live in areas of the city where other Caribbean migrants currently reside. This was not, however, the general pattern for all migrant's settlement choices. Most migrants with whom the formal interviews were conducted had been in Toronto for at least a decade, and even longer in some circumstances. In some cases, individuals had emigrated from the Caribbean to the United States or Britain in the first instance, but ended up in Toronto if their spouse had shifted for employment reasons.

From the fieldwork, two broad themes emerged demonstrating that there is an interface between return visits and return migration: (1) the need to facilitate social ties so that certain relationships are socially meaningful upon permanent return, and thus the return visit acts to facilitate that return; and (2) the functional nature of the return visit, such that changes are measured and benchmarked against what is remembered by the migrant. Taken together, these two themes represent how return visits function with respect to return migration. They demonstrate how a particular transnational activity may be used – potentially and actually – to facilitate return migration.

The Return Visit as a Conduit of Social Agency

For many of the Caribbean migrant informants encountered during the fieldwork, the social connection afforded through return visits forms a mechanism or strategy for maintaining social ties back home. As subjective as the notion of home is, it is important to point out that transnational connections are not only solidified with former homelands. Just as Byron (1999) found that social networks incorporated migrants in Britain as well the Caribbean, migrants in Canada recognised the importance of facilitating links and friendships with other migrants. To some extent, this counters the suggestion that transnational tendencies exist *in place of* assimilationist activities (see Waldinger and Fitzgerald 2004). Indeed, Portes (2003) is quite correct in stating that not all migrants are transnationals. Many migrant informants, who more characteristically fit a broad transnational classification, suggested that maintaining local (Toronto) social networks was often just as important as retaining social ties back in the external homeland. This was certainly accomplished using technological means such as email and the telephone, but more physical means of connection or retention were also sought.

For those considering returning permanently, it became apparent that making an effort to get involved in the home community in which they finally intend to re-settle is important. This is made possible, from their perspective, through periodic return visits. In fact, to do otherwise may serve to potentially jeopardize one's social standing back home. One female migrant from St. Lucia suggested that her reason for returning was to 'keep the network alive'. A Dominican male migrant, one who suggested that he was of 'some regard' in his village, stressed the importance of making trips back to his home:

> 'People used to look to me as their leader, knowing that I used
> to share information with them, and I brought information
> back, and up to this day, when I go back, they have that level
> of respect for me'.

Similar sentiments have been recognized in the literature. Hinds' study (1966) provides an interesting assessment of a fellow named 'Frank' who decided to return home to Jamaica after living in Britain for ten years. Upon arrival in Jamaica, he found that it had changed:

> 'The dream of returning was not so romantic when I reached
> the airport. It is a fact that the place is not very elegant. But
> returning migrants are considered tourists. The taxi drivers are
> out to squeeze everybody they can get hold of. You have to
> stick out or they would "thief" your eyes out of their sockets'.
> (Hinds 1966:191)

'Frank' experienced Jamaica in much the same way a foreign tourist would, without any social or familial ties to the overseas destination. What this suggests is that regular connection in the form of physical contact can often be critical. A migrant from Dominica living in Toronto noted that '… by going back on a regular note, they [locals] treat you as one of them …'. The familiarity is maintained and even enhanced.

This level of re-integration is above and beyond the level of the individual family unit. Instead, it speaks to conscious recognition of how returning migrants are viewed within wider Caribbean society. During fieldwork, informants emphasized that if overseas migrants do not build prior periodic return visits into their planned return migration, their degree of acceptance back into their home community is severely limited. One migrant from St. Vincent made this point quite clearly:

> 'What I find … [is that] individuals do not plan their move
> back process properly. People get someone to look for a piece
> of land in an area they consider suitable, and they build a
> house, and they set a date to move back. And while this house
> is building, they make a few trips back to check on it, and then
> they done with it and move back. I believe the correct
> approach should be you have to be focused on how do I
> become a member of this community. So while this house be a
> part of the community, some sportin' community or
> organization or school projects or function, get involved in
> these as well while you are visiting or when you just go back.
> But you cannot expect to just move back, live in your house,
> go to the city, buy your stuff, go in your house, and just be
> accepted'.

Another migrant from Dominica remarked:

> '… we have people coming home now from England, who
> have left for many years. They did contribute financially to the
> economic part of the island, but they themselves were absent.
> Okay, so when they get their retirement and they coming back,
> they still bring economic resources to the economy, they still
> treated as foreigners, because they don't know them. Where
> they been all those years?'

Importantly, while the eventual return of emigrants may be facilitated by periodic return visits, there is no automatic guarantee that the transition after return migration is made easier by such visits. Reintegration, and its problems and prospects, is situation specific. As such, if an individual concentrates his or her return visits on their former home community, yet returns permanently to a distinctly separate community, one might expect subtle differences to be experienced. In the words of one migrant from St. Vincent:

> '... a significantly increasing number of people who have returned in the last ten years, especially the last five years, and built nice homes and living well. But the generation they live with don't know them, because very often these people live in communities where they did not live before. So for example, if I return home, and I want to live in [Village A], where I grew up, those people who do not know me at least knew my father and my nephews and nieces and sisters and so. So, when they enquire "who's that guy" there's a connection because I grew up there. But if I were to go live in Kingstown [the capital of St. Vincent and the Grenadines], they would not know me. So, not only do you become a stranger to them, but then they would tap on the other element that these people have come to hate so much, namely a foreign element'.

What this suggests is that, while the return visit acts as a conduit through which migrants are able to maintain 'social visibility' in their external homeland, there may exist a degree of transnational connectedness that depends on the level of affiliation that a migrant may have in a locality. Some may feel nationalistic ties, while others may identify, first and foremost, with a smaller unit of association – a village, district, part of town, or a family home and neighborhood.

Regardless of the geographical variation in home 'ties that bind', the perception of returning migrants in the Caribbean is of particular interest here. During the course of the fieldwork frequent stories were relayed to the author about the practice by overseas migrants of regularly sending remittances back, sizeable portions of which are directed toward the construction of a retirement home. Problematic in this, however, is that some of migrants I spoke with may or may not have the financial ability to make a return visit in person, and their understanding of others' financial situations was such that return visits were also not possible. On the one hand, those who make repeated return visits have the benefit of maintaining visibility, and benefit from being on hand to oversee the building progress of their remittance investments. On the other hand, those who don't visit must rely more on their transnational social networks and the social obligations remittance flows tend to reinforce. 'Maintaining visibility' can be said to help attempts to 'return for good', since the return visitor is not as likely to be seen as a foreigner or individual for whom the place has no special social value. However, what might be the optimum (or minimum, for that matter) number of return visits needed in order to maintain transnational social networks is not clearly delineated. To some extent it will be dependent upon financial realities, but generalisable repetitive return visit impacts remain unknown. Among the migrants encountered during the fieldwork, some made frequent return visits (yearly and, in one case, twice annually), while others with lower net incomes would only make trips every three to five years. Indeed, some made no return visits, and their reasoning generally revolved around financial constraints, although some would speak of having no desire to retain ties. In essence, the reasons for non-return varied as widely as those given as reasons to return.

Identifying Change and Transformation

Another aspect of the return visit, which highlights its interface with return migration, is one that is somewhat more functionalist. While the overt social reasons for return visits have been discussed above, i.e. for maintaining networks of social relationships, some migrants make regular return trips in order to identify the tangible changes and transformation that have taken place in their homeland. Interestingly, despite technological advances in communication media such as electronic mail and satellite television, the return visit is still often seen as a preferable means of keeping pace with change.

One young male migrant from St. Lucia related a conversation he had with a friend who was about to return to her homeland in the Eastern Caribbean:

> '... last night we were talking and I said to her, "Yeah, you were up here, it was nice, you liked it, but going back home there is a lot of things you gotta t'ink about. The pace is too slow for you, the sun is too hot for you. You know, at 10 o'clock you might want eat pizza, you might pick up your phone and there's no pizza store open at that time. There's no delivery, there's no club, there's no Second Cup [Canadian coffee shop]. Where you can go and hang out? What you gonna do? Are you prepared for dat?"'.

In addition, identifying potential employment possibilities was a key consideration to the same young St. Lucian migrant for making a return visit:

> 'For me, I go to see the place and say "I probably need to do 'this' in college" or go to university and I might be able to make some money there... [or] "the island probably need this so why don't I try to make a research". I think these are things that we when we go home, these are things we should be looking for'.

The first quote from the male St. Lucian migrant highlights the relative importance of prior return visits when making the final decision to return. In essence, return visits point to a familiarity of social processes at home. In this particular case, it becomes a question of certain conveniences that may or may not be available. Consequently, migrants are genuinely aware of these 'conveniences' when planning to return. By keeping pace with change first-hand through a return visit, or two or more, migrants are able to strategically plan the final return. As evidenced in the second quote, this strategy can include taking advantage of educational training opportunities in Canada.

With much of the time spent on trips involving re-living of mental maps of village space, routes, and settlement patterns of friends and relatives, return visiting is effectively a process of re-familiarisation. Byron and Condon (1996), for example, note that measuring change during brief return visits might also be a way to assess the viability of a more permanent return: 'During visits to the region, migrants find that some people who remained in the Caribbean have built large houses and appear prosperous. To appear less successful would be unacceptable to most potential returnees' (Byron and Condon 1996:100).

The perception among many of those who migrated to Toronto in the 1950s and 1960s is that an entire generation has grown up in their absence, and many friends and relatives who have remained in the Eastern Caribbean have undergone some degree of personal change. While the length of time away from their natal home was often cited as a primary reason for going back, migrants certainly traveled home to see what had changed.

Discussion and Conclusion

Based on the brief case study presented above, migrants might well use return visits to measure change in their external homeland and assess the extent to which return is a viable option at a particular point in time. This temporal dimension is an important one, since it recognises that social and economic situations are indeed variable and, thus, the potential (or rather propensity) for return migration may vary considerably. What is important by way of conclusion, however, is that Caribbean migrants may well use the return visits as a means to maintaining visibility. To them, this results in a distinct advantage if a final return is eventually undertaken.

The utilisation of the return visit as a means by which return migration is facilitated is quite variable. The reason for this variation rests not only in the fact that the extent to which return visits are used by migrants themselves is variable. It is also because there are numerous and significant variables involved in decisions to either return visit or return migrate, not the least of which include the amount of social and economic capital raised and maintained in both the receiving country and the original homeland. Further, this will vary between and among first-generation migrants and second-generation migrants engaging in return visits or return migration (see, for example, the chapters by Potter, this volume).

Economic considerations of transnational living (Guarnizo 2003) could well determine how, and in what alternative ways, return visits are utilised by transnational migrants. We may even identify three distinct phases of the 'migrant experience' with respect to understanding how return visits may play a role in return migration. In the first phase, many new migrants, as a generalization, build both social and economic capital 'away from home'. Accordingly, return migration is theoretically less desirable in order to stay 'over there' and build a reputation back home as a successful migrant. In this first phase, the new migrant will work to build social capital in the receiving society, which may be manifested, for example, as social linkages with migrants and non-migrants, or in terms of the degree to which migrants become integrated (from an emic perspective) into their new place of residence. Similarly, during this phase the new migrant could well be working toward consolidating economic capital in the form of property acquired or even achieving employment and career gains. In effect, this first phase is representative of a period when, immediately following emigration, and during the formative years of immigrant adaptation, the propensity for return visits is greater than the propensity for return migration.

A second phase in the migrant experience could be characterised by the immigrant's duration of residence away from home and where stocks of social and economic capital accumulated are significant enough. In this phase it is expected that the propensity for return visits is roughly equal to the propensity to return migration. It is here where the migrant may consciously use the return visit as a means to maintain contacts, measure change and solidify any potential opportunities for return migration in the future, although the exact date and timing may not have been decided. This stage more or less reflects a period of 'migrant stasis' in order to reflect a somewhat more balanced migrant existence: they have established social networks in their new place of residence, they may have secured economic capital and investments, but at the same time they may still retain ties to their homeland. To some extent, this period is anything but one of stasis. In theory, migrants may be actively planning to return for good, but use the return visit as a means to extent or re-kindle their social networks back home. Eventually, or sooner or later, the third and final stage is characterised by a stronger propensity to return on the part of some migrants. In the face of (potentially) dwindling social and economic capital, with retirement, as 'empty-nesters', or as bereaved, widowed, separated, or disenchanted, the migrant may eventually decide that a final return is strategically advantageous. And, one final return visit may be all that is necessary to bring the final move back to fruition.

For the purposes of clarity, this conceptualisation of the return visit-return migration connection has been functional in that visiting helps the migrant strategically position him/herself in their personal post-migration environment. On the other hand, this is certainly not meant to imply that the return visit is the essential process affecting all migrants' return moves. To the contrary, I prefer to conceptualise the migration process in its entirety (thus, incorporating the return migration event). The inter-relationships between return visits and return migration requires a thorough understanding of the transnational nature of migrant behaviours and their interactions with migrant communities, both home and away. Return visits serve to bind migrant-sanctioned, transnational social fields that have been established and can therefore be characterised as a specific strategy (episode or repetitive episodes) within a wider migration process.

Overall, and as a demonstration of transnational agency, return visits can often aid in the successful reintegration of migrants who elect to re-migrate to their original homeland. They serve to link multiple 'homes' or spaces (Faist 2000) and bridge multiple identities that are manifested in these localities. In the context of Caribbean migrants, and as discussed above, it is as much economically strategic as social in its importance, since it can effectively help position and orientate the returnee to his or her former home. Without a doubt, and importantly, the return visit is one among many expressions of contemporary, transnational mobility. However, most importantly, this examination of the return visit-return migration connection is a valuable reminder that physical ties, person-to-person communication, and personal experiences are as essential a part of transnational lives as the international communication networks, the deterritorialised spaces, the institutional financial transfer systems, and the global theatre of information exchanges.

References

Anderson, W.W. (1993), *Caribbean immigrants: a socio-demographic profile*, Toronto: Canadian Scholars Press.

Anderson, A. and Grant, R. (1975), *The newcomers: problems and adjustments of West Indian immigrant children in Metro Toronto schools*, Toronto: York University.

Appadurai, A. (1991), 'Global ethnoscapes: notes and queries for a transnational anthropology', in Fox, R.G. (ed), *Recapturing anthropology: working in the present*, Santa Fe: School of American Research Press, 191-210.

Bailey, A.J. (2001), 'Turning transnational: notes on the theorization of international migration', *International Journal of Population Geography*, 7:413-428.

Baldassar, L. (2001), *Visits home: migration experiences between Italy and Australia*, Melbourne, Melbourne University Press.

Baldassar, L. (1998), 'The return visit as pilgrimage: secular redemption and cultural renewal in the migration process', in Richards, E. and Templeton, J. (eds), *The Australian immigrant in the 20th century: searching neglected sources*, Canberra: Research School of Social Sciences, The Australian National University, 127-156.

Baldassar, L. (1997), 'Home and away: migration, the return visit and 'transnational' identity', in Ang, I. and Symonds, M. (eds), *Communal plural: home, displacement, belonging*, Sydney: RCIS, 69-94.

Baldassar, L. (1995), 'Migration as transnational interaction: Italy re-visited', *ConVivio*, 1(2):114-126.

Basch, L., Glick Schiller, N. and Szanton Blanc, C. (1994), *Nations unbound: transnational projects and the deterritorialized nation-state*, New York: Gordon and Breach.

Bell, M. and Ward, G. (2000), 'Comparing temporary mobility with permanent migration', *Tourism Geographies*, 2(1):87-107.

Boyd, M. (1989), 'Family and personal networks in international migration: recent developments and new agendas', *International Migration Review*, 24(3):638-670.

Brah, A. (1996), *Cartographies of Diaspora: Contesting Identities*, London: Routledge.

Byron, M. (1999), 'The Caribbean-born population in the 1990s Britain: who will return?', *Journal of Ethnic and Migration Studies*, 25(2):285-301.

Byron, M. (1994), *Post-war Caribbean migration to Britain: the unfinished cycle*, Aldershot: Avebury.

Byron, M. and Condon, S. (1996), 'A comparative study of Caribbean return migration from Britain and France: towards a context-dependent explanation', *Transactions of the Institute of British Geographers*, 21:91-104.

Carter, B., Harris, C. and Joshi, S. (1993), 'The 1951-55 conservative government and the racialization of black immigration', in James, W. and Harris, C. (eds), *Inside Babylon: the Caribbean diaspora in Britain*, London: Verso, 55-72.

Chamberlain, M. (1997), 'Introduction', in Chamberlain, M. (ed), *Caribbean Migration: Globalised Identities*, London: Routledge, 1-20.

Coles, T.E. (2004), 'Diaspora, cultural capital and the production of tourism: lessons from enticing Jewish-Americans to Germany', in Coles, T.E. and Timothy, D.J. (eds), *Tourism, Diasporas and Space*, London: Routledge, 217-232.

Coles, T.E., Duval, D.T. and Hall, C.M. (2005), 'Tourism, Mobility and Global Communities: New Approaches to Theorising Tourism and Tourist Spaces', in Theobald, W. (ed), *Global Tourism: the next decade* (3rd Edition), Butterworth Heinemann, 463-481

Conway, D. (2000), 'Notions unbounded: A critical (re)read of transnationalism suggests that U.S.-Caribbean circuits tell the story, better', in Agozino, B. (ed), *Theoretical and Methodological Issues in Migration Research*, Brookfield: Ashgate, 171-190.

Duval, D.T. (2003), 'When hosts become guests: return visits and diasporic identities in a Commonwealth Eastern Caribbean community', *Current Issues in Tourism*, 6(4):267-308.

Duval, D.T. (2004), 'Linking return visits and return migration among Commonwealth Eastern Caribbean migrants in Toronto', *Global Networks: A Journal of Transnational Affairs*, 4(1):51-67.

Faist, T. (2000), 'Transnationalization in international migration: implications for the study of citizenship and culture', *Ethnic and Racial Studies*, 23(2):189-222.

Foner, N. (1997), 'Towards a comparative perspective on Caribbean migration', in Chamberlain, M. (ed), *Caribbean Migration: Globalised Identities*, London: Routledge, 47-62.

Fraser, P.D. (1990, 'Nineteenth-century West Indian migration to Britain', in Palmer, R.W. (ed), *In search of a better life: perspectives on migration from the Caribbean*, New York: Praeger, 19-39.

Glick Schiller, B. and G.E. Fouron (2001), *Georges Woke up Laughing: Long-distance Nationalism and the Search for Home*, Durham: Duke University Press.

Gmelch, G. (1980), 'Return migration', *Annual Review of Anthropology*, 9:135-159.

Gmelch, G. (1983), 'Who returns and why: return migration behavior in two North Atlantic societies', *Human Organization*, 42:46-54.

Gmelch, G. (1992), *Double Passage: The Lives of Caribbean Migrants Abroad and Back Home*, Ann Arbor: University of Michigan Press.

Goulbourne, H. (2002), *Caribbean Transnational Experience*, London: Pluto Press.

Harvey, D. (1989), *The Condition of Postmodernity*, Oxford: Blackwell.

Henry, F. (1994), *The Caribbean diaspora in Toronto: learning to live with racism*, Toronto: University of Toronto Press.

Hinds, D. (1966), *Journey to an Illusion: the West Indian in Britain*, London: Heinemann.

King, R. (1978), 'Return migration: a neglected aspect of population geography', *Area*, 10:175-182.

Kennedy, P. and V. Roudometof (2002), 'Transnationalism in a global age', in Kennedy, P. and Roudometof, V. (ed), *Communities across borders: new immigrants and transnational cultures*, London: Routledge, 1-26.

Larsen, J. (2001), 'Tourism mobilities and the travel glance: experiences of being on the move', *Scandinavian Journal of Hospitality and Tourism*, 1(2):80-98.

Longhurst, R. (1997), '(Dis)embodied geographies', *Progress in Human Geography*, 21(4):486-501.

Olwig, K.F. (1993), 'Defining the national in the transnational: cultural identity in the Afro-Caribbean diaspora', *Ethnos*, 58(3):361-376.

Olwig, K.F. and Sørensen, N.N. (2002), 'Mobile livelihoods: making a living in the world', in Sørensen, N.N. and Olwig, K.F. (eds), *Work and Migration: Life and Livelihoods in a Globalizing World*, London: Routledge, 1-19.

Parham, A.A. (2004), 'Diaspora, community and communication: internet use in transnational Haiti', *Global Networks: A Journal of Transnational Affairs*, 4(2):199-217.

Pessar, P. (1997), 'New approaches to Caribbean emigration and return', in Pessar, P. (ed), *Caribbean circuits: new directions in the study of Caribbean migration*, New York: Center for Migration Studies, 1-12.

Pool, G.R. (1989), 'Shifts in Grenadian migration: an historical perspective', *International Migration Review*, 23(2), 238-266.

Portes, A. (2003), 'Conclusion: theoretical convergencies and empirical evidence in the study of immigrant transnationalism', *The International Migration Review*, 37(3):874-892.

Portes, A., Guarnizo, L.E. and Landolt, P. (1999), 'The study of transnationalism: pitfalls and promise of an emergent research field', *Ethnic and Racial Studies*, 22(2):217-237.

Richardson, B.C. (1992), *The Caribbean in the wider world, 1492-1992,* New York: Cambridge University Press.

Richmond, A.H. and Mendoza, A. (1990), 'Education and qualifications of Caribbean immigrants and their children in Britain and Canada', in Palmer, R.W. (ed), *In search of a better life: perspectives on migration from the Caribbean*, New York: Praeger, 73-90.

Sørensen, N.N. (2003), '"Transnational" socio-cultural systems and ethnographic research: views from an extended field site', *The International Migration Review*, 37(3):787-811.

Sørensen, N.N. and Olwig, K.F. (2002)(eds), *Work and Migration: Life and Livelihoods in a Globalizing World*, London: Routledge.

Soto, I.M. (1987), 'West Indian child fostering: its role in migrant exchanges', in Sutton, C.R. and Chaney, E.M. (eds), *Caribbean Life in New York City: Sociocultural Dimensions*, New York: Center for Migration Studies of New York, Inc., 131-149.

Statistics Canada (2004), http://www.statcan.ca/english/Pgdb/demo35g.htm; accessed 7 June 2004.

Statistics Canada (2001a), www.statcan.ca/english/Pgdb/People/Population/demo35f.htm; accessed 26 January 2001.

Statistics Canada (2001b),www.statcan.ca:80/english/Pgdb/People/Population/demo28f.htm; accessed 26 January 2001.

Urry, J. (2003), 'Social networks, travel and talk', *British Journal of Sociology*, 53(2):155-175.

Vertovec, S. (2004), 'Cheap calls: the social glue of migrant transnationalism', *Global Networks: A Journal of Transnational Affairs*, 4(2):219-224.

Vertovec, S. (2001), 'Transnationalism and identity', *Journal of Ethnic and Migration Studies*, 27(4):573-582.

Vertovec, S. (1999), 'Conceiving and researching transnationalism', *Ethnic and Racial Studies*, 22(2):447-462.

Waldinger, R. and Fitzgerald, D. (2004), 'Transnationalism in question', *American Journal of Sociology*, 109(5):1177-1195.

Waldorf, B. (1995), 'Determinants of international return migration intentions', *The Professional Geographer*, 47(2):125-136.

Ward, R.G. (1997), 'Expanding worlds of Oceania: implications of migration', in K. Sudo and S. Yoshida (eds), *Contemporary Migration in Oceania: Diaspora and Network*, Osaka: The Japan Center for Area Studies, National Museum of Ethnology.

Yeoh, B.S.A., Lai, K.P.Y., Charney, M.W., Kiong, T.C. (2003), 'Approaching transnationalisms', in Yeoh, B.S.A., Charney, M.W., and Kiong, T.C. (eds), *Approaching Transnationalism: Studies on Transnational Societies, Multicultural Contacts, and Imaginings of Home*, Dordrecht: Kluwer Academic Publishers.

Chapter 13

Transnationalism and Return: 'Home' as an Enduring Fixture and 'Anchor'

Dennis Conway

Introduction: Transnationalism and Return Migration

New information technologies, a new international division of labor and the growing dominance of a rampant neoliberal capitalism are some of the interwoven yet fundamental 'global shifts' at work in today's globalizing world (Watson 1994). Emerging from current North American discourse on cross-border population movement in this new era of globalization has been the identification of 'transnational communities', where the people involved live between two worlds; their new migrant communities in the metropolitan North and their home communities in the South (Conway and Cohen 1998, 2003; Glick Schiller et al 1992; Massey et al 1998; Portes 1996). Transnationalism is often portrayed as a relatively new phenomenon; a product of contemporary globalization forces (Glick Schiller et al 1995). On the other hand, the idea that transnationalism is a new model of international mobility has been effectively questioned (Conway 2000; Foner 2001a). In the ensuing theoretical debate, alternative sociological constructs such as 'immigrant incorporation' (DeWind and Kasinitz 1997; Portes and Böröcz 1989) and 'selective assimilation' (Rumbaut 1999; Portes and Zhou 1993) have been advanced as more informative theoretical explanations of the contemporary experience of the new immigrant poor in North American cities.

Though perhaps not recognized, or labeled as such, transnational communities have existed from the time that immigrant enclave communities established themselves in the cities of colonial mother counties, and from the time of large scale international movements of people from Europe to the 'New Worlds' of the Americas – North and South (Conway 1998). As 'birds of passage', immigrants and their families brought with them their pre-migration cultural practices and social norms and values, while settling in and attempting to adapt to their new environments and ways of life (Piore 1979). Some fled oppression, destitution, landlessness, poverty and crises, with emigration to this new world across the Atlantic Ocean promising hope and a new, better life. Others, invariably the better endowed, who had pools of location-specific capital and strong social and familial ties with their source communities, preferred to retain the option for a possible return – what Rubenstein (1979) has labeled: a return mythology. Indeed, returns were common place (Kessner 1981), retention of old world values was the rule rather than the exception for many first generation immigrants, and only the duration of residence and inter-generational reproduction brought about changes in first and second generation immigrant perspectives and 'creolization' in family values (Foner 1997a).

The debate is multi-faceted, and the literature is burgeoning on transnationalism and transnational community formation (Faist 2000; Portes et al., 1999; Pries 1999; Yeoh et al., 2003). Difficulties lie in attempts to depict globalization's effects on peoples' mobility as a generalizable process when North American and European geographical and historical contexts differ so markedly, and structure-agency dynamics unfold in different ways. Even World Cities – like New York, London and Toronto – differ, having unique social and cultural milieus, while exhibiting similarities in terms of their international functions (Sassen 1996, 2000, 2001). As it happened, the first wave of literature announcing 'transnationalism' strongly suggested the emergence of a new process at large in the host cities of North America as they were re-structuring in response to globalization forces (Basch et al. 1994; Glick-Schiller et al 1992). There was a somewhat uncritical association of these 'transnational' processes with today's growing pre-eminence of globalization forces and the fast pace of integration of the capitalist world system (see Foner 1997b for a much-needed rebuttal). Globalization's rise to prominence as the pervasive super-ordinate set of macro-level forces responsible for the enlarging of the scale of world systems, responsible for geo-political re-alignments, and for the subordination of nation-state power to global capital, global industrial re-organization, and global intervention and mediation, certainly emerged as a new international force. And, there was some attempt to tie international movements of capital to international movements of labor under the rubric of this super-ordinate set of forces (Portes 1996, 1997; Sassen 1988). International (or transnational) migration, however, has been difficult to characterize as globalization's prodigy, in part because of the growing complexity of the international streams and their divergent characters in contemporary times (the 1970s-1990s), and in part because international mobility has long been part of social and structural changes through the eras of capitalism, beginning in the fifteenth century (Conway 1994, 2002; Massey 1988, 1990). Difference in nature and scope does not, however, constitute a new process, whether defined in terms of changing patterns of movement or enlarged systems of interaction (Gurak and Caces 1992).

Another uncritically accepted dictum, premature in my opinion, is that transnationality detaches people from their national or regional home, and conceives them as peripatetic movers interacting in a deterritorialized world, where they circulate among various home-places, become 'hydridized' by assuming multi-cultural identities, and become *de facto* 'global citizenry' (see Basch et al. 1994; Glick Schiller et al. 1992, 1995 as originators). Perhaps contributing to this globalizing conception has been the tendency for commentators and researchers to investigate the transnational migration behavior of target samples at only one location in their international trajectories; more often among enclave communities at their Core, metropolitan destinations, rather than their peripheral birth-places and 'home-places'. Both the metropolitan analyst-cum-commentator and the transnational population under the microscope are, at the time of examination, resident in the metropole, whether it is the United States – Basch et al (1994), Glick Schiller et al., (1995) Mahler (1995), Bailey et al., (2002) – or Britain – Chamberlain (1997, 1998), Goulbourne (2001). Theorizing about these transmigrant's experiences is therefore influenced by this geographical (co)incidence, and conceptualizing transnationalism is also framed in each particular metropolitan socio-cultural prism. To Nancy Foner (1997b), New York City transnationalism of West Indians is 'creolization'. To Mary Chamberlain (2001) and Stuart Hall (1996) London's Black British transnationalism is 'hybridization'.

Imposing a direction to the social and cultural molding of involved transnational migrants towards segmented assimilation or accommodation to the social milieu of only one locale of the life-spaces – the metropolitan host environment – seriously misrepresents the transnational process and infers an inevitability of modernist adoption that is oversimplified. Such a myopic focus on the migrant's socialization experiences and identification with the host culture misrepresents the abiding influence of the past, the memories of childhood, the mythical-to-real importance of family 'homes', the lasting power of attachments to birth-places and the emotional ties that bind people to home-places, land, territory, and localities in their peripheral societies (Douglas 1991; Mack 1993). Change in the strength and significance of these ties obviously occurs, because migrant travel experiences widen horizons, adds to their comparative knowledge and expands their *transnational geography* (see Duval 2002; Ellis et al., 1996). It may very well be the case that the notions of transnational communities as deterritorialized, *place-less* domains of interaction and of transnational identities as hegemonic cultural traits are more hypothetical constructs than tangible experiences. In part, this is because this global idea of transnational placelessness undervalues the importance of attachments to home-places and denies the time-space influences on peoples' senses of place. In part, it is a flawed notion because people's lives in and between two (or more) life-spaces – one their home, the other(s) accommodations 'away from home' – utilize these locales as their territorial anchors – the former an enduring haven, the latter being more temporary places. Furthermore, these transnational fields of circulation and time-space interludes of mobility and immobility will undergo change and alteration throughout the life-course, and even inter-generationally, thereby changing the multi-local transnational context in indeterminate directions.

More perceptively, Alison Mountz and Richard Wright (1996) characterize their Oaxacan (O) transnational community in Poughkeepsie (P), New York as a hybrid transnational locale 'OP', and conduct an illuminating and better-balanced ethnography of the changing practices therein. Metropolitan modernization is not so privileged, because changes are observed in both transnational spaces of OP, but the focus on individual life-histories, and only rarely families, limits considerations as to how transnational family life is changing in home places, or how transnational influences – circulating people, capital and information – are changing family structures in homes away from home. This criticism aside, Mountz and Wright's (1996) ethnographic account identifies five dissenting groups' antagonism and 'resistance' to the social rules and obligations of overseas enclave communities to maintain traditional transnational social practices. Among '*los irresponsables*' in Poughkeepsie are 'delinquent dads' or absentee non-remitters; 'independently-minded', female migrants; Seventh Day Adventist families; 'practical questioners' [after Bourdieu 1977]; and elopers, or wife-robbers. These dissenters appear to be using absence and distance to help them embrace modernization's messages of individualization, and more freely challenge social practices and communal and familial obligations in their Mexican/Oaxacan transnational homes.

'Hybridity' is evident among these metropolitan dissenters, but their impacts on the transnational family networks they leave behind, or set aside, is not evaluated in terms of the significance of their dissenting practices to community cohesion, or community evolution. These dissenters might embrace identity change, but the resilience of the transnational family structures and social practices does not appear to be undermined, nor is there much of a demonstration effect, apparently. The OP transnational system appears to be resilient enough to accommodate for such 'revisionist', or exit, strategies of these dissenting groups of individuals (Mountz and Wright 1996). Might it not be the case that the transnational system's consolidation and continuation is grounded in the conformists' behaviors, not the dissenters'? And, it is the transnational attachments to peripheral 'home-places' which remain among the conforming participants in 'away-places', that also consolidate the system through continuous time-space: simply, there would be no entrenched OP transnational space, without its Oaxacan home-land.

The 'Home as an Anchor' in Transnational Fields

A troubling omission in early perspectives on transnationalism is that the construct focuses on the bi- and/or multi- local nature of the international spaces in which the transnational migrants and their families live their daily lives, but gives little currency to the enduring influences of the socialization experiences 'back-home'. A transnational 'de-territorialized' space is instead conceived as an abstract simultaneously-occurring life-space 'between two worlds'. The conceptual focus is on the interactional field that lies across borders, links two places on either side of the border, and defines the backward and forward transfers of information, people, goods and services, political and social obligations and the like, as a transnational field or network of exchanges. Privileging metropolitan notions is one problem in this abstraction (Conway 2000). Denying, or diminishing, the importance of the migrant's source community and homeland, locational influences is another oversight. In short, the accumulated stocks of home-land attachments, obligations, social nets and individual identities are assumed away.

It is my thesis in this chapter that 'home' community social and cultural structures, social identities, obligations and networks and the social changes therein, are as influential a set of forces as the global, metropolitan structural forces for establishing the life-spaces within which transnational people move, make their life decisions, adapt to and/or resist. 'Home-places' provide the anchors for migrant's real world experiences. Home's opposite, namely 'homelessness' can not really be conceived as a voluntary livelihood option of any longevity, can it? The temporary accommodation a transnational migrant seeks when away from home is just that; a temporary shelter preferred over homelessness, but not comparable to a more permanent and secure livelihood option that a home provides. Home ties are an enduring fixture in people's life-courses, though there may be episodic stages when individuals prefer to be 'footloose and fancy free', there may be youthful episodes when alienation from one's home (and tyrannical fathers and mothers) is only too real, and there might well be life transitions wherein the ties to one's family/ancestral/island home may be exchanged for another of one's own making, or the children's making.

On the other hand, an entrenched, uninterrupted continuity of home ties is too stringent a requirement to be expected in the lives of immigrants and their children in today's transnational and global world. Flexibility (even impromptu opportunism) is as likely to feature in migrant's strategies as predictability and rationality. Ties to home may wax and wane, ties to another home may strengthen, ties to the migrant's birth place may weaken, be replaced, and so on. Yet, the sense of 'belonging' to a 'home as anchor' gives migrants the security that enables them to have flexible approaches to livelihood chances and options. In short, the connection to a home-place provides a much needed territorial fix, enabling/facilitating and even countering or reducing the risks and uncertainty inherent to an international mobility strategy, because it offers a return option as a fall-back strategy.

Transnationalism in Metropoles: 'Homes Away from Home'

Focusing on West Indian (Barbadian) migration to Britain, Mary Chamberlain (1995: 255-256) notes that '[t]he global dimension of migration, played out in international labor markets, and mediated by maneuvers of the host politics, engages with home-based social and cultural history which has furnished and continues to furnish Caribbean migrants with their own agenda. In this agenda, the family can be seen to play a significant role as both the end goal, and the means to achieve it'. Several important dimensions of the social process are suggested. 'First, the existence of a migration dynamic as a family dynamic which determined behavior and gave it meaning; second, the interplay between this migration dynamic and other family dynamics (such as color), and family goals (such as social mobility); third, the importance of the family in approving and enabling migration; and fourth, an ethos reflecting and reproducing a broader culture of migration which perhaps ran parallel with, but did not necessarily conform to the vagaries of international labor demands' (Chamberlain 1995: 256).

Furthermore, the creation of a new syncretic Caribbean culture in overseas enclaves, of increasing hybridity in identity formation, and the development and adaptation of older cultural patterns and social formations, occurs as migrants go about their day-to-day lives. There appears to be a 'creolization' (Foner 1997b, 1998) of the diaspora communities that is articulated in their material lives, within families, their workplaces, and their leisure spaces (Bailey 2001). It is within these micro-scale life-worlds that 'the points of similarity and difference, conformity and conflict are negotiated and resolved, where family values and cultural practices are transmitted, contested and transformed, and where national, transnational and/or hybrid identities evolve' (Chamberlain 1998: 8). Yet, this 'creolization' is not simply a metropolitan transformation or segmented assimilation, where the past socio-cultural traits are exchanged for the host society's. It is truly a hybridized existence in which both the 'home' and 'away from home' experiences are, relationally, enduring or involved frames of reference. Retention of the home-ties, maintenance of family practices, and referencing the two life-worlds as counterpoints, situates West Indians in between and within transnational networks and social spheres. Flexible and evolving, these spheres are comparative and inclusive, experiential and highly contextualized.

The resultant syncretic culture is diverse in its range, multi-cultural in its making, but most often generalized by reference to a common home-base in the Caribbean. Accordingly, British Jamaicans, Canadian Jamaicans, Afro-Jamaicans have common bonds as Jamaicans first and foremost, where points of departure are their hybrid metropolitan traits – their 'Britishness', their 'Americanisms', their Canadian affiliation. Their West Indian-ness distinguishes them while in their metropolitan milieu, while their British-ness or Canadian-ness might likewise provide them with distinction, or identity differentiation, in their island home place. And, significant identification with their island home can come about when deep-felt nationalism enters into the equation. For example, athlete Felix Sanchez, though born in Brooklyn, won the Athens 2004 Olympic gold medal running for his 'own country' in the 400 metres hurdles, the Dominican Republic. Eight years earlier in the 1996 Olympic Games, Donovan Bailey though winning the 100 metres final representing Canada, declared his gold medal was for Jamaica, too. On the other hand, Black-British immigrants, like Jamaican-born Linford Christie, won sprinting acclaim for his adopted host in the 1992 Barcelona Olympics, and captained the British track and field team from 1990 until 1997. 'Hybridity' has considerable flexibility in terms of self-identification, it appears.

Eric Swyngedouw (1997) appears to have it right when he identifies spatial scale as an arena and (space-time) moment that is socially produced, where power relations are contested and conflicts and compromises are negotiated. Unfortunately, Bailey's (2001) 'scaling up' of transnational identity-formation and relational ties of people to the communal territory, overlooks the deeper (and more meaningful) attachments (and detachments) of children, youths, young, middle-aged, old, elderly, aged-aged, women and men to home-places and to family-homes, and overlooks the gendered power-contests, negotiations and the like, that Chamberlain (1998, 2001) among others (Conway and Cohen 1998, 2003; Ellis et al. 1996; Folbre 1988, 1992; Katz 1991; Wheelock 1992) finds so significant at the familial and household levels. 'Geography matters' in this process too, because the repetitive 'circulations' of family members are movements of people bringing them together and then apart, moving individuals – children in their formative years, youth in their rebellious years, young adventurers in their risk-taking years, newly married adults in their successive life-course transitional phases, and so on. Such circulatory life experiences not only build transnational ties to home-places (and homes away from home), but their geographical configurations – distance traveled, durations-of-stay, repetitive cycles – and space-time life-worlds – local activity fields and experiences – in both home and away places also condition the transnational ties, social fields and networks.

'Class' matters too, partly because today's transnational families, circuits and communities are highly diversified in terms of their human capital stocks, their socio-economic standing, educational- and skill-levels, occupational and labor market experiences, family and kinship resources. In part, class power relations also matter as an essential feature of the transnational migrant's contextual environments at home and abroad, because under capitalism's mantle, both milieus have socially-constructed and gendered hierarchical structures which the migrants have to resist, negotiate, accommodate to, acquiesce, or escape. Class positions, acquired skills, knowledge-bases, familial and work experiences provide transnational migrants with accumulated 'bundles' of stocks of human and social capital, which they use to adapt to their changing circumstances and to achieve satisfactory levels of livelihood for themselves and their dependents. 'Class status' may even be a negotiating factor in its own right, if migrants' language-command, socialized patterns of dress, of manners and bearing can be utilized as effective symbols of power and prestige and of upper- (or even lower)-class distinctiveness (see Chapter 4 by Potter and Phillips in this collection).

Compared to previous eras of capitalism, and 'ages of international migration' (Castles and Miller 2003), a much higher proportion of today's transnational migrants originate from the upper, or privileged, classes of their sending societies; they have advanced education, professional skills, overseas experience, and some times substantial stocks of financial capital or family resources to ease their entry and adaptation to their new destination (Salt 1997). This privileged resource base and 'bundle' also facilitates their transnational communication and transnational transfers of knowledge, information and people, so that they are able 'to participate in the manner of modern-day cosmopolitans, in high-level institutions and enterprises here (in the US *[author's insertion]*) and in their home society' (Foner 2001a: 45). Clearly, the transnational experiences of these highly skilled migrants will differ markedly from those of the least-skilled. Indeed, we might opine that transnationalism privileges them even more, and rewards them more than the elitism they might have enjoyed in their former society. If we were to characterize them as global elites (see Martin 2001), experiencing considerable material and social returns from their cosmopolitan transnational identities and practices, we might not be far off the mark.

By contrast, the transnational experiences of those less well-endowed, and with lower class assignations, not only differ, but will be very much determined by the extent to which their social, familial and financial resource 'bundle' enables them to access transnational services, utilize transnational networks for sustenance and survival, rather than for further enrichment. A broad range of middle- to lower-class identities among transnational migrants should be expected, differentiated by previous class position, race and ethnic identities as 'others', and by their familial networks and human and capital resources. Those among the most unfortunate and least endowed – such as impoverished refugees, unauthorized migrants, the smuggled, enslaved and in bondage – might be least likely to adopt, or assume transnational identities, but such a *carte blanche* dismissal might have its limitations, because human resiliency and determination often accompany extreme hardship, not necessarily hopelessness.

Since 'hope' is an abiding feature of migration's promise, however, transnationalism is quite likely to emerge as a survival and sustenance strategy if modest resources come available, and if family or community ties 'back home' can be mobilized. In these most desperate of cases, and among the less well-endowed, the strength and resilience of transnational family networks, the growing involvement of more and more families and individuals in transnational circuits of people, capital, and information, and the cumulative growth and economic diversification of transnational societies both 'home and away', will build global-to-local linkages for the benefit of participants, mobile and immobile (Conway and Cohen 2003; Portes 1997; Smith and Guarnizo 1998). Wholesale 'new immigrant' upward social mobility can not be predicated as an outcome of this global movement of increasing transnationalism between the underdeveloped South and privileged North life-worlds. However, we might suggestively hypothesize that the small proportion of the world's 'poor and oppressed' who are using transnational paths are searching for the same avenues to self-advancement that earlier waves of international migrants sought, and quite a few reached (Foner 2001a; Morokvasic 1984; Piore 1979, 1986). Transnationalism and migrant upward social mobility is a productive project for many, that's for sure (Portes and Guarnizo 1991; Vertovec 2002).

Transnationalism at the Peripheries: 'Home-places', Real and Imaginary

Bonham Richardson (1983) evokes the historical traditions from slave times to present day, stressing the ecological limits of, and peoples' adaptations to the islands' fragile and denuded ecologies. 'Working local land has been half of the usual survival strategy of the people. The other half has been the world outside, which men and women of St. Kitts and Nevis have tapped for a century and a half' (Richardson, 1983: 170). 'Leaving in order to stay' was the metaphor coined by the late Bernie Neitschmann (1979) to characterize the migration of the Miskito of Nicaragua in the Far West Caribbean. Charles Carnegie (1982: 13-14) observed that 'Caribbean people routinely use cultural ideas which emphasize flexibility and the building of multiple option. Being flexible has at least two dimensions: adjusting rapidly to whatever comes along, and secondly the actual building of multiple options, potential capital as it were, to hedge against future insecurity'.

In an earlier attempt to generalize on the region's migration traditions, I characterized Caribbean circulators as opportunistic and individualistic, in large part because 'the incidence and durability of circulation is made possible by the flexibility of "home" residential environments, of intra-family relationships and of adult-children responsibilities' (Conway 1986: 4). Some time later, Murray Chapman (1991: 290) further characterized the contemporary movements of Pacific Islanders thus: 'At one level, the movement of Pacific Islanders can be conceived in terms of creative ambiguity and controlled paradox; at another, it constitutes regional interaction and socioeconomic inter-change among many and varied peoples'.

Harry Goulbourne (2001) observes that a fundamental feature of the value system of Caribbean societies is an extremely high level of individualism, in terms of people's actions and social acceptance. 'Consequently, whilst a person's colour/race is important, social action is judged according to a person's individual virtues rather than those purported to be exhibited by a whole group' (Goulbourne 2001: 237). And, while this leads to relatively high levels of social and political tolerance of racial/color and political difference in their island home-lands, it is an unwelcome contrast to the 'everyday racism' (Essed 1990) experienced by people of color in the Anglo-American metropoles of North America and Europe.

Individualism is only one identifiable trait of modernization among many, but as an autonomous attribute and foil to counter and 'resist' stereotyping and racist ascription by the 'white' host societies it has its pragmatic value for Caribbean women and men.

However, whether this generalization is as equally applicable to Caribbean women, as it is to men, remains an important unanswered question. We do know that many Caribbean women have been solely responsible for raising their children, and as heads of households have assumed the responsibilities of home management, children nurturing and raising and worked outside the home. Many have migrated overseas too, but a common strategy has been to leave dependent children with female members of extended families until they can be 'sent for' or 'returned to' (Ellis 2003). Both familism and individualism can be endearing qualities of Caribbean family members in the metropolitan North, because the former brings strength and cohesion to the migrant unit, while the latter provides members with the necessary self-reliance and self-confidence to effectively counter or reject racist pressures (Bryce-Laporte 1972, 1979; Dodgson 1984; Peach 1996).

Inserting 'Transnational Geography'

Given the growing complexity of the patterns of today's 'new age of migration' (Castles and Miller 2003), the multi-local, social contexts in which transnational migrants behave as risk-minimizers and 'satisficers', constitute a national-to-global set of nodes which are not simply (or solely) international. Rather, they can be local, regional and metropolitan in the periphery, as well as internationally varied in regional, hemispheric or global metropolitan. For example, recent research among Oaxaca valley transnational families, finds mobile family members in such multi-local venues as Oaxaca City, Mexico City, San Jose, California, Los Angeles (Conway and Cohen 2003). Karen Fog Olwig's (2001) Jamaican transnational family in New York City had family members in more than one transatlantic metropolitan, life-world – Britain, Canada or the United States – as well as islands other than the family's original birthplace – Barbados and the Virgin Islands. Intra-regional mobility of Caribbean migrants and their family members has long been a survival feature for many hopefuls, and the strategic flexibility of opportunistic individuals seeking work and a better life 'off the island' has bequeathed a transnational identity on many a family and its successive generations; some times by design, other times by chance, or 'happen-stance' (Carnegie 1982; Conway 1986).

Given the multi-local nature of life experiences, the complexity of movement options, the complex network of exchanges within a transnational circuit, or circuits, and last but by no means least, the enduring influence of, and ties to, the socio-cultural milieu of the transnational migrant's home-place, transnational identities will be forged from a combination of 'roots' and 'routes'. They should also be expected to change and evolve as sojourns and visits are undertaken, repeated and altered, as live-courses take unexpected turns, family relations change, ageing occurs, and life experiences widen (or narrow). Because continuous time is an overarching spatio-temporal framework for migration decision-making through the life course, acts of migration, or acts of immobility for that matter, may very well have different sets of determining factors and consequential outcomes.

'Home' and 'Away' Transnational Spaces: Different Geographies

Quite clearly, more than one locale, and more than one transnational space has to be examined (and its structural and agency interactions explained) in any conceptualization of transnationalism. It is the more complete case that transnational family and kinship networks, transnational circuits and transnational 'home' and 'away' communities, constitute interlocked, yet regionally differentiated, transnational social spaces (see Faist 2000). Transnational geographies and social capital linkages are the mutually-reinforcing socio-spatial frameworks of these transnational sub-systems; the former differentiating space(s) and determining local-to-global spatial interactions and relations, the latter the (social) organizational and institutional investments and involvements of people's time-space interactions with their locales.

Two micro-scale places/spaces of inter-personal, familial and social interaction need to be given explicit recognition – the family/kinship scale and the local community scale. The latter being a larger territory, envelops the former, while the family unit and its individual members constitute the more proximate decision-making unit(s). Extended family ties widen the net of transnational spaces, so that we should be wary of limiting the interaction space of family and 'home' to the nuclear unit. Community scales will be territorially determined by the family- and individual-level, 'every-day life-geographies'. Transnational social linkages will operate at both, sometimes independently, sometimes in co-ordination.

Different Growth Dynamics of Transnational Communities at 'Home' and 'Away'

Furthermore, an important distinction is that the genesis of many a transnational network originates at the familial scale and only through time and with the growth of transnational linkages, ties and interactions among families, their members and friends does the consolidation of a transnational identity to the home community occur. In contrast to the growth trajectory and consolidation of the 'home' transnational community in the periphery, the metropolitan, or 'away', transnational community takes shape and gains coherence through the adjustment experiences of individuals and family members to the host, metropolitan society; accommodating to, overcoming, or resisting the alienation pressures they experience. Indeed, the identity of this 'away' community is in large part given meaning and substance by the existence of a 'home' community, or more specifically a *common locality* of family home-places among absentee transnational migrants.

The two geographies of 'home' and 'away' are not only regionally distinctive in terms of their periphery or core contexts, but they are also situated at distant nodes in their transnational networks, or distant poles of the set of bi-national or multi-local transnational familial networks of migrants. In addition, the geographical processes of local movements, familial activity fields, household formation and consolidation, work regimes, leisure regimes and day-to-day material live-worlds in these dispersed communities, differ in their nature and their consequences. The two, three or more localities that make up the bi-national or multi-local networks, are, of course, influenced by the transitory through-flows of people, information and capital, but their social milieus are also strongly influenced by the contextual legacies and social codes of the past.

Transnational communities in metropolitan destinations, as 'homes away from home', are forever in states of flux and change, because movements of people in and out of such communities, in and out of family arrangements, in and out of communal and societal obligations, and exiting for good in some cases, bring about many changes in the family and community structures in such alien environments. Transnational communities in peripheral home-lands are likely to have more inertia, more social organizations, and more social capital investments, in large part because the home-places in these nodes within the networks, circuits and wider diasporas are the valued anchors for families. Migrant target earnings are therefore earmarked for marriage, family formation, family maintenance, family health, home building, maintaining, repairing and enlarging. Remittances are invested by recipients similarly, although providing the immediate necessities of life – food, clothing, medicine, school uniforms and books, etc – for family members is paramount in much of the day-to-day use of remittances. The bulk of the flows of remittances between 'home' and 'away' places are predominantly at the individual and family scales. Circuits of capital, information and people flow within family business ventures, and flows to home communities are mainly prompted by familial obligations, ties and personal linkages (Conway and Cohen 1998, 2003: Orozco 2002).

The Durability of Ties to Home

Today's transnational life-styles may be different in scale and scope than earlier mobility strategies. Innovations in communications and transportation technologies and logistical systems may shrink distance and facilitate transnational contact and interaction, but they do not fundamentally change people's territorial attachments and ties to a home-place, or home-land. 'Home' community social and cultural structures, social identities, obligations and networks and the social changes therein, are as influential a set of forces as the global, metropolitan structural forces for determining and circumscribing the life-spaces within which transnational people move, make their life decisions, adapt to and/or resist.

The 'home-place' is not only an anchor for local circulatory behavior as Chapman and Prothero (1985) so perceptively argue, but it is also an important anchor for migrants and their family members undertaking national and international circulation (Conway and Cohen 2003). It is not an unchanging 'home-place', obviously, because movement away changes the locales of interaction, and widens the scope of territorial experience for the migrant. Douglas (1991), while attempting to situate a home as an embryonic community, notes: '[h]ome is "here", or it is "not here" ... it is always a localizable idea ... a home is not only a space, it also has some structure in time; and because it is for people who are living in time and space, it has aesthetic and moral dimensions ... [w]hy some homes should have more complex orienting and bounding than others depends on the ideas that persons are carrying inside their heads about their lives in space and time. For the home is the realization of ideas' (Douglas 1991: 289-290).

When families migrate however, homes are exchanged, with one being left behind. Time spent in the new 'home away from home' deepens associations and attachments, and new social and spatial interactive fields are utilized in ways which imbue the new home with meaning and identity. Other family-related factors will re-value the new home while, devaluing the home left behind. If the extended family eventually leaves to rejoin their kin, then this lack of family 'back home' may very well reduce the obligations to return periodically. If or when lack of resources, or other obstacles, prevent circulation, delay, or preclude visits, then as the duration of staying away lengthens, devaluation of the ties to the former home is likely. Retention of ties to the previous home, however, can persist with the new home never supplanting the former in the minds of the migrants and family members as their 'real home', their 'island home'. This can come about because, regardless of the lack of personal visits, the feelings of alienation towards their new environment, the comparative lack of an appropriate 'homely' accommodation in the metropolitan destination, a resistance to identifying with the 'other man's world', and adherence to a target goal for the sojourn which is couched in temporary terms, and the durable strength of the family or kinship networks, which keep people in touch with their home-places: all these attachments to home, when combined to durable senses of alienation and distrust of the new place 'away from home', will ensure the retention of a return mythology (Byron 1999; King 2000; Rubenstein 1979).

Regardless of such changes, as life takes its twists and turns, and migrants move to and fro, it can still be claimed that a 'home' is an enduring territorial fixture; one that provides mobile people with a sense of place, a sense of belonging somewhere – a rootedness (Tuan 1980). For transnational migrants, the 'home away-from-home' can be merely accommodation – temporary, transitory, a 'no fixed abode'. The island home, on the other hand, can be both the migrant's constraining 'tyrannical' environment as it is her/his territorial haven (Fog Olwig 1998), but it is an enduring social construct held by mobile and/or immobile people; whether it is mythical or experiential; 'it all depends' (Hollander 1991).

Conclusion

Returning Home: A Long-held Intention, Realized Sooner, Later or Not at All

Mary Chamberlain's (1997) insightful examinations of British-Barbadian exiles' narratives and emigration stories, observes that 'transience and return' are two of the main themes of their experiences. She goes on to conclude that they 'suggest an "instability" in regard to national boundaries, and an implicit challenge to the idea of a nation-state (and the 'British' way of life) as the natural and only form of political and social organization'. She concludes: 'More particularly, the responses of Barbadians in Britain were historically conditioned, reflecting and creating a "structure of feeling" which was essentially international, and represented one resolution of a set of contradictory relationships. The narratives – of exile and return – are essentially stories of a journey, which is always a process and always relational' (Chamberlain 1997:88).

As other commentators on West Indian experiences in London, New York and Toronto – notably, Byron (1994), Foner (1997, 1998, 2001b), Henry (1994), Peach (1968, 1984), Philpott (1973, 1977) and Kasinitz (1992) – have also observed, the consensus is that an 'intention to return' is an essential feature of the migrant's calculus, and an essential part of their security-consciousness. First generation West Indian emigrants to Britain – the 'Windrush generation' – return home upon retirement, or at other equivalent life-course and/or community benchmarks, such as family reunions, baptisms and weddings, national celebrations (for example, Carnival in Trinidad, or Crop Over in Barbados), or after years of relatively permanent emigration in their colonial 'mother country'. Of course, some returned home much earlier when they failed to adapt to the new situation or achieve advantages from the initial international move (Davidson 1969; Dumon 1986; Peach 1968). Such returns home are to be expected in any long-distance movement, and accord with Ravenstein's (1885) long-held idea that every migration flow has its counter-flow.

However, for those who did adjust to their metropolitan situation, acquired material wealth, skills and experience during their working life, and successfully managed to build a family life as well, a rationalization of the situation, on or nearing retirement may very well induce an actualization of the perceived mythological intent to live the remainder of their days 'back home'. The actual home to which they retire may often be very different to the one left behind. The location may more likely be in an urban setting, or a site suitable for retirement and a more relaxed life-style (Tobago, for example). It may have been readied by remittances and investments during the time abroad, or it may be purchased by transfer payments, retirement savings, and investment monies, but it is likely to be suitably 'magnificent and opulent', or symbolic of the returnee's social standing and accomplishments. As a symbol of migration success it must be appropriately modern, a home to demonstrate that the returnees have made it (Connell and Conway 2000).

Children of the Windrush generation, as second-generation British-West Indians, may also have strong attachments to the island home of their parents (or both islands, if the parents have different regional origins). Family lore, family connections, family reminiscences are common social frames of reference within West Indian households, and when compared with the immediate metropolitan surroundings, the nostalgia for Caribbean life-styles, and Caribbean environments might very well be exaggerated, or mystified, but most certainly lauded and revered in many a case. Children brought up in West Indian families, therefore, echo the sentiments and views of their parents, often feel strong attachments to their parents, and find a common ground with their parents' 'senses of feeling'. They can build strong ties to their parents' island home within this family environment, but when and if return visits are made then their national ties might very well be strengthened further. Some West Indian parents deliberately foster this attachment by taking their children on repetitive return visits: 'to get them used to t'ings back home'. When the aging parents eventually return back home to realize their intention to finish out their days in their 'islands in the sun', their metropolitan-born children may very well be drawn to follow them, if the intra-family ties have been harmonious and deeply-felt. Familial support of grandchildren's day-care, support for single mothers with grandchildren, and strongly-felt family ties between mothers and daughters, can all be influential factors in the decision by second generation Black-British, Afro-Caribbean and American-West Indians to return 'home' to try and see whether things will work out.

First-, second-, and possibly third-generation West Indians retain double identities, that of their parents' and/or grand-parents' island home, and that of their metropolitan residence. Some hold to multi-local identities, because their life path has taken their parents and them from an island home, to Britain, then to Canada or the United States, where attachments to these places might grow with their durations of residence and life course experiences; marrying foreigners, having their children grow up in different places, visiting and revisiting these diasporic 'homes away from home'. Some, of course, turn away from the island home of their parents and through assimilation, adapt to, and adopt American, Canadian, or British identities.

The possibilities are relational and contingent on the individual's experiences, and a diversity of responses to transnational lives has to be expected. To predict otherwise, would deny West Indian and Caribbean people the strategic flexibility that generations have practiced. Equally important and influential, the home as an anchor and an enduring fixture are geographic and territorial identity-markers, which are so intertwined with the return intention to go 'back home' that they are essential features of West Indian and Caribbean transnational mobility and strategic flexibility. The resultant transnational geography is dynamic and diverse, complex and contingent. The regional diasporas can be expected to continue, and more and more of the peoples' global worlds will be transnational fields, experiences and livelihoods. Return migration is but one phase and one option in her/his life's pathways, and the intention to return is an essential ingredient in the transnational migrant's strategic praxis, whether it is realized, or postponed for ever.

References

Bailey, A.J. (2001), 'Turning Transnational: Notes on the Theorisation of International Migration', *International Journal of Population Geography*, 7: 413-428.

Bailey, A.J., Wright, R.A., Mountz, A. and I. Miyares (2002), '(Re)Producing Salvadoran Transnational Geographies', *Annals, Association of American Geographers*, 92(1): 125-144.

Basch, L., N. Glick Schiller and C. Szanton Blanc (1994), *Nations Unbound: Transnational Projects, Postcolonial Predicaments, and Deteritorrialized Nation-States*, OPA (Amsterdam) B.V. and Gordon & Breach.

Bourdieu, P. (1977), *Outline of a Theory in Practice,* Cambridge: Cambridge University Press.

Bryce-Laporte, R.S. (1972), 'Black Immigrants: The Experience of Invisibility and Inequality', *Journal of Black Studies*, 3: 29-56.

Bryce-Laporte, R.S. (1979), 'New York City and the New Caribbean Immigration: A Contextual Statement', *International Migration Review*, 13(1): 214-234.

Byron, M. (1994), *Post-War Caribbean Migration to Britain: The Unfinished Cycle*, Aldershot, Hong Kong, Singapore, Sydney: Avebury.

Carnegie, C.V. (1982), 'Strategic Flexibility in the West Indies: A Social Psychology of Caribbean Migration', *Caribbean Review*, 11(l):10-13, 54.

Castles, S. and M. Miller (2003), *The Age of Migration, 3rd Edition,* New York: Guildford.

Chamberlain, M. (1995), 'Family Narratives and Migration Dynamics: Barbadians to Britain', *New West Indies Guide/Nieuwe West-Indische Gids*, 69(3 & 4): 253-275.

Chamberlain, M. (1997), *Narratives of Exile and Return*, New York: St. Martin's Press.

Chamberlain, M. (1998), *Caribbean Migration: Globalised Identities*, London and New York: Routledge.

Chamberlain, M. (2001), 'Migration, the Caribbean and the Family', in H. Goulbourne and M. Chamberlain (eds), *Caribbean Families in Britain and the Trans-Atlantic World*, London and Oxford: MacMillan Press, Caribbean Studies, 32-47.

Chapman, M. (1991), 'Pacific Island Movement and Socioeconomic Change: Metaphors of Misunderstanding', *Population and Development Review*, 17(2), 263-292.

Chapman, M. and R.M. Prothero (1985), 'Circulation between "Home" and Other Places: Some Propositions', in M. Chapman and R.M. Prothero (eds), *Circulation in Population Movement: Substance and Concepts from the Melanesian Case*, New York: Routledge & Kegan Paul, 1-2.

Connell, J. and D. Conway (2000), 'Migration and Remittances in Island Microstates: A Comparative Perspective on the South Pacific and the Caribbean', *International Journal of Urban and Regional Research*, 24.1: 52-78.

Conway, D. (1986), *Caribbean Migrants: Opportunistic and Individualistic Sojourners* (Hanover, NH: University Field Staff International Report), No. 24, 15.

Conway, D. (1994), 'The Complexity of Caribbean Migration', *Caribbean Affairs*, 7(4): 96-19.

Conway, D. (1998), 'International Migration and Refugees in Latin America', in Jack W. Hopkins (ed), *Latin America: Perspectives on a Region, 2nd Edition*, New York: Holmes & Meier, 264-280.

Conway, D. (2000), 'Notions Unbounded: A Critical (Re)read of Transnationalism Suggests that U.S.-Caribbean Circuits Tell the Story Better', in *Theoretical and Methodological Issues in Migration Research: Interdisciplinary, Intergenerational and International Perspectives*, Biko Agozino (ed), Ashgate Publishers, Aldershot, UK, and Brookfield, USA, 03-226.

Conway, D. (2002), 'Gettin' there, despite the odds: Caribbean migration to the U.S. in the 1990s', *Journal of Eastern Caribbean Studies*, 27(4): 100-134.

Conway, D. and J.H. Cohen (1998), 'Consequences of Migration and Remittances for Mexican Transnational Communities', *Economic Geography*, 74(1): 26-44.

Conway, D. and J.H. Cohen (2003), 'Local Dynamics in Multi-local, Transnational Spaces of Rural Mexico: Oaxacan Experiences', *International Journal of Population Geography*, 9(1): 141-161.

Davidson, B. (1969), '"No Place Back Home": A Study of Jamaicans Returning to Kingston, Jamaica', *Race*, 9(4): 499-509.

DeWind, J. and P. Kasinitz. (1997), 'Everything Old is New Again? Processes and Theories of Immigrant Incorporation', *International Migration Review*, 31(4): 1096-1111.

Dodgson, E. (1984), *Motherland: West Indian Women to Britain in the 1950s*, London: Heinemann.

Douglas, M. (1991), 'The Idea of a Home: A Kind of Space', *Social Research,* 58(1): 288-307.

Dumon, W. (1986), 'Problems faced by Migrants and their Family Members, particularly Second Generation Migrants in Returning to and Reintegrating into their Countries of Origin', *International Migration*, 24(1): 113-128.

Duval, D.T. (2002), 'The Return Visit-Return Migration Connection', in A.M. Williams and C.M. Hall (eds), *Tourism and Migration: New Relationships between Production and Consumption,* Dordrecht, Boston and London: Kluwer Academic, 257-276.

Ellis, M., D. Conway and A.J. Bailey (1996), 'The Circular Migration of Puerto Rican Women: Towards a Gendered Explanation', *International Migration (Geneva)*, 34(1): 31-64.

Ellis, P. (2003), *Women, Gender and Development in the Caribbean: Reflections and Projections,* London & New York and Kingston, Jamaica: Zed Books and Ian Randle.

Essed, P. (1990), *Everyday Racism*, Claremont, CA: Hunter House.

Faist, T. (2000), *The Volume and Dynamics of International Migration and Transnational Social Spaces,* Oxford: Clarendon.

Fog, Olwig K. (1998), 'Epilogue: Contested Homes: Home-making and the Making of Anthropology', in N. Rapport and A. Dawson (eds), *Migrants of Identity: Perceptions of Home in a World of Movement*, Oxford and New York: Berg, 225-236.

Fog, Olwig K. (2001), 'New York as a Locality in a Global Family Network', in N. Foner (ed), *Islands in the City: West Indian Migration to New York*, Berkeley, Los Angeles and London: University of California Press, 142-160.

Folbre, N. (1988), 'The Black Four of Hearts: towards a New Paradigm of Household Economics', in D. Dwyer and J. Bruce (eds), *A Home Divided: Women and Income in the Third World*, Stanford: Stanford University Press, 248-262.

Folbre, N. (1992), 'Introduction: The Feminist Sphinx', in N. Folbre, B. Bergmann, B. Agarwal and M. Floro (eds), *Women's work in the World Economy*, New York: New York University Press, xxiii-xxx.

Foner, N. (1997a), 'The Immigrant Family: Cultural Legacies and Cultural Changes', *International Migration Review*, 31(4): 961-974.

Foner, N. (1997b), 'What's New about Transnationalism? New York Immigrants Today and at the Turn of the Century', *Diaspora*, 6(3): 355-375.

Foner, N. (1998), 'West Indian Identity in the Diaspora: Comparative and Historical Perspectives', *Latin American Perspectives*, 25(3): 173-188.

Foner, N. (2001a), 'Transnationalism Then and Now: New York Immigrants Today and at the Turn of the Twentieth Century', in R.C. Smith, H.R. Cordero-Guzman and R. Grosfuguel (eds), *Migration, Transnationalization and Race in New York City,* Philadelphia: Temple University Press, 35-57.

Foner, N. (2001b), *Islands in the City: West Indian Migration to New York*, Berkeley, Los Angeles and London: University of California Press.

Glick-Schiller, N., L. Basch and C. Blanc-Szanton (1992), *Towards a Transnational Perspective on Migration: Race, Class, Ethnicity, and Nationalism*, Annals of the New York Academy of Sciences, Volume 645, New York: The New York Academy of Sciences.

Glick-Schiller, N., L. Basch and C. Szanton Blanc (1995), 'From Immigrant to Transmigrant: Theorizing Transnational Migration', *Anthropological Quarterly*, 68, : 48-63.

Goulbourne, H. (2001), 'Trans-Atlantic Caribbean Futures', in H. Goulbourne and M. Chamberlain (eds), *Caribbean Families in Britain and the Trans-Atlantic World*, London and Oxford: Macmillan Caribbean, 234-242.

Gurak, D. and F. Caces (1992), 'Migration Networks and the Shaping of Migration Systems', in M.M. Kritz, L.L. Lim and H. Zlotnik (eds), *International Migration Systems: A Global Approach*, Oxford: Clarendon Press, 150-176.

Hall, S. (1996), 'Who needs Identity?', in S. Hal and P. du Guy (eds), *Questions of Cultural Identity*, London: Sage.

Henry, F. (1994), *The Caribbean Diaspora in Toronto: Learning to Live with Racism*, Toronto, Buffalo and London: University of Toronto Press.

Hollander, J. (1991), 'It All Depends', *Social Research*, 58(1): 31-49.

Kasinitz, P. (1992), *Caribbean New York: Black Immigrants and the Politics of Race*, Ithaca and London: Cornell University Press.

Katz, E. (1991), 'Breaking the Myth of Harmony: Theoretical and Methodological Guidelines to the Study of Rural Third World Households', *Review of Radical Political Economics*, 23(3 & 4): 37-56.

Kessner, T. (1981), *Repatriation in American History*, Appendix A to the Staff Report of the Select Commission on Immigration and Refugee Policy: Papers on U.S. Immigration History, April.

King, R. (2000), 'Generalizations from the History of Return Migration', in B. Ghosh (ed), *Return Migration: Journey of Hope or Despair?* Geneva: United Nations IOM, 7-55.

Mack, A. (1993), *Home: A Place in the World*, New York and London: New York University Press.

Mahler, S. (1995), *American Dreaming: Immigrant Life on the Margins*, Princeton, NJ: Princeton University Press.

Martin, S. (2001), 'Global Migration Trends and Asylum', *The Journal of Humanitarian Assistance: New Issues in Refugee Research, WP. 41* http://www.jha.ac/articles/u41.htm

Massey, D.S. (1988), 'Economic Development and International Migration in Comparative Perspective', *Population and Development Review*, 14(3): 383-413.

Massey, D.S. (1990), 'Social Structure, Household Strategies, and the Cumulative Causation of Migration', *Population Index*, 56(1): 3-26.

Massey, D.S., Arango J., Hugo G., Kouaouci A., Pellegrino A., and J. Edward Taylor, (1998), *Worlds in Motion: Understanding International Migration at the End of the Millennium*, Oxford: Clarendon Press.

Morokvasic, M. (1984), 'Birds of Passage are also Women ....', *International Migration Review*, 28(4): 886-907.

Mountz, A. and R.A. Wright (1996), 'Daily Life in the Transnational Migrant Community of San Agustin, Oaxaca, and Poughkeepsie, New York', *Diaspora*, 5(3): 403-428.

Nietschmann, B. (1979), 'Ecological change, inflation, and migration in the far Western Caribbean', *Geographical Review*, 69(1), 1-24.

Orozco, M. (2002), 'Globalization and Migration: The Impact of Family Remittances in Latin America', *Latin American Politics and Society*, 44(2): 41-67.

Peach, C. (1968), *West Indian Migration to Britain: A Social Geography*, Oxford University Press for the Institute of Race Relations.

Peach, C. (1984), 'The force of West Indian island identity in Britain', in Clarke, C., D. Lay and C. Peach (eds), *Geography and Ethnic Pluralism*, London: George Allen and Unwin, 214-230.

Philpott, S. (1973), *West Indian Migration: The Montserrat Case*, London: Athlone Press.

Philpott, S. (1977), 'The Montserratians: Migration Dependency and Maintenance of Island Ties in England', in J.L. Watson (ed), *Between Two Cultures: Migrants and Minorities in Britain*, Oxford: Basil Blackwell, 90-119.

Piore, M.J. (1979), *Birds of Passage: Migrant Labor and Industrial Societies*, New York: Cambridge University Press.

Piore, M.J. (1986), 'The Shifting Grounds for Immigration', *Annals, American Academy of Political and Social Sciences*, Volume 485: From Foreign workers to Settlers?: Transnational Migration and the Emergence of New Minorities, 23-33.

Portes, A. (1996), 'Transnational Communities: Their Emergence and Significance in the Contemporary World-system', in R.P. Korzeniewicz and W.C. Smith (eds), *Latin America in the World-Economy*. Westport, Connecticut: Praeger, 151-168.

Portes, A. (1997), 'Globalization from Below: The Rise of Transnational Communities', Working Paper in the WPTC Series (WPTC-97-01): Transnational Communities/ESRC Research Programme, Oxford University, Oxford, UK.

Portes, A. and J. Böröcz (1989), 'Contemporary Immigration: Theoretical Perspectives on Its Determinants and Modes of Incorporation', *International Migration Review*, 23: 606-630.

Portes, A. and L.E. Guarnizo (1991), 'Tropical Capitalists: U.S.-Bound Immigration and Small Enterprise Development in the Dominican Republic', in S. Diaz-Briquets and S. Weintraub (eds), *Migration, Remittances and Small Business Development*, Boulder: Westview, 101-131.

Portes, A. and M. Zhou (1993), 'The New Second Generation: Segmented Assimilation and Its Variants among Post-1965 Immigrant Youth', *Annals of the American Academy of Political and Social Science*, 530: 74-98.

Portes, A., Guarnizo, L. and P. Landolt (1999), 'The study of transnationalism: pitfalls and promise of an emergent research field', *Ethnic and Racial Studies*, 217-237.

Pries, L. (1999), *Migration and Transnational Social Spaces*, Aldershot UK, Brookfield, USA, Singapore and Sydney, Australia: Ashgate.

Ravenstein, E.G. (1885), 'The Laws of Migration', *Journal of the Royal Statistical Society*, 48, Part 2: 167-235.

Richardson, B.C. (1983), *Caribbean Migrants: Environment and Human Survival in St. Kitts and Nevis*, Knoxville, TN: University of Tennessee Press.

Rubenstein, H. (1979), 'The Return Ideology in West Indian Migration', in R.E. Rhoades (ed), *The Anthropology of Return Migration*, Papers in Anthropology, 20: 330-337.

Rumbaut, R. (1999), 'Assimilation and its Discontents: Ironies and Paradoxes', in C. Hirschman, P. Kasinitz and J. De Wind (eds), *Handbook of International Migration: The American Experience*, New York: Russell Sage Foundation, 172-195.

Salt, J. (1997), *International Movements of the Highly Skilled,* Paris: Organization of Economic Co-operation and Development, Occasional Paper, No. 3. ACDE/GD(97)169.

Sassen, S. (1988), *The Mobility of Labor and Capital*, Cambridge: Cambridge University Press.

Sassen, S. (1996), *Losing Control? Sovereignty in an Age of Globalization*, New York: Columbia Press.

Sassen, S. (2000), *Cities in a World Economy*, Thousand Oaks, CA: Pine Forge/Sage Press.

Sassen, S. (2001), *The Global City: New York, London Tokyo*, Princeton, NJ: Princeton University Press.

Smith, M.P. and L.E. Guarnizo (1998), *Transnationalism from Below,* New Brunswick, NJ: Transaction Books.

Swyngedouw, E. (1997), 'Neither Global or Local: "Glocalization" and the Politics of Scale', in K.R. Cox (ed), *Spaces of Globalization*, New York: Guildford, 137-166.

Tuan, Y.-F. (1980), 'Rootedness versus Sense of Place', *Landscape,* 24(1): 3-8.

Vertovec, S. (2002), *Transnational Networks and Skilled Labour Migration*, ESRC Transnational Communities Programme, Oxford, WPTC-02-02.

Watson, H.A. (1994), *The Caribbean in the Global Political Economy*, Boulder & London: Lynne Reiner.

Wheelock, J. (1992), 'The Household in the Total Economy', in P. Ekins and M. Max-Neef (eds) *Real-Life Economics: Understanding Wealth Creation*, London and New York: Routledge, 123-136.

Yeoh, B.S.A., Charney, M.W. and T.C. Kiong (2003), *Approaching Transnationalisms: Studies on Transnational Societies, Multicultural Contacts, and Imaginings of Home,* Boston, MA: Kluwer.

Chapter 14

Experiencing Return: Societal Contributions, Adaptations and Frustrations

Robert B. Potter and Dennis Conway

To this point, extant scholarship has commonly focused attention on the issues of adjustment faced by returning nationals of retirement age, together with their development implications. In other words, the focus has been on those who have consummated their long-held desires and intentions to return to the Caribbean islands and territories of their youth, and who have left metropolitan Britain, Europe or North America for good. But retiree returnees are generally not considered effective 'agents of change', in part because of their age, their exit from the formal labour market and the adaptation problems they face due to their lengthy sojourns overseas.

Indeed, these older returnees will have left a very different island setting in the 1950s, 1960s or 1970s. Some are all too likely to be characterized by intransigence and reactionary attitudes towards the changes that have occurred during the intervening years in which they have been away. On the other hand, this cohort is the agency for the injection of large amounts of capital and pensions into the local economy. For example, as noted in the case of Barbados, it is estimated by the Central Bank that in 1996 remittances from nationals overseas and returnees amounted to US $62.5 million (see Conway, 1985; 1993 for a wider discussion). In addition, these older retirees may well have time on their hands and can make valuable additions to social capital in the form of reinforcing and cementing traditional Caribbean family values, providing child minding and child care. Notably, this was witnessed among those returnees who were consulted as part of the research projects discussed in this volume, who specifically mentioned their retiree parents' return migration as a strong factor in prompting their decision to migrate to the Caribbean. Such a situation stresses the importance of the family in Caribbean migration (Chamberlain, 1995, 2001).

Young and younger return migrants, who have variously been referred to in the chapters making up this volume as second generation returnees, foreign-born returning nationals and citizens of decent, have rarely been the focus of specific academic attention until comparatively recently (see Potter 2001a, 2001b, Plaza, 2002, Potter, 2003, 2005). This group has remained relatively invisible as a return migration cohort, largely on account of the relatively small numbers involved, although they have occasionally formed part of wider samples of returnees

(DeSouza, 1998; Byron, 2000). However, as the first generation migrants reach retirement age and beyond, the second and third generations are poised to become an ever more important group; perhaps serving as a transnational cohort, who forge further links between the Caribbean and Europe-North America.

Thus, as the chapters in this research anthology have ably demonstrated, in today's global-transnational world, return migrants and those repetitive circulators who are undertaking more temporary sojourns or visits are no longer insignificant demographic cohorts. Despite their numerically small size, many are demonstrating that they can act as influential 'agents of change'. No longer consisting mainly of returning retirees, today's return cohort is becoming noticeably more diverse, in respect of age and family-life cycle characteristics, class, social and gendered positions, family networks and migration histories.

As we might expect, the research studies presented in this volume stress the importance of pull factors in promoting return migration. There also may well have been negative experiences in the metropolitan societies of origin, such as being made to feel like second-class citizens, the occurrence of racial discrimination and harassment. But, undoubtedly it is the pull factors of the Caribbean region, including the climate, returning to family roots, and the availability of opportunities – albeit selective and promissory – that seem to be of vital significance in promoting return. It is axiomatic that the chance to improve standards of living and quality of life is likely to be central to the returnee's calculus. Several of the studies included in this volume have shown that a prime motive is that the returnees feel that the Caribbean is the best place to bring up children and that opportunities in education, in particular, are likely to be better for their offspring. As first generation parents have often returned on retirement, the core of an extended family is also on hand to help with the care of children and wider family support networks in times of need. Others talked of the need to return in order to fulfill family duties, such as providing care for ageing parents. In a few instances, the agency of national pride and wanting to do something for the 'Motherland' was revealed as a motivational factor prompting return migration.

As the chapters in this collection have also served to demonstrate, return migrant experiences differ according to island context, the metropolitan backgrounds of returnees, and the extent to which returnees are well-supported by transnational networks of extended and nuclear families, while others are not so socially embedded in such transatlantic or pan-American fields. But on the other hand, the chapters in this volume suggest that return migrants of whatever age or background face a range of common adjustments and ultimately, in some instances, strongly perceived frustrations in coming to terms with their new island homes.

As was the case with the first generation (retiree) returnees, the length of time the migrants have spent away from their birthplaces may pose problems. Accordingly, the accents, dress codes and learned behaviours of those born overseas of Caribbean parents can all too easily represent a tangible badge of difference, making for real difficulties of transition. One of the contributors to this volume refers to 'the suffering of return' when describing the difficulties involved in adjustment. Returning nationals are not always easily accepted, as some of the chapters here have shown. Frequently, this is due to competition in the workplace, and for women in particular, 'competition for men' emerges as an interesting aspect of social distancing between returnees and stay-at-homes. In several cases, this has also been shown to be closely linked to returning women's difficulties in making female friends. In extreme instances, this appears to have led to a degree of 'othering', that is the manifest marginalizing of the returnees as outsiders who are fundamentally different from the indigenous population. For example, in several instances cited in this volume, issues of national identity rather that racial identity loom to the fore in the day-to-day lives of the returnees. But such occurrences clearly vary from place to place, person-to-person and time-to-time. The 'madness accusation' encountered in one of the chapters in this collection would seem to be an extreme example of this process of 'othering'. What is more certain is that a sizeable number of returnees, young and old, are characterized by their 'inbetween-ness'; that is, their existence and identification with two 'worlds within worlds', European and New World, black and white. Some returnees also feel frustrations at the poorer facilities, limited shopping facilities and higher prices that they experience in their new Caribbean homes. In other cases, power cuts, water shortages and domestic tasks being harder to complete, pose major problems on a day-to-day basis.

Gender is another major dimension of distinction, with many female return migrants reporting the loss of the gendered gains made in metropolitan societies. This has been expressed in terms of being more dependent on husbands for income and for transport and the like. In some instances, female returning nationals have been shown as being less likely to work, due to societal expectations and constraints. The corollary is more time spent on domestic work, child-care, food preparation, with women assuming even more domestic responsibilities. Another theme expressed by female returning nationals was the need to conform to local gender norms, especially those connected with female respectability (Wilson, 1969; see also Pratt and Yeoh, 2003). In those instances where younger returnees were participating in the labour market, the behaviour of 'macho' colleagues and bosses and the exercise of patriarchal power within society as a whole were commented upon in a number of cases. As several of the case studies demonstrate, issues relating to race relations and the continued operation of the colour class system (Potter, Barker, Conway and Klak, 2004), came as real surprises to many returnees, amounting to a further aspect of a distinct 'culture shock'.

We can conclude that many return migrants inevitably experience frustration while they adapt, whilst others adapt more easily and quickly. By definition, those who remain are the ones who have developed niches from which they can build island contacts, make new circles of friends, and generally participate in the social and economic fields associated with their professions and businesses in the island homes of their parents, now their homes and nations. Where this is happening, as the comparative studies included in this volume have demonstrated, dual citizenship and multiple, linked identities are becoming the norm, rather that the exception, as returnees of different ages choose island life, and decide to live, work and play in the Caribbean island 'homes' rather than struggle to make their livelihoods in the metropolitan centers of their youth. They do not, however, generally sever their ties with the metropolitan centers of their past, but rather retain property, keep in close touch with family and friends there, and by and large, adhere to transnational strategies to live in and between two worlds, or sometimes three or more if their family's international reach is multi-local.

Most saliently, the research in this volume has pointed towards the new migration experiences of the Caribbean at the beginning of the twenty-first century. Movements have been shown to involve a wider group of returnees. As well as the expected counter-stream of returning first generation retirees, more youthful and younger migrants of working age, in their 30s and 40s, have decided to give it a try 'back home'. These young returning nationals, however, have more information available at their finger-tips via the internet, telephone and other global networked systems, and as a direct consequence, are more directly aware of the opportunities that are open to them, and also perhaps to the adjustments they may have to make in their new environments. They are likely to be more skilled than their migratory counterparts were in the past. Some might very well be moving back and forth, trying out strategies to see if they can enhance their standards of living and the quality of their lives. And it is scarcely surprising that Caribbean tourism ensembles, which present so many opportunities in the service sector, are ideal locales for such individuals to search for, and acquire, appropriate jobs.

But the previous links developed in the metropolitan society are likely to remain very important to such migrants, and they may update and enhance these by means of regular return visits. Some will be renting out property in the metropole while renting, building or buying a property in their new Caribbean home. In these circumstances, 'home' is likely to be a multi-dimensional construct of territorial attachment. Such migrants will be characterized by multi-local, transnational networks, with this amounting to a clear expression of the new migration circumstances that are evolving at the beginning of the twenty-first century. By definition, such peripatetic migrants are likely to hold dual nationality and may hold dual passports. These migrants are likely to be very much part of the Transatlantic or pan-American worlds (Goulbourne, 2000). It seems highly unlikely that the present migratory location is seen as the final one, with the 'migration option' remaining a distinct possibility for the future. Contributors to this volume talk of 'transnational migrants', and of 'nations on the move'. Others have commented that what is occurring is not so much an emigration with a return intention, but rather a more transitory 'phenomenon of circulation'.

However, as we have previously summarized in this chapter, such circulation is neither occurring in a frictionless global economic space, nor in a homogeneous cultural realm. Even in these days of globalisation and transnationalism, the processes involved in this variant of so-called 'hypermobility' are far from seamless and costless, when viewed in terms of the dimensions of adjustment that have been so clearly outlined and discussed in the essays making up this volume. Thus, while skilled young migrants may be economically advantaged and find it easy to gain and change jobs, others are bemoaning the severity of their day-to-day domestic struggles and their loss of gendered autonomy. All of this reflects the essential conclusion that transnationality and mobility are part and parcel of the increasingly unequal neo-liberal world order, one that is giving rise to further heterogeneity and difference, rather than any semblance of homogeneity and uniformity.

References

Byron, M. (2000), 'Return migration to the eastern Caribbean: comparative experiences and policy implications', *Social and Economic Studies*, 47, 155-188.

Chamberlain, M. (1995), 'Family narratives and migration dynamics: Barbadians in Britain', *Nieuwe West Indische Gids*, 69, 253-275.

Chamberlain, M. (2001), 'Migration, the Caribbean and the family', in Goulbourne, H. and Chamberlain, M. (eds), *Caribbean Families in Britain and the Trans-Atlantic World*, Macmillan: Oxford and London: Macmillan.

Conway, D. (1985), 'Remittance impacts on development in the eastern Caribbean', *Bulletin of Eastern Caribbean Affairs*, 11, 31-40.

Conway, D. (1993), 'Rethinking the consequences of remittances for eastern Caribbean development', *Caribbean Geography*, 4, 116-130.

De Souza, R-M. (1998), 'The spell of the cascadura: West Indian return migration', in Klak, T. (ed), *Globalisation and Neoliberalism: the Caribbean context*, Rowman and Littleford: London, 227-253.

Goulbourne, H. (2002), *Caribbean Transnational Experience*, Pluto Press: London and Arawak Publications: Kingston.

Chamberlain, M. and Goulbourne, H. (2001), *Caribbean Families in Britain and the Trans-Atlantic World*, Macmillan: London and Oxford.

Plaza, D. (2002), 'The socio-cultural adjustments of second generation British-Caribbean "return" migrants to Barbados and Jamaica', *Journal of Eastern Caribbean Studies*, 27, 135-160.

Potter, R.B. (2001a), '"Tales of two societies": young return migrants to St Lucia and Barbados', *Caribbean Geography*, 12, 24-43.

Potter, R.B. (2001b), 'Narratives of socio-cultural adjustment among young return migrants to St Lucia and Barbados', *Caribbean Geography*, 12, 70-89.

Potter, R.B. (2003), 'Foreign-born and young returning nationals to Barbados: results of a pilot study', *Reading Geographical Papers*, 166, 40pp.

Potter, R.B. (2005), '"Citizens of descent": "foreign-born" and "young" returning nationals to St Lucia', *Journal of Eastern Caribbean Studies*, 30` (in press).

Potter, R.B., Barker, D., Conway, D. and Klak, T. (2004), *The Contemporary Caribbean*, Pearson-Prentice Hall: London and New York.

Pratt, G. and Yeoh, B. (2003), 'Transnational (counter) topographies', *Gender, Place and Space*, 10, 159-166.

Wilson, P.J. (1969), 'Reputation and respectability: a suggestion for Caribbean ethnology', *Man*, 70-84.

Index